Anthocyanins in Subtropical Fruits

Anthocyanins are one of the powerful antioxidants that can alleviate several lifestyle diseases such as heart diseases and hypertension. They can reduce cancer by protecting cells against damage. Several subtropical fruits, including berries, plums, black grapes, apricots, and peaches, among others, are a rich source of anthocyanin. Consumption of these fruits will prolong the longevity of consumers; this is ascribed to the curative effects of anthocyanins present in those fruits. *Anthocyanins in Subtropical Fruits: Chemical Properties, Processing, and Health Benefits* discusses novel techniques adopted for the extraction of anthocyanins from various subtropical fruits.

In this book, experts in the field examine solutions for efficiently extracting anthocyanins from subtropical fruits with higher yield. Protocols for the commercial production of anthocyanins from various subtropical fruits with their applications are also discussed in detail.

Additional features:

- Addresses chemical properties, classification, and stability of anthocyanins during processing and storage
- Discusses the benefits of using both thermal and non-thermal processing methods for extraction of anthocyanins from various subtropical fruits
- Explains the applications of synthetic and natural anthocyanins in foods and their regulatory aspects

Providing comprehensive information on extraction techniques as well as the chemical and health properties of anthocyanins from various subtropical fruits, this book is a valuable resource for academic students, research scholars, and food scientists.

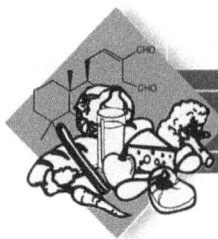

FUNCTIONAL FOODS AND NUTRACEUTICALS SERIES

Series Editor
John Shi, Ph.D.
Guelph Food Research Center, Canada

Anthocyanins in Subtropical Fruits

Chemical Properties, Processing, and Health Benefits

Edited by
M. Selvamuthukumaran

CRC Press
Taylor & Francis Group
Boca Raton London New York

CRC Press is an imprint of the
Taylor & Francis Group, an **informa** business

First edition published 2023
by CRC Press
6000 Broken Sound Parkway NW, Suite 300, Boca Raton, FL 33487-2742

and by CRC Press
4 Park Square, Milton Park, Abingdon, Oxon, OX14 4RN

CRC Press is an imprint of Taylor & Francis Group, LLC

Library of Congress Cataloging-in-Publication Data
Names: Selvamuthukumaran, M., editor.
Title: Anthocyanins in subtropical fruits : chemical properties, processing,
and health benefits / edited by M. Selvamuthukumaran.
Description: First edition. | Boca Raton : CRC Press, 2023. |
Series: Functional foods and nutraceuticals |
Includes bibliographical references and index.
Identifiers: LCCN 2022034771 (print) | LCCN 2022034772 (ebook) |
ISBN 9781032127958 (hardback) | ISBN 9781032151175 (paperback) |
ISBN 9781003242598 (ebook)
Subjects: LCSH: Anthocyanins–Health aspects. | Anthocyanins–
Physiological effect. | Tropical fruit–Health aspects.
Classification: LCC QK898.A55 A643 2023 (print) |
LCC QK898.A55 (ebook) | DDC 582.13–dc23/eng/20221209
LC record available at https://lccn.loc.gov/2022034771
LC ebook record available at https://lccn.loc.gov/2022034772

ISBN: 978-1-032-12795-8 (hbk)
ISBN: 978-1-032-15117-5 (pbk)
ISBN: 978-1-003-24259-8 (ebk)

DOI: 10.1201/9781003242598

Typeset in Times
by Newgen Publishing UK

I profoundly thank
The Almighty
My family
My friends
and
Everybody
Who have constantly encouraged and helped me to complete this book successfully

Selvamuthukumaran .M

Contents

Preface

Anthocyanins are among the powerful antioxidants that can alleviate several lifestyle diseases such as heart diseases and hypertension. They can reduce cancer by protecting cells against damage. They are water-soluble pigments that exhibit different colors such as purple, red, and blue depending upon their pH value. Several subtropical fruits are a rich source of anthocyanins, including berries, plums, black grapes, apricots, peaches, dragon fruits, pomegranates, kiwis, and lychees. Consumption of these fruits will prolong the longevity of consumers; this is ascribed to the curative effects of anthocyanins present in those fruits.

This book describes various anthocyanins sourced from tropical fruits of different continents, i.e., North and South America, Africa, Oceania, Europe, Asia. It explains the benefits of using both thermal and non-thermal processing methods for the extraction of anthocyanins from various subtropical fruits and portrays the commercial production methodology for manufacturing anthocyanins from various subtropical fruits. The book also addresses the chemical properties, classification, and stability of anthocyanins during processing and storage. It explains the applications of synthetic and natural anthocyanins in foods together with their regulatory aspects and the significant health benefits of consuming anthocyanins from the sole source of subtropical fruits.

I would like to express my sincere thanks to all the contributors; without their continuous support this book would not have seen daylight. We would also like to express our gratitude to Steve Zollo and all other CRC Press people, who have made every continuous cooperative effort to make this book a great standard publication at a global level.

About the Editor

M. Selvamuthukumaran is Professor, Department of Food Science & Technology, Hamelmalo Agricultural College, Eritrea. He was a visiting professor at Haramaya University, School of Food Science & Postharvest Technology, Institute of Technology, Dire Dawa, Ethiopia. He received his PhD in Food Science from Defence Food Research Laboratory affiliated to University of Mysore, India. His core area of research is processing of underutilized fruits for the development of antioxidant-rich functional food products. He has transferred several technologies to Indian firms as an outcome of his research work. He has received several awards and citations for his research work. He has published several international papers and book chapters in the area of antioxidants and functional foods. He has guided several national and international postgraduate students in the area of food science and technology.

Contributors

Asmita Acharya
Department of Food Science and Technology, Shivaji University, Kolhapur, Maharashtra, India

Weibin Bai
Department of Food Science and Engineering, Institute of Food Safety and Nutrition, Guangdong Engineering Technology Center of Food Safety Molecular Rapid Detection, Jinan University, Guangzhou, China

Khalid Bashir
Department of Food Technology, Jamia Hamdard, New Delhi, India

Shraddha Bhat
Department of Food Science and Technology, Shivaji University, Kolhapur, Maharashtra, India

Shivani Bisht
Food Science & Technology, College of Agriculture, Govind Ballabh Pant University of Agriculture &Technology, Pantnagar, U.S. Nagar, Uttarakhand, India

Dongbao Cai
Department of Food Science and Engineering, Institute of Food Safety and Nutrition, Guangdong Engineering Technology Center of Food Safety Molecular Rapid Detection, Jinan University, Guangzhou, China

Abishek Chandra
Department of Food Engineering, National Institute if Food Technology Entrepreneurship and Management, Haryana, India

Ajay Chinchkar
Department of Food Science and Technology, National Institute of Food Technology Entrepreneurship and Management, Haryana, India

Rohini Dhenge
Department of Food and Drug, University of Parma, Italy

Mehvish Habib
Department of Food Technology, School of Interdisciplinary Sciences & Technology, Jamia Hamdard, New Delhi, India

Kulsum Jan
Department of Food Technology, Jamia Hamdard, New Delhi, India

Shumaila Jan
Department of Food Science and Technology, National Institute of Food Technology Entrepreneurship and Management, Sonipat, Haryana, India

Aman Kaushik
Department of Basic and Applied Sciences, National Institute of Food Technology Entrepreneurship and Management, Haryana, India

Sidra Kazmi
Department of Food Technology, Jamia Hamdard, New Delhi, India

Sachin Kumar
Department of Food Engineering, National Institute of Food Technology Entrepreneurship and Management, Haryana, India

Sourabh Kumar
Department of Food Engineering, National Institute of Food Technology Entrepreneurship and Management, Haryana, India

Swati Mitharwal
Department of Food Science and Technology, National Institute of Food Technology Entrepreneurship and Management, Haryana, India

Anil S. Nandane
Department of Food Processing Technology, A.D. Patel Institute of Technology, New Vallabh Vidyanagar, Anand, Gujarat, India

Mohd. Nazim
Food Science & Technology, College of Agriculture, Govind Ballabh Pant University of Agriculture &Technology, Pantnagar, U.S. Nagar, Uttarakhand, India

Qianqian Qi
Tobacco Research Institute of Chinese Academy of Agricultural Sciences, Qingdao, China; Graduate School of Chinese Academy of Agricultural Sciences, Beijing, China

Sweta Rai
Food Science & Technology, College of Agriculture, Govind Ballabh Pant University of Agriculture &Technology, Pantnagar, U.S. Nagar, Uttarakhand, India

Savita Rani
Department of Life Science, Sharda University, Greater Noida, India

Rahul C. Ranveer
Department of PHM of Meat, Poultry and Fish, PG Institute of Post-Harvest Management, Killa-Roha, Dr BSKKV, Dapoli, Dist Raigad, India

Sabbu Sangeeta
Food Science & Technology, College of Agriculture, Govind Ballabh Pant University of Agriculture &Technology, Pantnagar, U.S. Nagar, Uttarakhand, India

John Shi
Guelph Food Research Centre, Agriculture and Agri-Food Canada, Guelph, Ontario N1G 5C9, Canada

Ning Yan
Tobacco Research Institute of Chinese Academy of Agricultural Sciences, Qingdao, China

Xiuting Yu
Tobacco Research Institute of Chinese Academy of Agricultural Sciences, Qingdao, China; Graduate School of Chinese Academy of Agricultural Sciences, Beijing, China

1 Anthocyanins in Subtropical Fruits

*Qianqian Qi, Dongbao Cai, Xiuting Yu, John Shi,
Weibin Bai, and Ning Yan*

CONTENTS

1.1 INTRODUCTION

China is rich in a wide variety of subtropical fruit resources with high nutritional value. The climate of the south of the Yangtze River in China is warm and humid. The natural conditions here are superior to those of other regions, and thus, the region is suitable for planting subtropical fruits. In recent years, due to the high economic benefits of planting subtropical fruits, with the deepening of poverty alleviation and prosperity in rural and mountainous areas in China, the planting area of subtropical fruits continues to expand, and output has increased significantly. For example, the world-famous litchi, longan, bayberry, orange, grape, mulberry, pitaya, banana, mango, and their processed products are also very popular in the domestic market. These fruits are rich in carbohydrates, vitamins, minerals, organic acids, and celluloses. They not only supply the heat energy and various nutrients needed by

DOI: 10.1201/9781003242598-1

the human body and regulate the normal physiological functions of the human body, but also promote appetite and help digestion. They have become one of the indispensable foods in people's life. Therefore, their high production and demand are due to their proven health benefits, in addition to their flavour and exotic character. Besides their contribution to macronutrient supply, subtropical fruits are characterised as a natural source of bioactive compounds with a high antioxidant capacity associated with the prevention of diseases related to oxidative stress, ageing, and chronic inflammation, indicating the great interest of the food, pharmaceutical, and cosmeceutical industries in these fruits (Rinaldo, 2020). Among the bioactive compounds in these fruits, phenolic compounds must be highlighted, especially phenolic acids, flavonoids, flavan-3-ols, anthocyanidins, coumarins, and isoflavonoids (Faria et al., 2021; Pereira-Netto, 2018; Septembre-Malaterre et al., 2016).

Anthocyanins are natural water-soluble pigments that are commonly found in the sap of plant cells, and they are the main factors underlying the red, blue, and purple colours of certain vegetables, fruits, and grains (Burton-Freeman et al., 2019). The content of anthocyanidin (the aglycone form) is particularly high in fruits, such as grapes and blueberries. Anthocyanidin not only gives plants a unique flavour and colour but also helps them to attract insects to complete the pollination process. Currently, over 550 types of anthocyanidins are recognised, of which six are widely distributed in plants, namely, cyanidin, delphinidin, pelargonidin, peonidin, malvidin, and petunidin, accounting for approximately 50%, 12%, 12%, 12%, 7%, and 7% of the anthocyanidin content in nature, respectively (Castañeda-Ovando et al., 2009; Gowd et al., 2017; You et al., 2011). Anthocyanidins are classified as flavonoids and exhibit diverse pharmacological activities that are primarily manifested in their anti-cancer, antioxidant, and anti-inflammatory properties (Reis et al., 2016). Therefore, this chapter reviews anthocyanins in subtropical fruits to provide a reference value for the research of subtropical fruits and anthocyanins, so as to improve the utilisation rate of subtropical fruits.

1.2 CHEMICAL STRUCTURES AND PROPERTIES OF ANTHOCYANINS

Anthocyanins belong to the flavonoid group of phenolic compounds. The basic structure of their parent nucleus is a highly conjugated 2-phenylbenzopyran cation. The two benzene rings are connected by three carbon atoms to form a C6–C3–C6 skeleton, which is the anthocyanin motif. Alternate substituents on the carbon positions of the two benzene rings lead to the formation of anthocyanidins with diverse structures; the six most common types are cyanidin, pelargonidin, delphinidin, peonidin, petunidin, and malvidin (Figure 1.1). Cyanidin-3-O-glucoside (C3G) is the most common, stable, and easily available anthocyanin in nature and has the strongest antioxidant effect among the six common plant pigments. Free anthocyanidin is rare under natural conditions and it often exists in the form of glycosides in fruits. Anthocyanidin is usually combined with one or more glucose, galactose, arabinose, or rhamnose moieties to form anthocyanins through glycosidic bonds (Manolescu et al., 2019). Moreover, the glycosidic bonds in anthocyanins can also form acylated anthocyanins with an organic acid of one or more molecules through ester bonds (Fernandes et al., 2014). Glycosyl derivatives are esterified by aromatics or fatty acids. The different numbers of substitutions, sites of hydroxylation on different rings, patterns of methylation, and number of glycosylation sites are the primary differences responsible for the various anthocyanidin chemical structures and colours found in nature (Manolescu et al., 2019).

The stability and colour of anthocyanidins are affected by internal and external factors, primarily the chemical structure and the presence of oxygen, illumination, temperature, pH, metal ions, and enzymes, respectively. Generally, anthocyanidins are easily degraded under light conditions and can readily oxidise and fade under heating or high temperature conditions, but are more stable under low temperature and acidic conditions (Fernandes et al., 2014). Anthocyanidins change colour as the pH value of the cell sap changes, being reddish at pH < 7, purple between pH 7 and 8, and blue at pH > 11 (Zhao et al., 2017). In addition, the anthocyanidin colour is affected by the number and

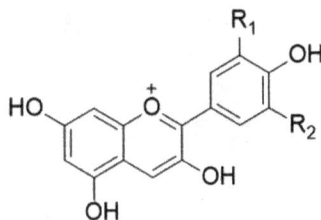

Cyanidin: R_1=OH, R_2=H
Pelargonidin: R_1=H, R_2=H
Delphinidin: R_1=OH, R_2=OH
Peonidin: R_1=OMe, R_2=H
Petunidin: R_1=OMe, R_2=OH
Malvidin: R_1=OMe, R_2=OMe

FIGURE 1.1 Common chemical structures of anthocyanins.

position of the hydroxyl, methoxy, and aglycone moieties. Moreover, the colour-rendering effect of anthocyanidin is enhanced under acidic conditions, and the colour can be stabilised via a complex reaction with metal ions. The anthocyanidin molecule contains both an acidic group and a basic group and is readily soluble in polar solvents, such as water and alcohol compounds (e.g., methanol and ethanol), but it is almost insoluble in non-polar solvents (e.g., chloroform). Finally, anthocyanidin has two main absorption wavelength ranges, with one in the visible region (465–560 nm) and the other in the ultraviolet region (270–280 nm) (Shi et al., 2021).

1.3 SUBTROPICAL FRUITS CONTAINING ANTHOCYANINS

Anthocyanins are usually found in subtropical fruits, although their content and composition can vary considerably among different plants. Commonly consumed foods with higher anthocyanins content include grapes, pitayas, mulberries, bayberries, litchis, and blood oranges (Figure 1.2). Representative fruits are briefly introduced in the following sections.

1.3.1 Grapes

Grapes constitute a rich source of anthocyanins, whereas grape seeds contain high quantities of proanthocyanidins. The colour of grape peel is mainly determined by the composition and content of anthocyanin glycosides. The different types of anthocyanins found in the outermost 3–4 cells of grape peel are responsible for the green, purple, red, and other colours of the grape. Grapes are rich in nutrients, containing various fruit acids, vitamins, minerals, flavonoids, and essential amino acids required by the human body. These agents can effectively scavenge free radicals and prevent senility, protect against cardiovascular and cerebrovascular disease, relieve fatigue, promote digestion, infection resistance, and blood circulation, and improve neurasthenia (Wang et al., 2010b). In particular, the anthocyanins and resveratrol found in grapes endow grape-derived products with health-promoting functions, such as antioxidation, enhancement of blood vessel elasticity, and improvement of skin radiance (Wang et al., 2010b).

Grape seeds have a higher level and richer variety of proanthocyanidins. The proanthocyanidins extracted from grape seeds comprise oligomers formed from varying amounts of epicatechin or catechin, usually in the form of glycosides (Rauf et al., 2019). Numerous *in vitro* and *in vivo* studies have also shown that grape seed proanthocyanidins can exert antioxidative, anti-microbial, anti-obesity, anti-diabetic, anti-neurodegenerative, anti-osteoarthritic, anti-cancer, and heart and eye protective effects,

FIGURE 1.2 Subtropical fruits with high anthocyanin content.

along with various other pharmacological effects (Unusan, 2020). Grapes consequently offer considerable benefits for the market economy; moreover, the majority of proanthocyanidin products currently on the market are derived from grape seeds. Consumer education has also increased the appreciation for wine, grape seed oil, and grape products as being rich in anthocyanins and proanthocyanidins.

1.3.2 Pitaya

Pitaya (*Hylocereus undulatus* Britt.), also known as red dragon fruit, fairy honey fruit, lover fruit, is one of the main tropical and subtropical fruits. Pitaya is a perennial climbing succulent plant, which likes light, fertiliser, heat, and drought. It mainly grows in tropical and subtropical areas, such as Taiwan, Guangxi, and Hainan. According to the colour of its peel and pulp, pitaya can be divided into three varieties: red skin and white meat, red skin and red meat, yellow skin and white meat. Among them, red-meat pitaya is rich in natural anthocyanins, and its colour from the skin to the meat is rose red to purple red. It is a good source for natural pigment extraction and processing. Pitaya fruit has high nutritional value. It is rich in protein, dietary fibre, vitamins, calcium, magnesium, potassium, iron, and other mineral elements (Xu et al., 2016). In particular, the pulp contains a high concentration of anthocyanins and almost no fructose and sucrose (Xu et al., 2016). It is widely used in the development of pitaya fruit wine, vinegar, jam, and health foods. In addition, pitaya fruit has medicinal effects, such as antioxidation, detoxification, stomach protection, whitening, and beauty effects. The anthocyanin content in pitaya fruit is rich and stable, and the content in the pulp is higher than that in the peel. Extracting anthocyanins from the peel can not only improve the utilisation value of pitaya fruit, but also reduce the environmental pollution caused by the peel.

1.3.3 Mulberry

Mulberry is a perennial woody plant native to central China. It has a wide planting range and a long cultivation history. China ranks first worldwide in terms of the variety of mulberry trees and yields (Jiang

& Nie, 2015). Mulberry is the fruit of the mulberry tree and can be consumed as a sweet and juicy fruit; it is also used in Chinese medicine. It was included in the first list of medicinal and edible plants published by the National Health Commission of the People's Republic of China (Jiang & Nie, 2015). Mulberry is not only rich in organic acids, vitamins, essential amino acids, and other nutrients, but also contains bioactive components, such as anthocyanins and phenols (Lin & Tang, 2007). Traditional Chinese medicine believes that mulberry is helpful for the treatment of conditions such as insomnia, neurasthenia, hypertension, and diabetes. It has also been reported to exert anti-fatigue effects and help prevent constipation (Özgen et al., 2009). Studies have also found that anthocyanins have physiological effects, such as free radical scavenging, anti-cancer, and fat-reducing activities (Chang et al., 2013; Huang et al., 2011a; Krishna et al., 2018; Raman et al., 2016). At present, solvent extraction is widely used to extract mulberry anthocyanins in China and overseas, and the amount of anthocyanins extracted using 70% ethanol as the extractant is 170.5 mg/100 g (Suh et al., 2003). Studies have shown that ripe mulberries are abundant in anthocyanins, predominantly C3G and cyanidin-3-rutinoside (Kabi & Bareeba, 2008; Wu et al., 2011). Mulberry is currently developed and utilised predominantly by food industries in products such as fruit juice, vinegar, wine, and jam, as well as in other industries for natural dye and cosmetic production. The medicinal value of the anthocyanins in mulberry, however, has been less widely developed and utilised (Liang et al., 2012). Moreover, studies have found that mulberry anthocyanins have a strong sensitivity to factors such as temperature, light, pH, and enzymes, owing to their strong activity and low levels of stability. This limits, to a certain extent, the production of mulberry anthocyanins (Dastmalchi et al., 2007; Jia et al., 1999). Therefore, optimising the extraction process conditions for anthocyanins is key to enable their wider use.

1.3.4 BAYBERRY

Bayberry is a subtropical evergreen tree that produces an economically important fruit in southern China (Chen et al., 2004). Its ripe berries can be purple, red, pink, or white, depending on the variety (Cheng et al., 2015). Bayberry is sweet and sour, with abundant juice and a high nutritive value. A high consumer demand exists for this fruit both in China and overseas. It can be eaten fresh or used to make products such as juices, preserves, jams, and wines (Cheng et al., 2015). Bayberry is utilised in medicines as it is rich in nutrients, such as vitamins and dietary fibres, as well as eight types of amino acids and 17 types of mineral elements, which are necessary for the human body (Bao et al., 2005). Bayberry also contains phenolic compounds, such as anthocyanins, phenolic acids, and flavonols, which are involved in functions such as antioxidation, cancer prevention, and hyperglycaemia alleviation (Fang et al., 2007; Sun et al., 2012, , 2013). The anthocyanins in bayberry are predominantly C3G (Zhang et al., 2008). The content of anthocyanins in bayberry is as high as 76.2 mg/100 g (in terms of fresh fruits). The C3G content of bayberry (64.8 mg/100 g) accounts for at least 85% of the anthocyanins in the fruit, which is similar to that of blackberries, and is markedly higher than the proportion found in cranberries and blueberries (Bao et al., 2005; Zhang et al., 2008). In addition, bayberry leaves and bark are rich in compounds such as flavonoids and proanthocyanidins, and they have a strong oxidation resistance (Zhang et al., 2016b). Researchers studying natural berries, such as blueberries, strawberries, cranberries, mulberries, and bayberries, found that berry anthocyanins have a significant α-glucosidase inhibitory activity, which is significant for controlling blood sugar after meals (Xu et al., 2019).

1.3.5 LITCHI

Litchi (*Litchi chinensis* Sonn.) is an evergreen tree of the genus *Liriodendron*, which is an important southern subtropical fruit tree in China. It serves as a germplasm resource; its fruit colour, aroma, and taste are good and it is widely loved by consumers. Litchi is the largest fruit tree in the south, with a cultivation area of about 900,000 hectares and an output of about 3.5 million tons. The cultivation area and output account for 80% and 60% of the world's total, both of which are the highest in the world. China is the main producing country of litchi, mainly distributed in Guangdong, Guangxi, Fujian, Hainan, Taiwan, Yunnan, and Sichuan. The colour of litchi fruit is an important index of its appearance quality. When the litchi is mature, the coloration of the peel is mainly caused by the accumulation of anthocyanins. A red colour of litchi fruit is caused by anthocyanins. With the development of litchi fruit, the chlorophyll in the litchi pericarp gradually degrades, the background colour fades, and anthocyanin accumulates continuously, showing a red surface colour (Zhang et al., 2004). Even if the litchi fruit is from a highly flavoured variety, it will be more attractive and competitive in the market only when it shows its typical colour. A study found that the anthocyanin in the peel of early-maturing varieties was mainly cyanidin-3-glucoside, while that in the peel of middle and late-ripening varieties was mainly C3G (Wei et al., 2011).

1.3.6   BLOOD ORANGE

Blood orange is widely loved by consumers because of its rich nutrition, colour, and unique flavour. Among citrus species, anthocyanins are not common in all varieties, but only in a few red and purple varieties, such as purple pomelo and blood orange (Butelli et al., 2017). Blood orange is a typical variety coloured by anthocyanins. The red flesh of blood orange is caused by glycosides formed by the combination of anthocyanins and sugar. Tarocco blood orange, known as the king of blood orange, is the main variety in Italy. The cultivation amount accounts for more than 80% of all blood oranges. It is planted in several provinces (cities and districts), such as Guangdong, Hunan, Guangxi, Sichuan, and Chongqing. There are two types of pigments responsible for the red colour of citrus: fat-soluble carotenoids, such as lycopene and β-carotene, and water-soluble anthocyanins. Anthocyanins were reportedly found in the blood orange flower, bud, and fruit. In the blood orange fruit, the most important anthocyanin components are C3G and cyanidin-3-(6'-malonyl glucoside), accounting for about 25.40–45.01% and 22.35–46.81% of the total anthocyanin content of the fruit, respectively (Fabroni et al., 2016).

1.4   BIOSYNTHESIS OF ANTHOCYANINS

The synthesis of anthocyanins is mainly carried out on the surface of the endoplasmic reticulum. The biosynthesis pathway of anthocyanin glycosides in plants is basically clear and is a branch of the flavonoid synthesis pathway. Phenylalanine is a direct precursor for the biosynthesis of anthocyanins and other flavonoids, and the process from phenylalanine to anthocyanins roughly goes through three stages. The biosynthesis of anthocyanins is shown in Figure 1.3. The first stage is from phenylalanine to coumaroyl-coenzyme A (CoA); this step is regulated by the phenylalanine ammonia-lyase (PAL) gene. Firstly, PAL catalyses the formation of cinnamate from phenylalanine, and then cinnamic acid 4-hydroxylase and 4-coumarate CoA ligase catalyse the formation of 4-coumaryl-CoA. The second stage is from coumaroyl-CoA to dihydroflavonol. Coumaroyl-CoA and malonyl-CoA are catalysed by chalcone synthase (CHS) to produce chalcone, and then chalcone isomerase (CHI). Flavanone-3-hydroxylase (F3H) catalyses the production of dihydroflavonols flavonoid 3'-hydroxylase (F3'H) and flavonoid 3',5'-hydroxylase (flavonoid 3'5'-hydroxylase, F3'5' H) to generate dihydroquercetin and dihydromyricetin, which are catalysed by dihydroflavonol-4-reductase (DFR) to generate leucocyanidin.

The third stage is the synthesis of various anthocyanin glycosides. Colourless anthocyanin glycosides are converted into coloured anthocyanin glycosides by oxygenation under the action of anthocyanin synthase (ANS), and then various anthocyanin glycosides with different colours are catalysed by glycosyl transferase, methyl transferase, and acyltransferase (Hichri et al., 2011; Zhao et al., 2015).

The biosynthesis of anthocyanins is controlled by two sets of genes, namely structural genes and regulatory genes. Structural genes are shared by different plants and directly encode enzymes in the anthocyanin biosynthetic pathway. Structural genes include two important gene groups, namely upstream gene groups (PAL, CHS, CHI, and F3H) and downstream gene groups (F3'H, F3'5'H, DFR,

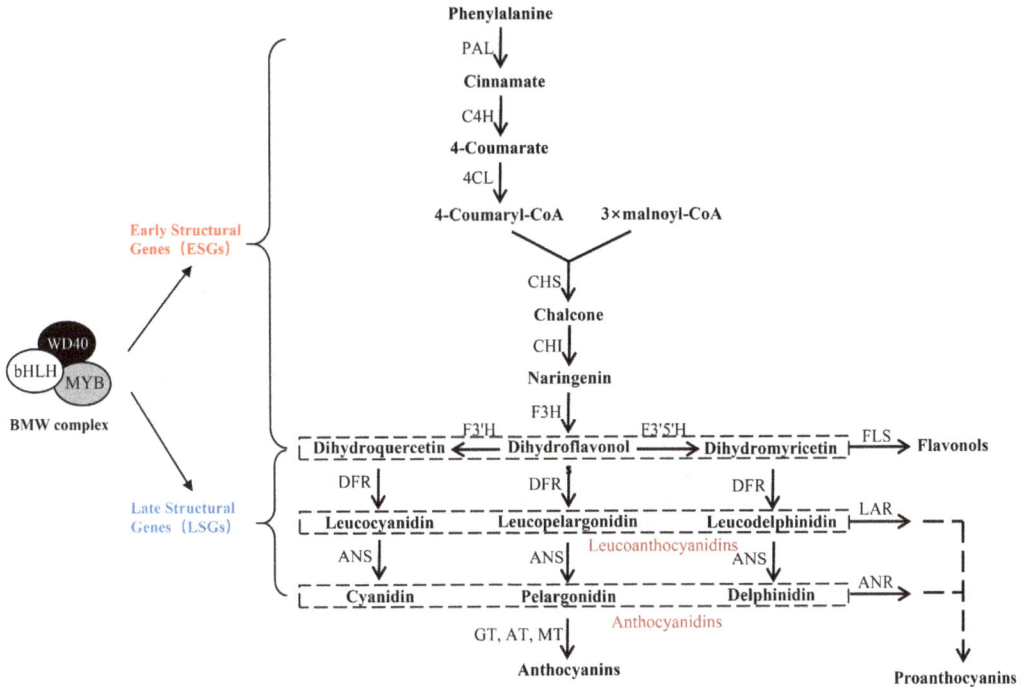

FIGURE 1.3 Biosynthetic pathway of anthocyanins.

ANS, and UDP-glucose flavonoid 3-O-glucosyltransferase (UFGT)). The upstream gene group is relatively conserved and found in lower bryophytes. They usually encode the enzymes involved in the synthesis of flavonoids and flavonols at the starting point of the pathway, and their expression is related to the process of fruit development (Honda et al., 2002; Kumar & Ellis, 2001; Lo Piero et al., 2005; Muir et al., 2001). The downstream gene group mainly encodes enzymes for synthesising anthocyanins and proanthocyanidins. Generally, when a fruit presents bright colours, high expression levels of these enzyme genes are observed (Han et al., 2012; Honda et al., 2002; Kobayashi et al., 2001; Lo Piero et al., 2005; Wang et al., 2010a). Regulatory genes control the intensity and mode of action of structural gene expression. Regulatory genes mainly include genes encoding three types of transcription factors: MYB, bHLH, and WD40. They regulate the expression of anthocyanin biosynthesis-related genes by binding to corresponding *cis*-acting elements in structural gene promoters. The three types of transcription factors can interact to form the MBW (MYB–BHLH–WD40) complex to jointly regulate anthocyanin biosynthesis (Xu et al., 2015). In the MBW complex, MYB is generally the dominant regulator (Hichri et al., 2011). Transcription factor MYB is a superfamily that plays a very important role in plant secondary metabolism, cell morphology, and response to stress. bHLH (basic helix–loop–helix) is another family of transcription factors that can regulate anthocyanin biosynthesis. WD40 regulates gene expression mainly by modifying histones to regulate chromatin remodelling (Suganuma et al., 2008). From the molecular point of view, the accumulation of anthocyanins in fruits is the result of the efficient expression of various key enzymes in their biosynthesis, and the expression of these key enzymes is closely related to the role of regulatory genes.

1.5 BIOAVAILABILITY OF ANTHOCYANINS

It should be noted that the resulting biological activities of anthocyanins usually depend on their bioavailabilities It is known that anthocyanins are responsible for many plant colours, and the change or loss of native anthocyanin colours is an important quality indicator for anthocyanins and anthocyanin-rich food stability, which further influences the bioavailability of anthocyanins.

Anthocyanins with a high bioavailability efficiently reduce cellular lipid peroxidation, hence reducing the risk of many diseases. As a nutraceutical, the bioavailability of anthocyanins is the key factor for maintaining good health and preventing diseases. Nevertheless, the use of anthocyanins in the food industry is restricted due to rapid degradation under different pH, temperature, oxygen, light, and metal ion conditions, among others. Additionally, the bioavailability of anthocyanins is also very low due to degradation by the gastrointestinal digestion system (Khoo et al., 2017; Salah et al., 2020; Tena et al., 2020). Recent bioavailability studies have demonstrated that anthocyanins are quickly absorbed in the stomach (Passamonti et al., 2003; Talavera et al., 2003) and small intestine (about 5–20 min) (Miyazawa et al., 1999), that they appear in plasma and urine in their parental form or as methylated, glucuronidated, or sulphated compounds (Felgines et al., 2003; Kay et al., 2004, 2005), and that most absorbed anthocyanins are in the intact glycoside form (McGhie & Walton, 2007). Additionally, anthocyanins are metabolised by gut microbiota to generate a series of phenolic acids (Aura et al., 2005; Azzini et al., 2010; Czank et al., 2013). Some studies reported that the bioavailability of anthocyanins ranged from 0.26 to 1.80% (Borges et al., 2007). An *in vitro* study using a gastrointestinal digestion model to investigate the bio-accessibility of anthocyanin-rich pomegranate showed that about 89% of anthocyanins were stable in gastric conditions, whereas this ratio was just 38% after pancreatic digestion, and only 12% was available in the serum (Sengul et al., 2014). As a consequence, only a small amount of anthocyanins can be absorbed in their intact form and enter the systemic circulation. The low bioavailability of anthocyanins causes low absorption of these compounds into the blood circulatory system and a high excretion via the urine and faeces, thus reducing the efficacy of anthocyanins in scavenging free radicals.

The relative bioavailability of red-wine anthocyanins has been reported previously, where the glycosides of peonidin had the highest relative bioavailability, followed by those of cyanidin, malvidin, delphinidin, and petunidin (Frank et al., 2003). Additionally, an *in vivo* study showed that the bioavailability of pelargonidin-3-O-rutinoside (1.13%) was fourfold higher than that of pelargonidin-3-O-glucoside (0.28%), demonstrating that anthocyanins linked to rutinoside may exhibit a higher bioavailability than those linked to glucoside (Xu et al., 2021).

The bioavailability of anthocyanins may also be affected by other food components (food matrix), where anthocyanins and other phytochemicals or vitamins may interact with each other synergistically or antagonistically (Eker et al., 2019; Yang et al., 2011). A study using red wine or red grape juice showed that anthocyanins in red grape juice were more readily absorbed than those in red wine (urinary excretion 0.23% vs. 0.18%). The decrease in anthocyanin absorption from red wine may be due to the presence of alcohol in red wine (Bitsch et al., 2004; Frank et al., 2003). A study on hyperlipidaemic rabbits showed a higher absorption rate of anthocyanins from blackcurrant juice than from aqueous citric acid solution of purified anthocyanins, suggesting a possible increase in bioavailability due to the presence of anthocyanins in the food matrix (Nielsen et al., 2003).Therefore, it has been found that the stability and bioavailability of anthocyanins during *in vitro* digestion were significantly improved using desolvated β-Lg nanoparticles, which could be further utilised in various food and pharmaceutical matrices (Salah et al., 2020). In addition, to overcome anthocyanins degradation and improve bioactivities, researchers are using several micro-/nano-encapsulation systems (Rashwan et al., 2021).

1.6 BIOACTIVITIES OF ANTHOCYANINS

Anthocyanins have attracted considerable attention owing to their notable health benefits, which involve a wide range of pharmacological activities. In addition to their potential applications in the medical field, anthocyanins have promising developmental prospects in other fields, such as the food and health product industries and in the cosmetics industry as natural pigments to replace artificial synthetic pigments (Teng et al., 2017). The bioactive effects and clinical trials of anthocyanins are summarised in Table 1.1.

TABLE 1.1
Bioactivity of Anthocyanins

Bioactivity	Research Design	Conclusion	Reference
Antioxidant activity	*In vivo* experiment, *Lycium ruthenicum* anthocyanins, mice	Improved activity of antioxidant enzymes and efficiency of free radical scavenging; reduced radiation damage	Duan et al. (2015)
	In vivo experiment, red bean and black bean anthocyanins, rats	Anthocyanin-3-glucoside improved liver ischemia–reperfusion effects in rats induced by oxidative stress	Tsuda et al. (2000)
	In vivo experiment, blackcurrant skin anthocyanins, rats	Induced upregulation of the antioxidant response element pathway	Thoppil et al. (2012)
	Clinical trial, red mixed berry juice anthocyanins	DNA oxidative damage decreased in human subjects and showed significant increases in reduced glutathione	Weisel et al. (2006)
	Clinical trial, *Aristotelia chilensis* anthocyanins	Improved the oxidative status (oxidized low-density lipoprotein (Ox-LDL) and F_2-isoprostanes) of healthy adults, overweight adults, and adult smokers	Davinelli et al. (2015)
Anti-cancer activity	*In vivo* experiment, black bean skin anthocyanins, *Drosophila* model of malignant tumour	Inhibited autonomous and non-autonomous autophagy of tumour cells, affected the tumour growth environment, and inhibited tumour growth and invasion	Wei et al. (2021)
	In vivo experiment, black raspberry anthocyanins, mice with oesophageal cancer	Regulated the expression of cyclooxygenase 2, inducible nitric oxide synthase, and nuclear factor activated B cells to inhibit the tumour inflammatory response and tumour cell proliferation	Shi et al. (2016)
	Clinical trial, strawberry anthocyanins	Reduced the progression of precancerous lesions and cell proliferation in subjects affected by oesophageal dysplastic lesions	Chen et al. (2012)
	Clinical trial, anthocyanin-enriched fruits	Reduced breast cancer recurrence	Butalla et al. (2012)
Prevention of cardiovascular disease	Clinical trial, blueberry anthocyanins and strawberry anthocyanins, healthy menopausal women	Reduced morbidity and mortality of coronary heart disease and myocardial infarction	Cassidy et al. (2011)
	In vivo experiment, grape anthocyanins and bilberry anthocyanins, rats	Reduced levels of total cholesterol and triglycerides	Seeram et al. (2003)
	In vivo experiment, bilberry anthocyanins and blueberry anthocyanins, mice	Reduced the formation of atherosclerotic lesions	Mauray et al. (2009)
	Clinical trial, strawberry anthocyanins	Decreased levels of cholesterol, LDL, and triglyceride	Alvarez-Suarez et al. (2014)

(continued)

TABLE 1.1 (Continued)
Bioactivity of Anthocyanins

Bioactivity	Research Design	Conclusion	Reference
	Clinical trial, blueberry anthocyanins	Showed a more significant reduction in blood pressure, Ox-LDL, and serum malondialdehyde and hydroxynonenal	Basu et al. (2010)
Bacteriostatic and anti-inflammatory effects	*In vivo* experiment, strawberry anthocyanins, rats	Inhibited the expression of inflammatory factors in mice and reduced the production of inflammatory substances	Winter et al. (2018)
	In vivo experiment, *Lycium ruthenicum* anthocyanins, rats	Reduced the content of monosodium urate and helped prevent gouty arthritis and colitis	Zhang et al. (2019a)
	Clinical trial, blueberry anthocyanins	Inhibited the expression of cyclooxygenase-2 and NF-κB	Xu et al. (2016)
Neuroprotective effects	*In vivo* experiment, black rice anthocyanins, *Drosophila*	Inhibited paraquat-induced neurodegeneration in *Drosophila*	Zuo et al. (2012)
	In vivo experiment, black bean anthocyanins, mice	Reduced AβO-induced reactive oxygen species and oxidative stress and helped to prevent cell apoptosis and neurodegeneration	Ali et al. (2018)
	In vivo experiment, blueberry anthocyanins, mice and rats	Increased short-term memory and motivation, and reversed the process of neuronal and behavioural ageing	Papandreou et al. (2009)
	Clinical trial, cherry anthocyanins, patients with Alzheimer's disease	Significantly improved memory and cognitive ability	Kent et al. (2017)
	Clinical trial, blueberry anthocyanins, elderly patients with mild cognitive impairment	Enhanced nerve response and relieved nervous system diseases	Boespflug et al. (2018)
Anti-diabetic effect	*In vivo* experiment, blueberry anthocyanins, hyperglycaemic mice	Lowered blood sugar levels	Grace et al. (2009)
	In vitro experiment, blueberry anthocyanins	Inhibited the activity of glucosidase, prevented the process of glucose metabolism in the body, and reduced blood sugar	Pranprawit et al. (2015)
	In vivo experiment, blueberry anthocyanins, diabetic mice	Relieved the adverse symptoms of diseased mice, such as polydipsia, hyperphagia, and polyuria; exerted equivalent blood sugar-lowering effects to those of acarbose	Johnson et al. (2011)
	In vivo experiment, black bean seed coat anthocyanins, diabetic mice	Improved the blood sugar and insulin sensitivity of diabetic mice; the effect was related to the activation of adenylate-activated protein kinase	Kurimoto et al. (2013)
	In vivo experiment, mulberry anthocyanins, diabetic mice	Relieved the high blood sugar of diabetic mice, reduced triglyceride accumulation and cholesterol levels, and increased adiponectin levels	Yan et al. (2016)

Relief of intestinal diseases	Clinical trial, blackberries anthocyanins	Increased fat oxidation and improved insulin sensitivity	Solverson et al. (2018)
	Clinical trial, red raspberry anthocyanins	Reduced peak and postprandial (2-h) glucose concentrations	Xiao et al. (2019)
	In vitro experiment, black rice anthocyanins	Increased the number of bifidobacteria and lactobacilli	Zhu et al. (2018)
	In vivo experiment, black raspberry anthocyanins, rats	Promoted the growth of Lactobacillus, Faecalibacterium prausnitzii, and Eubacterium rectale and could also inhibit the growth of harmful bacteria, such as Enterococcus spp.	Chen et al. (2018)
	In vitro experiment, blueberry anthocyanins	Increased diversity and abundance of human gut microbes	Zhou et al. (2020)
	In vitro experiment, purple sweet potato anthocyanins, faecal samples of healthy volunteers	Significantly increased the Bifidobacterium and Lactobacillus-Enterococcus spp. populations and short-chain fatty acid production	Zhang et al. (2016a)
	In vivo experiment, blueberry anthocyanins, rats	Promoted reverse cholesterol transport mediated by macrophage intestinal flora metabolites and reversed intestinal atherosclerosis lesions	Lee et al. (2018)
	Clinical trial, blueberry anthocyanins	Modulated cytokine expression in the intestine of patients with chronic colitis	Roth et al. (2016)

1.6.1 ANTIOXIDANT ACTIVITY

Anthocyanins are natural antioxidants. Moreover, anthocyanins are the most effective, safe, and natural water-soluble free radical scavengers identified to date (Fernandes et al., 2014). Their antioxidant activity is primarily related to their chemical structures, and generally, the simpler the chemical structure, the stronger the antioxidant activity. Furthermore, their health and therapeutic effects are primarily related to their antioxidant activity, which can be improved by methoxylation, hydroxylation, and glycosylation of the B-ring structure. The antioxidant activity of anthocyanins is comparable to that of vitamin C and is unmatched in the body compared with other antioxidants. This capacity can enhance the human body's immune system and may play an important role in future developments of modern medicine. In particular, studies have found that the anthocyanins in black goji berries can effectively increase the activity of antioxidant enzymes in mice and improve the efficiency of free radical scavenging, thereby preventing radiation-induced damage (Duan et al., 2015). Anthocyanin-3-glucoside also relieves the effects of liver ischemia–reperfusion induced by oxidative stress in rats (Tsuda et al., 2000); moreover, anthocyanin, delphinidin, and malvidin-induced antioxidant enzymes can induce the upregulation of the antioxidant response element pathway in these animals (Thoppil et al., 2012). Clinical trials found that human subjects who had consumed a juice rich in anthocyanins exhibited decreased DNA oxidative damage and showed significant increases in reduced glutathione compared with those in the control group (Weisel et al., 2006). The results of clinical studies suggest that anthocyanin diet intervention may improve the oxidative status (Ox-LDL and F_2-isoprostanes) of healthy adults, overweight adults, and adult smokers (Davinelli et al., 2015).

1.6.2 ANTI-CANCER ACTIVITY

Cancer remains a serious threat to human health. Cancer cells can spread and proliferate rapidly without limits, causing considerable harm to the human body and sometimes death. It is noteworthy that lung, colorectal, stomach, liver, and breast cancer were ranked in the top five for mortality (Bray et al., 2018). As both chemotherapy and radiation therapy have adverse side effects on human health, natural anti-cancer substances that effectively treat cancer without causing side effects are playing increasingly important roles in cancer treatment. Anthocyanins as a group of flavonoids, widely exists in fruits and vegetables, and possesses various bioactivities for cancer prevention and management. The anti-cancer activity of anthocyanins is primarily attributed to the catechol structure on the Bring structure (Marko et al., 2004). Notably, the free radical-scavenging ability of anthocyanins can prevent the diffusion of cancer cells, thereby reducing the likelihood of metastasis. Moreover, the research found that the addition of anthocyanin extracts in specific proportions to the daily feed of mice could inhibit the inflammatory reactions of tumours by regulating the expression of cyclooxygenase-2 (COX-2), inducible nitric oxide synthase (NOS), and nuclear factor kappa-B (NF-κB), thereby inhibiting tumour cell proliferation (Shi et al., 2016). Therefore, antioxidation and anti-inflammation were the main potential mechanisms of anthocyanins for cancer prevention. Donating electron, Nrf2 activation, and greater hydroxylation are dominant approaches to preventing cancer with anthocyanin supplement. Moreover, anthocyanin treatments have the capacity to achieve cancer prevention by inhibiting pro-inflammatory cytokines, and reducing pro-inflammatory gut bacteria (Figure 1.4) (Chen et al., 2021). Aimed as an antioxidant for cancer prevention, anthocyanins performed strong antioxidant capacity via scavenging free radicals, thereby reducing DNA damage, followed by preventing the tumorigenesis (Kalemba-Drożdż et al., 2020). Similarly, anthocyanin regulated the nuclear factor erythroid 2 (Nrf2) signalling pathway in colorectal cancer cells (Lee et al., 2020). In regard to an anti-inflammatory effect for cancer prevention, anthocyanins could down-regulate the expression levels of pro-inflammatory cytokines, including tumour necrosis factor-α (TNF-α), interleukin-1β (IL-1β), IL-6, and NF-κB in HT-29 colorectal cell and colorectal cancer cell (Chen et al., 2018; Venancio et al., 2017). Therefore, the potential

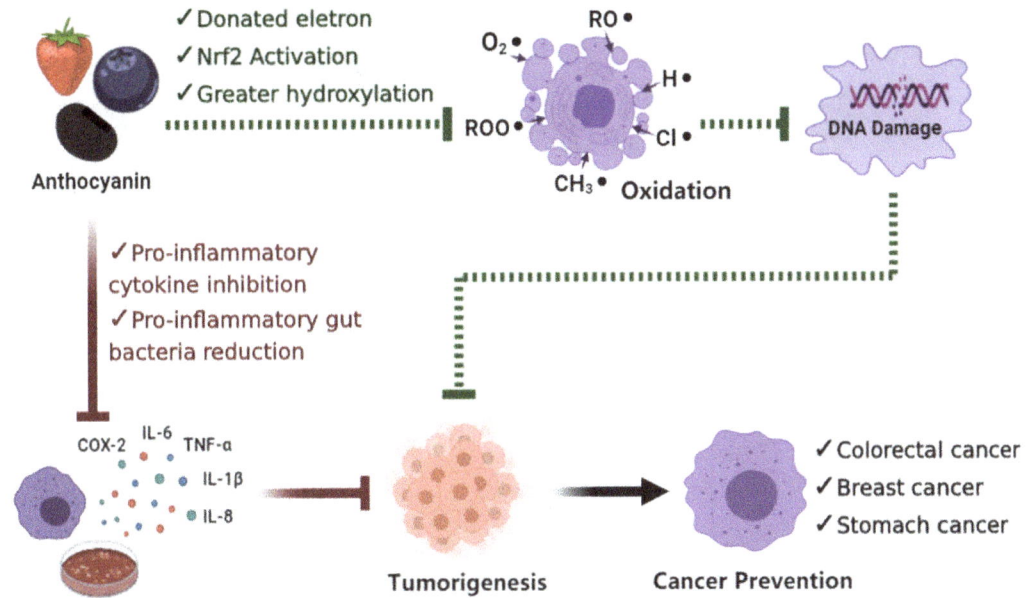

FIGURE 1.4 The main potential mechanisms of the cancer prevention of anthocyanins (Chen et al., 2021).

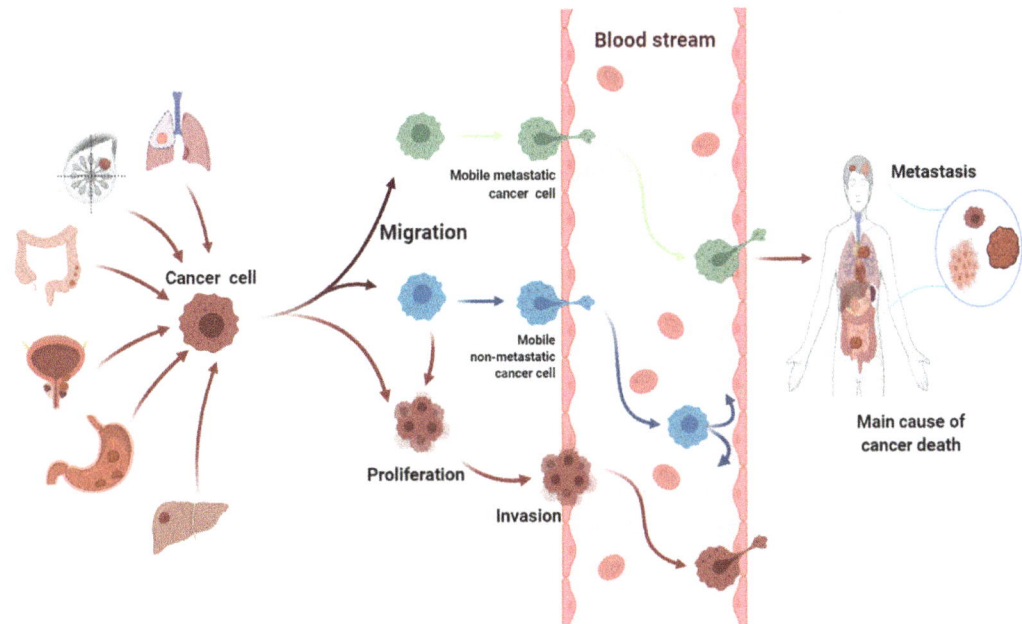

FIGURE 1.5 The potential mechanism of anthocyanin to inhibit cancer death (Chen et al., 2021).

anti-cancer activities of anthocyanins may inhibit the migration, invasion, and proliferation of corresponding cancer cells via antioxidation and anti-inflammation (Figure 1.5) (Chen et al., 2021). In particular, the combined use of C3G and chloroquine was highly effective in inhibiting tumour growth and metastasis (Wei et al., 2021). Epidemiologic studies showed that cancers ranked in the top five for incidence were lung, breast, colorectal, prostate, and stomach cancer; as regards

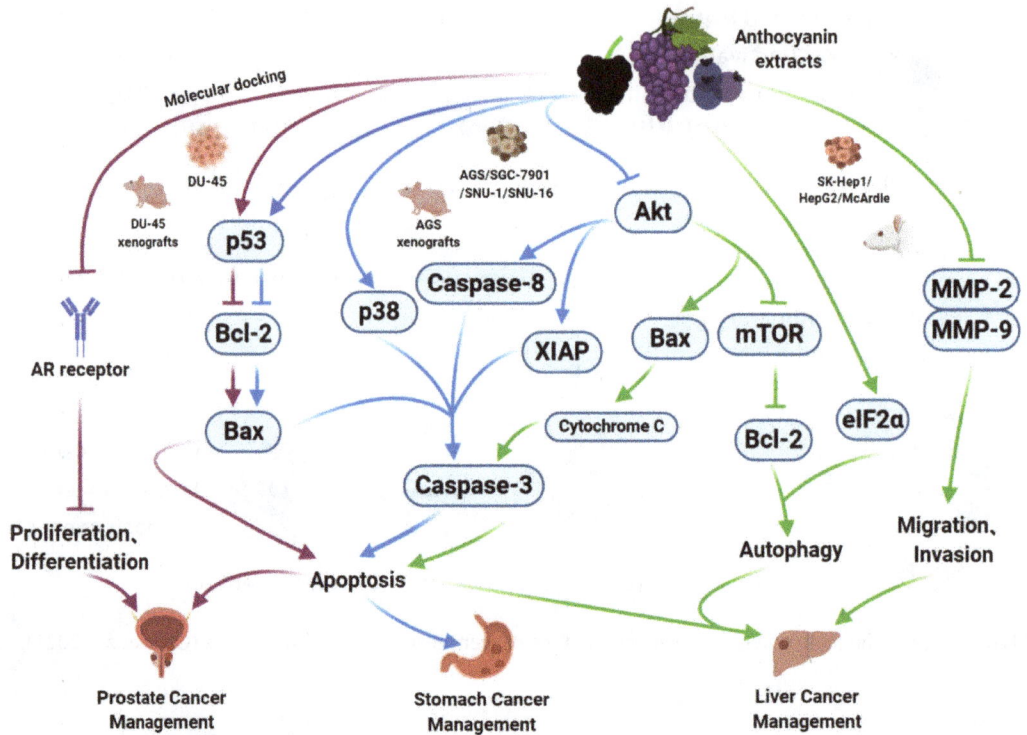

FIGURE 1.6 The main potential mechanisms of anthocyanins, e.g., prostate, stomach, and liver cancer (Chen et al., 2021).

mortality, the cancers ranked in the top five included lung, colorectal, stomach, liver, and breast cancer. The interaction and corresponding mechanisms of anthocyanins for cancer prevention are mainly attributed to down-regulating PI3/AKT/mTOR, Wnt, Notch, or NF-κB signalling pathways (Figure 1.6) (Chen et al., 2021). More precisely, anthocyanins had an anti-tumour effect via promoting apoptosis of A549 or H1299 cells through Notch and WNT pathways (Kausar et al., 2012). Additionally, migration and proliferation of A549 cells can be inhibited by down-regulating the expression level of matrix metalloproteinase-2 (MMP-2), matrix metalloproteinase-9 (MMP-9), COX-2, C-myc, and cyclin D1 (Aqil et al., 2016). Besides, anthocyanins can significantly reduce the proliferation and induce apoptosis by inhibiting Akt/mTOR and Sirt1/survivin pathways against breast cancer cells (Lage et al., 2020).

In another clinical trial, daily intake of anthocyanin-enriched fruits attenuated cancer metabolism and reduced breast cancer recurrence (Butalla et al., 2012). On account of other cancers, stomach, prostate, and liver cancer cells can be suppressed by anthocyanins through adjusting the apoptosis-related mediators (e.g., p53, Bcl-2, Bax, cytochrome C, Akt, and Caspase-3). Furthermore, anthocyanins reduced invasion of prostate cancer and liver cancer by targeting AR receptor, and regulating the expression levels of mTOR, Bcl-2, eIF2α, MMP-2, and MMP-9 (Figure 1.5). Black raspberries (*Rubus occidentalis* L.) exerted beneficial effects on the prevention of prostate cancer in men enrolled in a clinical trial (Gu et al., 2014). Taken together, anthocyanins could thus be promising medicines to help inhibit the proliferation and spread of cancer cells.

1.6.3 PREVENTION OF CARDIOVASCULAR DISEASE

Cardiovascular and cerebrovascular diseases are a particularly serious threat for the elderly, as incidence rates generally increase in a rapidly ageing society. Among these, cardiovascular disease is still the leading cause of morbidity and mortality in analyses of all death causes. It is predicted that nearly 28 million people will die of cardiovascular disease every year by 2030 (Wallace et al., 2016). Multiple studies have shown that anthocyanins are beneficial against cardiovascular disease, as they can inhibit the inflammatory process, endothelial dysfunction, and the production of nitric oxide (NO) (Hori & Nishida, 2009; Juránek & Bezek, 2005). Anthocyanins can also maintain the stability of blood pressure by regulating the contraction of blood vessels in the body, thereby reducing the incidence of cardiovascular disease (Cutler et al., 2017). Moreover, anthocyanins can also protect the cardiovascular wall and inhibit platelet aggregation. In addition, the incidence rate of myocardial infarction is significantly reduced, with the probability of myocardial infarction reduced by up to 32% (Cassidy et al., 2013).

Endothelial cells, which cover the luminal surface of all blood vessels, play a critical role in the control of vascular homeostasis. Healthy endothelium regulates vascular tone, and is the basis for relaxing or constricting the vessel. Anthocyanins reduced or even reversed risk factors for cardiovascular disease via maintaining the function of endothelial cells, including alleviating endothelial dysfunction and apoptosis, attenuating oxidative stress in endothelial cells, inhibiting cytokine secretion from endothelial cells, improving insulin resistance and NO release in endothelial cells (Jiang et al., 2019). Specifically, C3G as a principal type of anthocyanin could restore endothelial cell dysfunction by inhibiting the increase of gLDL, GRP78/94, XBP1, and CHOP (Zhao et al., 2015). Anthocyanins promoted the efflux of cholesterol from endothelial cells via regulating the expression of ABCG1 and ABCG5/8 pathways. For example, C3G upregulated ATP-binding ABCG1 and ABCA1 expression in oxidized sterol-stimulated human aortic endothelial cells, and thereafter promoted the efflux of 7-ketocholesterol (7-KC) from the cells (Wang et al., 2012a, b). Meanwhile, C3G up-regulated ABCG1 and 7-KC efflux from HAECs was dependent on activation of LXR-α (Wang et al., 2012a). Anthocyanins had an anti-apoptosis effect on endothelial cells, induced by angiotensin II, ox-LDL, high glucose, cluster of differentiation 40 ligand (CD40L). For example, anthocyanins can inhibit the oxidation of low-density lipoproteins (Muñoz-Espada & Watkins, 2006). Research also confirmed the anti-atherosclerosis effects of grape and bilberry juices that are rich in anthocyanins, which can reduce the level of total cholesterol and triglycerides (Graf et al., 2013). Moreover, in a study, Apo-E-deficient mice fed anthocyanins for 16 weeks exhibited reduced atherosclerosis lesions regardless of whether the anthocyanins were derived from bilberries or blueberries (Mauray et al., 2009). Oxidative stress induces endothelial dysfunction that plays a central role in the pathogenesis of vascular diseases. Therefore, inhibition of vascular reactive oxygen species (ROS)-producing enzymes (NOX1, NOX4) is a critical approach to prevent the development of cardiovascular disease. Notably, regular consumption of foods rich in anthocyanins can increase the antioxidant capacity of the serum. Particularly, C3G reduced the level of ROS and suppressed NOX protein level, including NOX1 and NOX4 in vascular endothelial cells (Xie et al., 2012). C3G also attenuated oxidative stress by activating Nrf2 signalling pathways, elevating endogenous antioxidant enzymes, and reducing the expression of iNOS (Huang et al., 2016; Ma, 2013; Sivasinprasasn et al., 2016). The expression of VCAM-1 and ICAM-1 induced by proinflammatory cytokines could recruit monocytes to the intima of blood vessels, thereby exacerbating cardiovascular disease. Anthocyanins suppressed the cytokines and lessened monocyte adhesion to human umbilical endothelial cells due to the mediation of I kappa B and NF-κB pathway (Del Bo et al., 2016; Pantan et al., 2016; Sivasinprasasn et al., 2016). In addition to the three mechanisms mentioned above, anthocyanins improved insulin resistance and NO release in endothelial cells. Insulin resistance not only happens in obesity, diabetes, and fatty liver disease but is also recognized as an independent risk factor for cardiovascular disease according to epidemiological studies (Mita et al., 2010). Moreover, insulin mediated the generation of vasodilator NO in endothelium, which was essential for vasodilation and neovascularisation (Hamed et al., 2011). NO is a ubiquitous free radical gaseous messenger synthesised from L-arginine in a reaction catalysed by NOS, including neuronal NOS (nNOS), endothelial NOS (eNOS), and inducible NOS (iNOS). Among these, eNOS generates NO in blood vessels and is involved in regulating vascular function. Abnormal vascular NO production and transport result in endothelial dysfunction (Rochette et al., 2013). Surprisingly, C3G effectively reversed the PI3K/Akt axis, restored eNOS expression, and promoted NO release (Fratantonio et al., 2017). In general, anthocyanin possesses favourable capacity of preventing cardiovascular disease.

For example, research spanning over 16 years in healthy menopausal women revealed that the incidence and mortality rates of coronary heart disease were generally reduced for women who regularly ingested foods containing anthocyanins. Researchers treated healthy humans with fresh anthocyanin-rich fruit (Alvarez-Suarez et al., 2014). The participants were subjected to cycles of dietary consumption of strawberries and times when they were advised to avoid strawberries and other polyphenols. After 10 days, the patients received the strawberry for 30 days at 500 g daily, and blood and urinary samples were collected. At the end of the 30 days, the patients were recommended to avoid strawberries for 15 further days. After this period, blood and urinary samples were collected again for analysis. The results demonstrated that strawberry-supplemented patients presented decreased levels of cholesterol, LDL, and triglycerides. These parameters returned to baseline levels after 15 days of strawberry supplementation. The study also showed lower levels of spontaneous and oxidative haemolysis.

In another study on blueberry anthocyanins, researchers evaluated 66 obese patients with metabolic syndrome. The patients were treated for 8 weeks with evaluation after each 4-week period (Basu et al., 2010). This evaluation consisted of anthropometric and blood pressure measurements, assessment of dietary intake, and fasting blood draws. The data showed a more significant reduction in blood pressure, Ox-LDL, and serum malondialdehyde (MDA) and hydroxynonenal; serum glucose concentration and lipid profiles showed no significant alteration. These results suggest that blueberries have a correlation with cardiovascular disease markers and improve aspects of metabolic syndrome. Thus, these authors concluded that a strawberry-rich diet could partially prevent cardiovascular disease.

1.6.4 Bacteriostatic and Anti-Inflammatory Effects

Polyphenols play an important role in plant resistance to bacterial pathogen invasion. Consequently, anthocyanins can also be used as preservatives and bacteriostatic agents to help improve the colour and lustre of food, as natural colorants, and to extend shelf-life (Yoshimoto et al., 2001). Anthocyanins, as natural functional substances, can also reduce inflammatory reactions, which involve a series of complex physiological responses that occur when tissues are damaged, including fever, swelling, and headache. Specifically, anthocyanins function by inhibiting cytokinins and histamines via anti-inflammatory effects, resulting from the inhibition of cyclooxygenase, and bacteriostatic effects, inferred from their minimum inhibitory concentration and minimum bactericidal concentration against bacteria, fungus, and mildew. For example, in a study evaluating the effects of dietary anthocyanins on acute inflammatory responses in rats, researchers found that the anti-inflammatory effects became more evident as the concentration of anthocyanins increased (Winter et al., 2018). Notably, anthocyanins can inhibit the expression of inflammatory factors in rats and reduce the production of inflammatory substances. Moreover, *Lycium ruthenicum* anthocyanins extract and petunidin-3-glucoside, the anthocyanin with the highest content, could both effectively reduce the content of monosodium urate and prevent gouty arthritis in rats. Furthermore, both agents exhibited intestinal anti-inflammatory effects in a mouse model of colitis (Zhang et al., 2019a). Further, researchers found that blueberry anthocyanins significantly inhibited the expression of COX-2 and NF-κB, further confirming that these anthocyanins exert anti-inflammatory effects via an NF-κB mechanism (Xu et al., 2016).

1.6.5 Neuroprotective Effects

Neurodegenerative diseases constitute a major threat to the elderly, and although their incidence is currently increasing (Li et al., 2021), only a few effective methods are available for their prevention or treatment. Dietary intervention constitutes a feasible strategy for the treatment of neurodegenerative diseases. Notably, anthocyanins can relieve the symptoms of neurodegenerative diseases and could thus be useful as dietary supplements. Notably, anthocyanins can protect neurons, glial cells, and hippocampal nerve cells from being damaged by β-amyloid (Aβ), glutamate, and lipopolysaccharide, thereby reducing nerve injury-related disorders, such as cognitive and memory impairment (Li et al., 2021). In the central nervous system, anthocyanins and its major component C3G have been reported to exhibit preventive and/or therapeutic activities in a wide range of disorders, such as cerebral ischaemia, Alzheimer's disease, Parkinson's disease, multiple sclerosis, and glioblastoma. Both anthocyanins and C3G can also affect important processes in ageing, including neuronal apoptosis and death as well as learning and memory impairment (Zhang et al., 2019b).

To be more specific, the mechanisms of anthocyanins regulating neurodegenerative diseases consist of alleviation of oxidative stress, ameliorating inflammation, inhibition of caspase-mediated apoptosis activation, and regulation of key enzymes (Li et al., 2021). Brain that is rich in fatty acids generates oxidative stress, causing neuronal cell death. Anthocyanin possesses the favourable antioxidant capacity of increasing the expression of antioxidant enzymes and suppressing oxidative stress in neuronal cytopathic changes (Figure 1.7) (Sukprasansap et al., 2017). Similarly, accumulation of inflammatory cytokines plays an important role in neurodegenerative disease. The

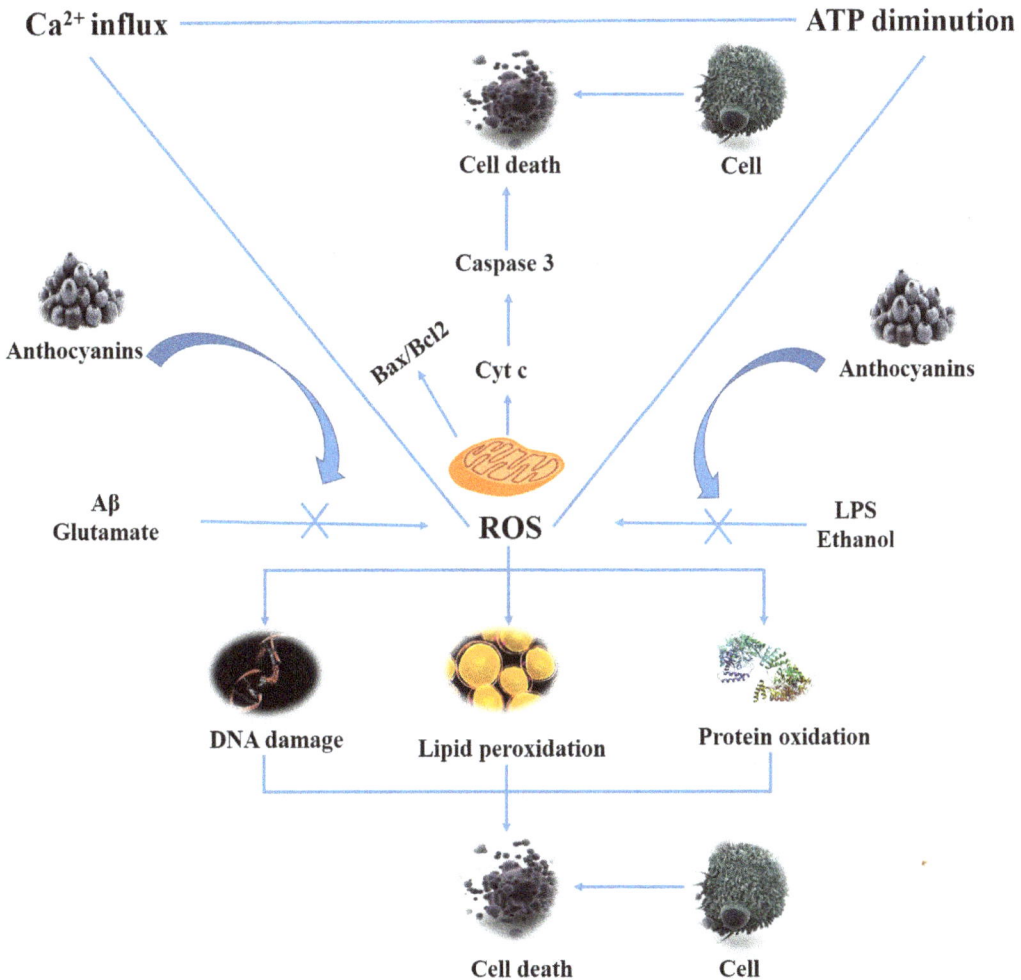

FIGURE 1.7 Schematic representation of the protective effect of anthocyanins on pathways involving reactive oxygen species (ROS), mitochondria, and elevated intracellular calcium in a neuron (Li et al., 2021).

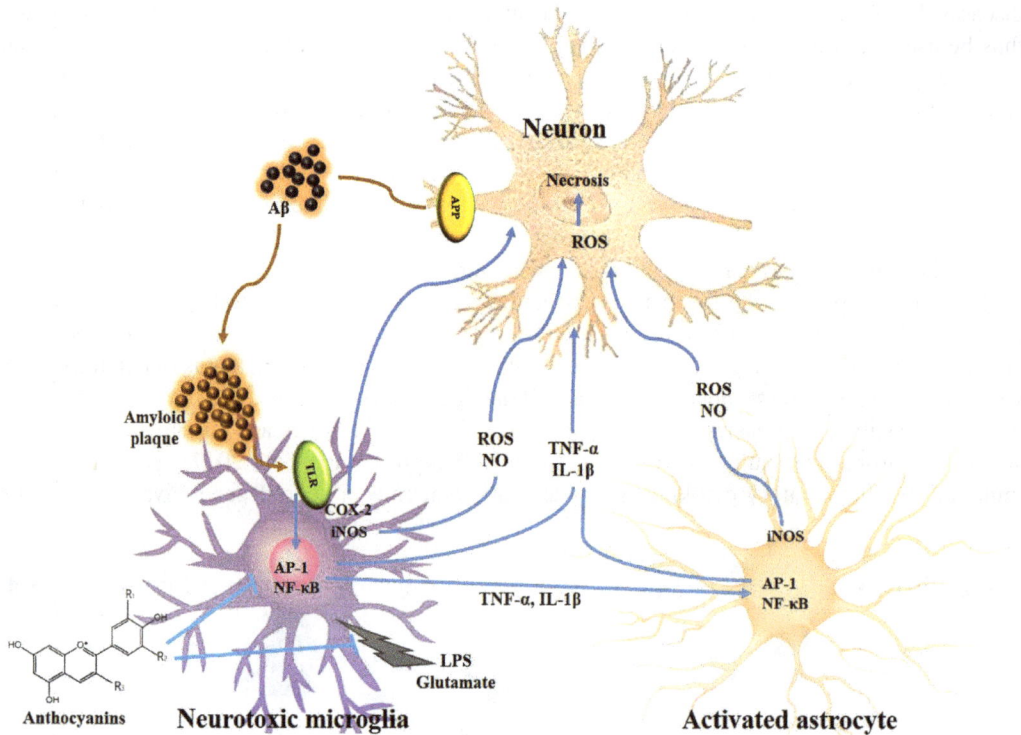

FIGURE 1.8   Anthocyanins ameliorate inflammation (Li et al., 2021).

aggregation of Aβ and the remains derived from dead cells can activate microglia and astrocytes through the Tol-like receptor (TLR)-dependent pathway, leading to local inflammation and amplifying neuronal death (Glass et al., 2010). However, anthocyanin reduced the expression of inflammation cytokines, phagocytoses and cleared excess Aβ and tau oligomers via NF-κB/JNK/ GSK3beta signalling pathway (Figure 1.8) (Kim et al., 2017; Thummayot et al., 2014). Moreover, anthocyanins inhibited apoptosis via the AMPK pathway, presented as the activation of caspase-3, decrease in Bax, cytochrome C, and increase in Bcl-2 (Ullah et al., 2014). Aimed at the key enzymes in the nervous system, anthocyanins modulate expression to maintain normal levels. Among these, acetylcholinesterase is the key enzyme regulating neurotransmitter levels, in order to keep cognitive ability (Moseley et al., 2007). Na^+, K^+-ATPase and Ca^{2+}-ATPase are applied to maintain ion and synaptic transmission, thereby ensuring memory ability. Anthocyanins restored homeostasis of acetylcholinesterase, Na^+, K^+-ATPase and Ca^{2+}-ATPase, thereby improving the abilities of learning and memory. Therefore, anthocyanins could be the potential dietary supplement to prevent against neurodegenerative disease.

For example, researchers showed that feeding hypertoxic paraquat-induced *Drosophila* with black rice anthocyanin extracts could inhibit disease progression and reduce the paraquat lethality rate (Zuo et al., 2012). Researchers found that black bean anthocyanins could reduce the ROS and oxidative stress induced by Aβ protein oligomers, thereby preventing apoptosis and neurodegeneration in a mouse model of Alzheimer's disease (Ali et al., 2018). In addition, anthocyanins could reverse age-related cognitive deficits and neurological problems caused by neurodegenerative diseases (Mattioli et al., 2020; Shishtar et al., 2020). For example, adding blueberry anthocyanins to the diet of mice or rats could increase short-term memory and motivation (Papandreou et al., 2009) and reverse the process of neuronal and behavioural ageing (Ramirez et al., 2005). Accumulating evidence shows that anthocyanins could be potentially applied for clinical therapy for neurodegenerative diseases (Cásedas et al., 2017). In clinical trials, patients with Alzheimer's disease were administered daily with supplemental cherry juice rich in anthocyanins, leading to significant improvements in their memory and cognitive ability after 12 weeks of treatment (Kent et al., 2017). In addition, after elderly patients with mild cognitive impairment received blueberry anthocyanins for 16 weeks, the blood oxygen level in their brain increased, indicating that blueberry anthocyanins could alleviate nervous system disease (Boespflug et al., 2018). Therefore, anthocyanins could be useful in future pharmaceutical applications for the treatment of neuronal diseases.

1.6.6 ANTI-DIABETIC EFFECTS

Diabetes is a non-infectious severe endocrine and metabolic disease that results from insufficient insulin secretion or insulin resistance owing to the destruction of islet cells. Diabetes is primarily characterised by chronic hyperglycaemia (Zhang et al., 2012) and can induce complications in various organs, leading to heart disease, kidney failure, liver dysfunction, nerve injury, stroke, and numerous other diseases (Kam et al., 2016; Manna et al., 2010). Studies have shown that anthocyanins can regulate the glucose and lipid metabolism levels in the body and lower blood sugar; they can also reduce lipotoxicity-induced endothelial dysfunction, thereby minimising diabetes-related complications. For example, researchers found that the blood sugar level of hyperglycaemic mice subjected to gavage with blueberry anthocyanins was reduced by 33–51%, whereas that of mice subjected to gavage with the hypoglycaemic drug metformin hydrochloride was reduced by only 27% (Grace et al., 2009). Researchers further demonstrated that blueberry anthocyanins could inhibit the activity of glucosidase and reduce blood sugar levels by preventing the process of glucose metabolism in the body (Pranprawit et al., 2015). Researchers also found that blueberry anthocyanins could not only reduce the blood sugar level of diabetic mice but also relieve adverse symptoms, such as polydipsia, hyperphagia, and polyuria, with the blood sugar-lowering effect of blueberry anthocyanins being equivalent to that of acarbose (Johnson et al., 2011). Existing studies showed that diabetes was closely associated with obesity. In particular, the increasing size and proliferation of adipocyte cells could secrete inflammatory factor to promote inflammation, leading to changes in insulin resistance and increasing the risk of type 2 diabetes. Anthocyanins markedly reduced the hypertrophy of the adipocytes via affecting PPARγ and PPARα-mediated energy substrate metabolism and inflammation (Seymour et al., 2011). Anthocyanins affected adipokine secretion; adipokines mediated many signalling cascades in target tissues, including insulin sensitivity and other procedures of energy homeostasis (Klöting & Blüher, 2014). For example, anthocyanin supplements reversely shifted the elevation of IL-6 and monocyte chemoattractant protein-1 (MCP-1) in diabetes mice (Sasaki et al., 2007). It is noteworthy that anthocyanins ameliorate the insulin resistance in adipocytes. Insulin resistance as the dominating feature of type 2 diabetes results from inflammatory cytokines, adipokines, and free fatty acid that disrupt insulin signalling pathways (Bastard et al., 2006). For example, anthocyanins could improve the blood sugar and insulin sensitivity of diabetic mice via a mechanism related to the activation of adenylate-activated protein kinase (Kurimoto et al., 2013). Other studies have confirmed that anthocyanins were effective inhibitors of α-amylase and α-glucosidase and that they helped regulate diabetes (Adisakwattana et al., 2004; Sui et al., 2016).

Finally, researchers demonstrated that mulberry anthocyanins relieved the hyperglycaemic symptoms of diabetic mice, reduced the accumulation of triglycerides and cholesterol, and increased adiponectin levels (Yan et al., 2016). In a clinical trial wherein 27 overweight or obese men were enrolled in a randomised, placebo-controlled crossover study and fed a high-fat diet containing either 600 g/day blackberries or a calorie- and carbohydrate-matched amount of gelatin for 7 days (Solverson et al., 2018), blackberries consumption promoted increased fat oxidation and improved insulin sensitivity in the overweight or obese males fed the high-fat diet (Solverson et al., 2018). In individuals with pre-diabetes and insulin resistance, intake of 250 g of frozen red raspberries (~2 cups) in a breakfast meal significantly reduced peak and postprandial (2-h) glucose concentrations when compared with the control group (Xiao et al., 2019). Overall, anthocyanins are potential compounds to protect against diabetes.

1.6.7 EFFECTS AGAINST INTESTINAL DISEASES

Intestinal health is currently a hot topic in both research and public health arenas as intestinal functions play a central role in host health (Staudacher & Loughman, 2021). Intestinal cells consist of a cell barrier, which prevents the gut microbiota and some toxicants migrating into blood circulation. Anthocyanins can protect the intestinal barrier. For example, C3G as a typical anthocyanin has considerable benefits for the protection of intestinal health, including repairing the intestinal mucosal barrier by regulating the levels of colitis-related indicators and signalling pathways, thereby alleviating inflammatory bowel diseases (Ferrari et al., 2017; Gan et al., 2020; Xia et al., 2019). Additionally, the interaction between C3G and the intestinal mucosal immune system represents the core mechanism by which C3G impacts intestinal diseases (Cheng et al., 2021). Gut microbiota, considered as a large metabolic "organ" in our body, can be regulated by the immune system which is crucial to defend against infection by pathogenic microbes. To our delight, anthocyanin target gut microbiota, and promote the growth of beneficial bacteria. Researchers found that black rice anthocyanins and C3G significantly increased the number of bifidobacteria and lactobacilli (Zhu et al., 2018). Other researchers found that black raspberry anthocyanins can promote the growth of *Lactobacillus*, *Faecalibacterium prausnitzii*, and *Eubacterium rectale* and that they can also inhibit the growth of harmful bacteria, such as *Enterococcus* spp. (Chen et al., 2018). Another study showed that blueberry anthocyanins can increase the diversity and abundance of intestinal microorganisms in humans (Zhou et al., 2020). Consistent with this, the results of *in vitro* experiments (Zhang et al., 2016a) evaluating the regulatory effects of purple sweet potato anthocyanins on human intestinal flora suggested that ingesting purple sweet potato rich in anthocyanins was beneficial for intestinal flora and synergistically improved host health. Notably, epidemiological studies suggest that with consumption of foods rich in anthocyanins, the resultant metabolites may contribute to health benefits. Anthocyanins exerted health-promoting effects mainly through upper gastrointestinal absorption and direct metabolism by endogenous enzymes, and colonic absorption of metabolites produced by microbiota after metabolism of anthocyanins (Figure 1.9) (Tian et al., 2019). Given their low bioavailability, only a small proportion of anthocyanins is directly absorbed through the small intestine, and up to 65% is not absorbed after passing through the small intestine, and consequently, most must be further metabolised (Zhang et al., 2016a). Therefore, the specific mechanism of anthocyanins and its metabolites to adjust and even balance the gut microbia is still to be investigated. Nevertheless, consumption of a diet rich in anthocyanin-containing plant foods contributes to the prevention of gastrointestinal disorders and intestinal diseases.

1.7 ANTHOCYANIN PRODUCT DEVELOPMENT

1.7.1 FOOD AND HEALTH PRODUCTS

Fruits and vegetables are generally rich in anthocyanins, proanthocyanidins, vitamins, and other nutrients, suggesting that their consumption will result in health benefits, such as reducing the occurrence of some chronic diseases. During food processing and production, the addition of edible pigments is generally also required to improve sensory properties to stimulate the desire of consumers to purchase products, thus increasing sales. At present, the majority of pigments used in

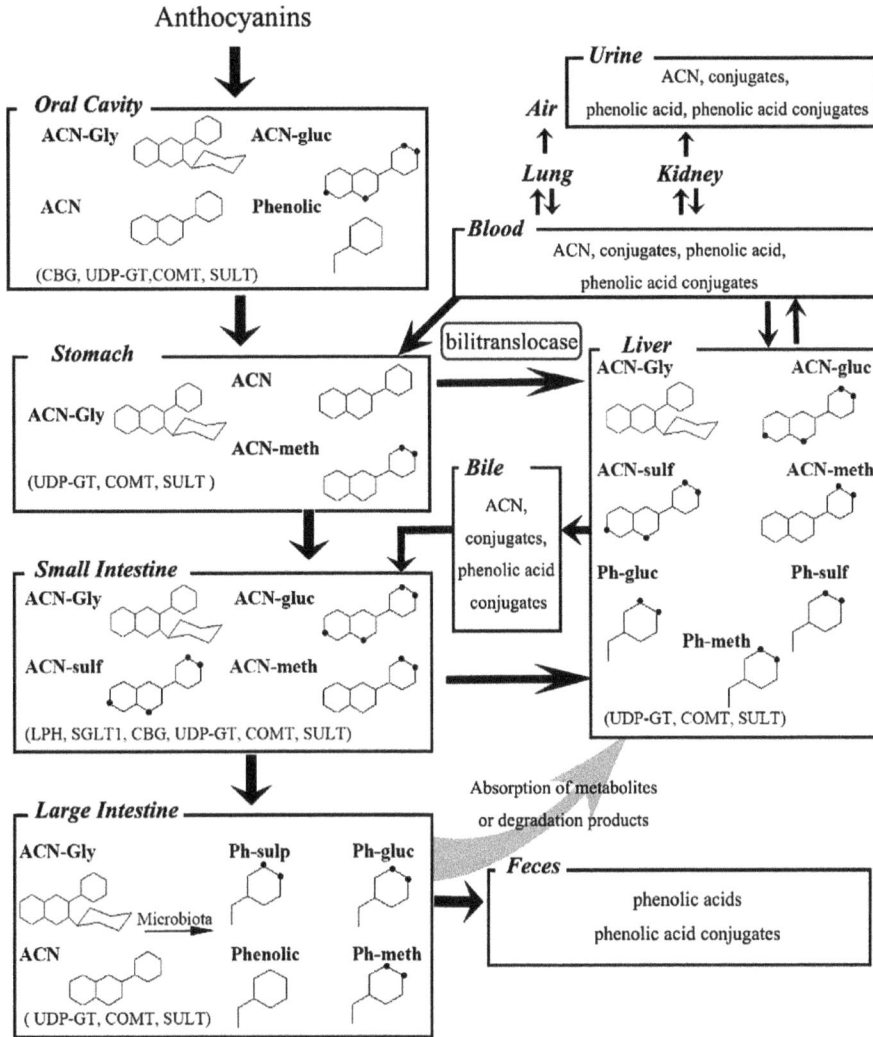

FIGURE 1.9 Summary of absorption, distribution, first-pass metabolism, and excretion (Tian et al., 2019).

the food industry are chemically synthesised, cheap, easy to obtain, and have a certain toxicity (Poorniammal et al., 2021). Compared with synthetic pigments, natural pigments are safer and may even have some medicinal value (Huang et al., 2011b). With the general improvement of living standards and the deepening awareness of the side effects associated with synthetic chemical additives and the knowledge that long-term consumption will harm human health and may even cause cancer, consumers are increasingly pursuing natural, green, and healthy lifestyles. Therefore, there is a growing need for research and development regarding natural pigments. As anthocyanins can be utilised as safe and non-toxic food additives and nutrition enhancers, they are expected to be widely used in the food industry in the future. Anthocyanins can also be used in the research and development of health foods with age-delaying, blood pressure- and blood lipid-regulating, anti-tumour, and brain-strengthening effects. Numerous health foods in the domestic and foreign markets primarily utilise the biological activities of anthocyanins to prevent and treat diseases.

1.7.2 Cosmetics

Consumer requirements for cosmetics are ever increasing, requiring the ability not only to protect and improve the skin but also to contain safe and reliable ingredients. Anthocyanins are known as "oral skin cosmetics" in Europe. They are strong absorbers of both visible and ultraviolet light and can reduce the damage from ultraviolet rays to the skin, which is helpful for antioxidation and improving skin inflammation to a certain extent. Thus, anthocyanins have the potential for wide use in cosmetics as well as in food industries, such as for sunscreen, hair dye, lipstick, and rouge as a replacement for chemically synthesised pigments (Huang et al., 2011b). Moreover, as anthocyanins are natural antioxidants with an antioxidant capacity far exceeding that of vitamin C and they exhibit strong abilities to scavenge free radicals and inhibit lipid peroxidation, they meet the prerequisites for widespread application in cosmetics. Anthocyanins can also help delay skin ageing, whiten skin, provide anti-wrinkle moisturisation, and maintain skin elasticity. Consequently, they have become a focus for research and development in the related cosmetics fields.

1.7.3 Medicine

Anthocyanins exhibit strong antioxidation and free radical-scavenging abilities and have medicinal value for purposes such as regulating lipid metabolism and tumour resistance, treatment of eye diseases, immunoregulation, treatment of diabetes, lowering blood pressure, and inducing anti-inflammatory functions. Moreover, the pharmaceutical industry is also investigating the use of anthocyanins as a replacement for chemically synthesised pigments, such as carmine, lemon yellow, and indigo, which are often used as drug dyes to help distinguish and identify them, as their regular consumption has adverse effects on human health. It is therefore expected that the application of anthocyanins as natural colorants in different products will increase significantly in the coming years. Moreover, demand for the development of new natural bacteriostatic agents is increasing sharply, as the side effects associated with drug-resistant bacterial strains and conventional antibiotic drugs are now seriously endangering human safety. In this context, anthocyanins can also be used as effective drugs to prevent various diseases and can inhibit inflammation and control obesity. Compared with drugs, anthocyanins are effective alternatives; however, further research is needed prior to their application in humans, especially to determine the exact dosages required to achieve the desired effects (Gomes et al., 2019).

1.8 CONCLUSIONS AND FUTURE TRENDS

In summary, the chemical structures of anthocyanins determine their chemical properties. Extensive research has indicated that anthocyanins exhibit strong biological activities and health-promoting effects, which are relevant for a wide range of experimental and practical applications. In-depth research related to the mechanisms and functions of anthocyanins has led to new breakthroughs and has revealed new challenges in the fields of medicine, food, and cosmetics. Anthocyanins are present in various fruits, vegetables, grains, and other foods that are consumed daily. However, additional research and exploration are warranted to understand better their bioactivities owing to their relative instability, low extraction rate, and inability to exert their bioactivity fully in the digestive pathways of the body. If subtropical fruits containing anthocyanins can be fully utilised, it would not only increase the income of growers but also provide overall social and ecological benefits. Moreover, it is anticipated that the advances made during the numerous years of exploration and research to improve our understanding of the characteristics and bioactivities of plant anthocyanins and their considerable application potentials will stimulate researchers and investors to conduct more in-depth studies and exploration to help promote their utilisation and relevant product development.

REFERENCES

Adisakwattana, S., Ngamrojanavanich, N., Kalampakorn, K., Tiravanit, W., Roengsumran, S., & Yibchok-Anun, S. (2004). Inhibitory activity of cyanidin-3-rutinoside on α-glucosidase. *Journal of Enzyme Inhibition and Medicinal Chemistry, 19*(4), 313–316.

Ali, T., Kim, T., Rehman, S. U., Khan, M. S., Amin, F. U., Khan, M., ... & Kim, M. O. (2018). Natural dietary supplementation of anthocyanins via PI3K/Akt/Nrf2/HO-1 pathways mitigate oxidative stress, neurodegeneration, and memory impairment in a mouse model of Alzheimer's disease. *Molecular Neurobiology, 55*(7), 6076–6093.

Alvarez-Suarez, J. M., Giampieri, F., Tulipani, S., Casoli, T., Di Stefano, G., González-Paramás, A. M., ... & Battino, M. (2014). One-month strawberry-rich anthocyanin supplementation ameliorates cardiovascular risk, oxidative stress markers and platelet activation in humans. *The Journal of Nutritional Biochemistry, 25*(3), 289–294.

Aqil, F., Jeyabalan, J., Kausar, H., Munagala, R., Singh, I. P., & Gupta, R. (2016). Lung cancer inhibitory activity of dietary berries and berry polyphenolics. *Journal of Berry Research, 6*(2), 105–114.

Aura, A. M., Martin-Lopez, P., O'Leary, K. A., Williamson, G., Oksman-Caldentey, K. M., Poutanen, K., & Santos-Buelga, C. (2005). *In vitro* metabolism of anthocyanins by human gut microflora. *European Journal of Nutrition, 44*(3), 133–142.

Azzini, E., Vitaglione, P., Intorre, F., Napolitano, A., Durazzo, A., Foddai, M. S., ... & Maiani, G. (2010). Bioavailability of strawberry antioxidants in human subjects. *British Journal of Nutrition, 104*(8), 1165–1173.

Bao, J., Cai, Y., Sun, M., Wang, G., & Corke, H. (2005). Anthocyanins, flavonols, and free radical scavenging activity of Chinese bayberry (*Myrica rubra*) extracts and their color properties and stability. *Journal of Agricultural and Food Chemistry, 53*(6), 2327–2332.

Bastard, J. P., Maachi, M., Lagathu, C., Kim, M. J., Caron, M., Vidal, H., ... & Feve, B. (2006). Recent advances in the relationship between obesity, inflammation, and insulin resistance. *European Cytokine Network, 17*(1), 4–12.

Basu, A., Du, M., Leyva, M. J., Sanchez, K., Betts, N. M., Wu, M., ... & Lyons, T. J. (2010). Blueberries decrease cardiovascular risk factors in obese men and women with metabolic syndrome. *The Journal of Nutrition, 140*(9), 1582–1587.

Bitsch, R., Netzel, M., Frank, T., Strass, G., & Bitsch, I. (2004). Bioavailability and biokinetics of anthocyanins from red grape juice and red wine. *Journal of Biomedicine and Biotechnology, 2004*(5), 293–298.

Boespflug, E. L., Eliassen, J. C., Dudley, J. A., Shidler, M. D., Kalt, W., Summer, S. S., ... & Krikorian, R. (2018). Enhanced neural activation with blueberry supplementation in mild cognitive impairment. *Nutritional Neuroscience, 21*(4), 297–305.

Borges, G., Roowi, S., Rouanet, J. M., Duthie, G. G., Lean, M. E., & Crozier, A. (2007). The bioavailability of raspberry anthocyanins and ellagitannins in rats. *Molecular Nutrition & Food Research, 51*(6), 714–725.

Bray, F., Ferlay, J., Soerjomataram, I., Siegel, R. L., Torre, L. A., & Jemal, A. (2018). Global cancer statistics 2018: GLOBOCAN estimates of incidence and mortality worldwide for 36 cancers in 185 countries. *CA: A Cancer Journal for Clinicians, 68*(6), 394–424.

Burton-Freeman, B., Brzeziński, M., Park, E., Sandhu, A., Xiao, D., & Edirisinghe, I. (2019). A selective role of dietary anthocyanins and flavan-3-ols in reducing the risk of type 2 diabetes mellitus: A review of recent evidence. *Nutrients, 11*(4), 841.

Butalla, A. C., Crane, T. E., Patil, B., Wertheim, B. C., Thompson, P., & Thomson, C. A. (2012). Effects of a carrot juice intervention on plasma carotenoids, oxidative stress, and inflammation in overweight breast cancer survivors. *Nutrition and Cancer, 64*(2), 331–341.

Butelli, E., Garcia-Lor, A., Licciardello, C., Las Casas, G., Hill, L., Recupero, G. R., ... & Martin, C. (2017). Changes in anthocyanin production during domestication of *Citrus*. *Plant Physiology, 173*(4), 2225–2242.

Cásedas, G., Les, F., Gómez-Serranillos, M. P., Smith, C., & López, V. (2017). Anthocyanin profile, antioxidant activity and enzyme inhibiting properties of blueberry and cranberry juices: A comparative study. *Food & Function, 8*(11), 4187–4193.

Cassidy, A., O'Reilly, É. J., Kay, C., Sampson, L., Franz, M., Forman, J. P., ... & Rimm, E. B. (2011). Habitual intake of flavonoid subclasses and incident hypertension in adults. *The American Journal of Clinical Nutrition, 93*(2), 338–347.

Cassidy, A., Mukamal, K. J., Liu, L., Franz, M., Eliassen, A. H., & Rimm, E. B. (2013). High anthocyanin intake is associated with a reduced risk of myocardial infarction in young and middle-aged women. *Circulation*, *127*(2), 188–196.

Castañeda-Ovando, A., de Lourdes Pacheco-Hernández, M., Páez-Hernández, M. E., Rodríguez, J. A., & Galán-Vidal, C. A. (2009). Chemical studies of anthocyanins: A review. *Food Chemistry*, *113*(4), 859–871.

Chang, J. J., Hsu, M. J., Huang, H. P., Chung, D. J., Chang, Y. C., & Wang, C. J. (2013). Mulberry anthocyanins inhibit oleic acid induced lipid accumulation by reduction of lipogenesis and promotion of hepatic lipid clearance. *Journal of Agricultural and Food Chemistry*, *61*(25), 6069–6076.

Chen, K., Xu, C., Zhang, B., & Ferguson, I. B. (2004). Red bayberry: Botany and horticulture. *Horticultural Reviews*, *30*(83), 114.

Chen, T., Yan, F., Qian, J., Guo, M., Zhang, H., Tang, X., ... & Wang, X. (2012). Randomized phase II trial of lyophilized strawberries in patients with dysplastic precancerous lesions of the esophagus. *Cancer Prevention Research*, *5*(1), 41–50.

Chen, L., Jiang, B., Zhong, C., Guo, J., Zhang, L., Mu, T., ... & Bi, X. (2018). Chemoprevention of colorectal cancer by black raspberry anthocyanins involved the modulation of gut microbiota and SFRP2 demethylation. *Carcinogenesis*, *39*(3), 471–481.

Chen, J., Xu, B., Sun, J., Jiang, X., & Bai, W. (2022). Anthocyanin supplement as a dietary strategy in cancer prevention and management: A comprehensive review. *Critical Reviews in Food Science and Nutrition*, *62*(26), 7242–7254.

Cheng, H., Chen, J., Chen, S., Wu, D., Liu, D., & Ye, X. (2015). Characterization of aroma-active volatiles in three Chinese bayberry (*Myrica rubra*) cultivars using GC–MS–olfactometry and an electronic nose combined with principal component analysis. *Food Research International*, *72*, 8–15.

Cheng, Z., Si, X., Tan, H., Zang, Z., Tian, J., Shu, C., ... & Wang, Y. (2021). Cyanidin-3-O-glucoside and its phenolic metabolites ameliorate intestinal diseases via modulating intestinal mucosal immune system: Potential mechanisms and therapeutic strategies. *Critical Reviews in Food Science and Nutrition*, 1–19.

Cutler, B. R., Petersen, C., & Anandh Babu, P. V. (2017). Mechanistic insights into the vascular effects of blueberries: Evidence from recent studies. *Molecular Nutrition & Food Research*, *61*(6), 1600271.

Czank, C., Cassidy, A., Zhang, Q., Morrison, D. J., Preston, T., Kroon, P. A., ... & Kay, C. D. (2013). Human metabolism and elimination of the anthocyanin, cyanidin-3-glucoside: A ^{13}C-tracer study. *The American Journal of Clinical Nutrition*, *97*(5), 995–1003.

Dastmalchi, K., Dorman, H. D., Koşar, M., & Hiltunen, R. (2007). Chemical composition and *in vitro* antioxidant evaluation of a water-soluble Moldavian balm (*Dracocephalum moldavica* L.) extract. *LWT-Food Science and Technology*, *40*(2), 239–248.

Davinelli, S., Bertoglio, J. C., Zarrelli, A., Pina, R., & Scapagnini, G. (2015). A randomized clinical trial evaluating the efficacy of an anthocyanin–maqui berry extract (Delphinol) on oxidative stress biomarkers. *Journal of the American College of Nutrition*, *34*(1), 28–33.

Del Bo', C., Roursgaard, M., Porrini, M., Loft, S., Møller, P., & Riso, P. (2016). Different effects of anthocyanins and phenolic acids from wild blueberry (*Vaccinium angustifolium*) on monocytes adhesion to endothelial cells in a TNF-α stimulated proinflammatory environment. *Molecular Nutrition & Food Research*, *60*(11), 2355–2366.

Duan, Y., Chen, F., Yao, X., Zhu, J., Wang, C., Zhang, J., & Li, X. (2015). Protective effect of *Lycium ruthenicum* Murr. against radiation injury in mice. *International Journal of Environmental Research and Public Health*, *12*(7), 8332–8347.

Eker, M. E., Aaby, K., Budic-Leto, I., Rimac Brnčić, S., El, S. N., Karakaya, S., ... & de Pascual-Teresa, S. (2019). A review of factors affecting anthocyanin bioavailability: Possible implications for the inter-individual variability. *Foods*, *9*(1), 2.

Fabroni, S., Ballistreri, G., Amenta, M., & Rapisarda, P. (2016). Anthocyanins in different *Citrus* species: an UHPLC-PDA-ESI/MS n-assisted qualitative and quantitative investigation. *Journal of the Science of Food and Agriculture*, *96*(14), 4797–4808.

Fang, Z., Zhang, M., & Wang, L. (2007). HPLC-DAD-ESIMS analysis of phenolic compounds in bayberries (*Myrica rubra* Sieb. et Zucc.). *Food Chemistry*, *100*(2), 845–852.

Faria, J. V., Valido, I. H., Paz, W. H., da Silva, F. M., De Souza, A. D., Acho, L. R., ... & Bataglion, G. A. (2021). Comparative evaluation of chemical composition and biological activities of tropical fruits consumed in Manaus, central Amazonia, Brazil. *Food Research International*, *139*, 109836.

Felgines, C., Talavéra, S., Gonthier, M. P., Texier, O., Scalbert, A., Lamaison, J. L., & Rémésy, C. (2003). Strawberry anthocyanins are recovered in urine as glucuro-and sulfoconjugates in humans. *The Journal of Nutrition, 133*(5), 1296–1301.

Fernandes, I., Faria, A., Calhau, C., de Freitas, V., & Mateus, N. (2014). Bioavailability of anthocyanins and derivatives. *Journal of Functional Foods, 7*, 54–66.

Ferrari, D., Cimino, F., Fratantonio, D., Molonia, M. S., Bashllari, R., Busà, R., ... & Speciale, A. (2017). Cyanidin-3-O-glucoside modulates the *in vitro* inflammatory crosstalk between intestinal epithelial and endothelial cells. *Mediators of Inflammation, 2017*, 3454023.

Frank, T., Netzel, M., Strass, G., Bitsch, R., & Bitsch, I. (2003). Bioavailability of anthocyanidin-3-glucosides following consumption of red wine and red grape juice. *Canadian Journal of Physiology and Pharmacology, 81*(5), 423–435.

Fratantonio, D., Cimino, F., Molonia, M. S., Ferrari, D., Saija, A., Virgili, F., & Speciale, A. (2017). Cyanidin-3-O-glucoside ameliorates palmitate-induced insulin resistance by modulating IRS-1 phosphorylation and release of endothelial derived vasoactive factors. *Biochimica et Biophysica Acta (BBA)-Molecular and Cell Biology of Lipids, 1862*(3), 351–357.

Gan, Y., Fu, Y., Yang, L., Chen, J., Lei, H., & Liu, Q. (2020). Cyanidin-3-O-glucoside and cyanidin protect against intestinal barrier damage and 2, 4, 6-trinitrobenzenesulfonic acid-induced colitis. *Journal of Medicinal Food, 23*(1), 90–99.

Glass, C. K., Saijo, K., Winner, B., Marchetto, M. C., & Gage, F. H. (2010). Mechanisms underlying inflammation in neurodegeneration. *Cell, 140*(6), 918–934.

Gomes, J. V. P., Rigolon, T. C. B., da Silveira Souza, M. S., Alvarez-Leite, J. I., Della Lucia, C. M., Martino, H. S. D., & Rosa, C. D. O. B. (2019). Antiobesity effects of anthocyanins on mitochondrial biogenesis, inflammation, and oxidative stress: A systematic review. *Nutrition, 66*, 192–202.

Gowd, V., Jia, Z., & Chen, W. (2017). Anthocyanins as promising molecules and dietary bioactive components against diabetes – A review of recent advances. *Trends in Food Science & Technology, 68*, 1–13.

Grace, M. H., Ribnicky, D. M., Kuhn, P., Poulev, A., Logendra, S., Yousef, G. G., ... & Lila, M. A. (2009). Hypoglycemic activity of a novel anthocyanin-rich formulation from lowbush blueberry, *Vaccinium angustifolium* Aiton. *Phytomedicine, 16*(5), 406–415.

Graf, D., Seifert, S., Jaudszus, A., Bub, A., & Watzl, B. (2013). Anthocyanin-rich juice lowers serum cholesterol, leptin, and resistin and improves plasma fatty acid composition in Fischer rats. *PloS ONE, 8*(6), e66690.

Gu, J., Ahn-Jarvis, J. H., Riedl, K. M., Schwartz, S. J., Clinton, S. K., & Vodovotz, Y. (2014). Characterization of black raspberry functional food products for cancer prevention human clinical trials. *Journal of Agricultural and Food Chemistry, 62*(18), 3997–4006.

Hamed, S., Brenner, B., & Roguin, A. (2011). Nitric oxide: A key factor behind the dysfunctionality of endothelial progenitor cells in diabetes mellitus type-2. *Cardiovascular Research, 91*(1), 9–15.

Han, Y., Vimolmangkang, S., Soria-Guerra, R. E., & Korban, S. S. (2012). Introduction of apple ANR genes into tobacco inhibits expression of both CHI and DFR genes in flowers, leading to loss of anthocyanin. *Journal of Experimental Botany, 63*(7), 2437–2447.

Hichri, I., Barrieu, F., Bogs, J., Kappel, C., Delrot, S., & Lauvergeat, V. (2011). Recent advances in the transcriptional regulation of the flavonoid biosynthetic pathway. *Journal of Experimental Botany, 62*(8), 2465–2483.

Honda, C., Kotoda, N., Wada, M., Kondo, S., Kobayashi, S., Soejima, J., ... & Moriguchi, T. (2002). Anthocyanin biosynthetic genes are coordinately expressed during red coloration in apple skin. *Plant Physiology and Biochemistry, 40*(11), 955–962.

Hori, M., & Nishida, K. (2009). Oxidative stress and left ventricular remodelling after myocardial infarction. *Cardiovascular Research, 81*(3), 457–464.

Huang, H. P., Chang, Y. C., Wu, C. H., Hung, C. N., & Wang, C. J. (2011a). Anthocyanin-rich mulberry extract inhibits the gastric cancer cell growth *in vitro* and xenograft mice by inducing signals of p38/p53 and c-jun. *Food Chemistry, 129*(4), 1703–1709.

Huang, X. D., Liang, J. B., Tan, H. Y., Yahya, R., Long, R., & Ho, Y. W. (2011b). Protein-binding affinity of *Leucaena* condensed tannins of differing molecular weights. *Journal of Agricultural and Food Chemistry, 59*(19), 10677–10682.

Huang, W., Zhu, Y., Li, C., Sui, Z., & Min, W. (2016). Effect of blueberry anthocyanins malvidin and glycosides on the antioxidant properties in endothelial cells. *Oxidative Medicine and Cellular Longevity, 2016.*

Jia, Z. S., Tang, M. C., & Wu, J. M. (1999). The determination of flavonoid contents in mulberry and their scavenging effects on superoxide radicals. *Food Chemistry, 64*(4), 555–559.

Jiang, Y., & Nie, W. J. (2015). Chemical properties in fruits of mulberry species from the Xinjiang province of China. *Food Chemistry, 174*, 460–466.

Jiang, X., Li, X., Zhu, C., Sun, J., Tian, L., Chen, W., & Bai, W. (2019). The target cells of anthocyanins in metabolic syndrome. *Critical Reviews in Food Science and Nutrition, 59*(6), 921–946.

Johnson, M. H., Lucius, A., Meyer, T., & Gonzalez de Mejia, E. (2011). Cultivar evaluation and effect of fermentation on antioxidant capacity and *in vitro* inhibition of α-amylase and α-glucosidase by highbush blueberry (*Vaccinium corombosum*). *Journal of Agricultural and Food Chemistry, 59*(16), 8923–8930.

Juránek, I., & Bezek, S. (2005). Controversy of free radical hypothesis: Reactive oxygen species – cause or consequence of tissue injury?. *General Physiology and Biophysics, 24*(3), 263.

Kabi, F., & Bareeba, F. B. (2008). Herbage biomass production and nutritive value of mulberry (*Morus alba*) and *Calliandra calothyrsus* harvested at different cutting frequencies. *Animal Feed Science and Technology, 140*(1-2), 178–190.

Kalemba-Drożdż, M., Cierniak, A., & Cichoń, I. (2020). Berry fruit juices protect lymphocytes against DNA damage and ROS formation induced with heterocyclic aromatic amine PhIP. *Journal of Berry Research, 10*(1), 95–113.

Kam, J., Puranik, S., Yadav, R., Manwaring, H. R., Pierre, S., Srivastava, R. K., & Yadav, R. S. (2016). Dietary interventions for type 2 diabetes: How millet comes to help. *Frontiers in Plant Science, 7*, 1454.

Kausar, H., Jeyabalan, J., Aqil, F., Chabba, D., Sidana, J., Singh, I. P., & Gupta, R. C. (2012). Berry anthocyanidins synergistically suppress growth and invasive potential of human non-small-cell lung cancer cells. *Cancer Letters, 325*(1), 54–62.

Kay, C. D., Mazza, G., Holub, B. J., & Wang, J. (2004). Anthocyanin metabolites in human urine and serum. *British Journal of Nutrition, 91*(6), 933–942.

Kay, C. D., Mazza, G., & Holub, B. J. (2005). Anthocyanins exist in the circulation primarily as metabolites in adult men. *The Journal of Nutrition, 135*(11), 2582–2588.

Kent, K., Charlton, K., Roodenrys, S., Batterham, M., Potter, J., Traynor, V., ... & Richards, R. (2017). Consumption of anthocyanin-rich cherry juice for 12 weeks improves memory and cognition in older adults with mild-to-moderate dementia. *European Journal of Nutrition, 56*(1), 333–341.

Khoo, H. E., Azlan, A., Tang, S. T., & Lim, S. M. (2017). Anthocyanidins and anthocyanins: Colored pigments as food, pharmaceutical ingredients, and the potential health benefits. *Food & Nutrition Research, 61*(1), 1361779.

Kim, M. J., Rehman, S. U., Amin, F. U., & Kim, M. O. (2017). Enhanced neuroprotection of anthocyanin-loaded PEG-gold nanoparticles against Aβ1-42-induced neuroinflammation and neurodegeneration via the NF-KB/JNK/GSK3β signaling pathway. *Nanomedicine: Nanotechnology, Biology and Medicine, 13*(8), 2533–2544.

Klöting, N., & Blüher, M. (2014). Adipocyte dysfunction, inflammation and metabolic syndrome. *Reviews in Endocrine and Metabolic Disorders, 15*(4), 277–287.

Kobayashi, S., Ishimaru, M., Ding, C. K., Yakushiji, H., & Goto, N. (2001). Comparison of UDP-glucose flavonoid 3-O-glucosyltransferase (UFGT) gene sequences between white grapes (*Vitis vinifera*) and their sports with red skin. *Plant Science, 160*(3), 543–550.

Krishna, P. G. A., Sivakumar, T. R., Jin, C., Li, S. H., Weng, Y. J., Yin, J., ... & Gui, Z. Z. (2018). Antioxidant and hemolysis protective effects of polyphenol-rich extract from mulberry fruits. *Pharmacognosy Magazine, 14*(53), 103.

Kumar, A., & Ellis, B. E. (2001). The phenylalanine ammonia-lyase gene family in raspberry. Structure, expression, and evolution. *Plant Physiology, 127*(1), 230–239.

Kurimoto, Y., Shibayama, Y., Inoue, S., Soga, M., Takikawa, M., Ito, C., ... & Tsuda, T. (2013). Black soybean seed coat extract ameliorates hyperglycemia and insulin sensitivity via the activation of AMP-activated protein kinase in diabetic mice. *Journal of Agricultural and Food Chemistry, 61*(23), 5558–5564.

Lage, N. N., Layosa, M. A. A., Arbizu, S., Chew, B. P., Pedrosa, M. L., Mertens-Talcott, S., ... & Noratto, G. D. (2020). Dark sweet cherry (*Prunus avium*) phenolics enriched in anthocyanins exhibit enhanced activity against the most aggressive breast cancer subtypes without toxicity to normal breast cells. *Journal of Functional Foods*, *64*, 103710.

Lee, S., Keirsey, K. I., Kirkland, R., Grunewald, Z. I., Fischer, J. G., & de La Serre, C. B. (2018). Blueberry supplementation influences the gut microbiota, inflammation, and insulin resistance in high-fat-diet-fed rats. *The Journal of Nutrition*, *148*(2), 209–219.

Lee, D. Y., Yun, S. M., Song, M. Y., Jung, K., & Kim, E. H. (2020). Cyanidin chloride induces apoptosis by inhibiting NF-κB signaling through activation of Nrf2 in colorectal cancer cells. *Antioxidants*, *9*(4), 285.

Li, P., Feng, D., Yang, D., Li, X., Sun, J., Wang, G., ... & Bai, W. (2021). Protective effects of anthocyanins on neurodegenerative diseases. *Trends in Food Science & Technology*, *117*, 205–217.

Liang, L., Wu, X., Zhu, M., Zhao, W., Li, F., Zou, Y., & Yang, L. (2012). Chemical composition, nutritional value, and antioxidant activities of eight mulberry cultivars from China. *Pharmacognosy Magazine*, *8*(31), 215.

Lin, J. Y., & Tang, C. Y. (2007). Determination of total phenolic and flavonoid contents in selected fruits and vegetables, as well as their stimulatory effects on mouse splenocyte proliferation. *Food Chemistry*, *101*(1), 140–147.

Lo Piero, A. R., Puglisi, I., Rapisarda, P., & Petrone, G. (2005). Anthocyanins accumulation and related gene expression in red orange fruit induced by low temperature storage. *Journal of Agricultural and Food Chemistry*, *53*(23), 9083–9088.

Ma, Q. (2013). Role of nrf2 in oxidative stress and toxicity. *Annual Review of Pharmacology and Toxicology*, *53*, 401–426.

Manna, P., Das, J., Ghosh, J., & Sil, P. C. (2010). Contribution of type 1 diabetes to rat liver dysfunction and cellular damage via activation of NOS, PARP, IκBα/NF-κBκBα/NF-κB, MAPKs, and mitochondria-dependent pathways: Prophylactic role of arjunolic acid. *Free Radical Biology and Medicine*, *48*(11), 1465–1484.

Manolescu, B. N., Oprea, E., Mititelu, M., Ruta, L. L., & Farcasanu, I. C. (2019). Dietary anthocyanins and stroke: A review of pharmacokinetic and pharmacodynamic studies. *Nutrients*, *11*(7), 1479.

Marko, D., Puppel, N., Tjaden, Z., Jakobs, S., & Pahlke, G. (2004). The substitution pattern of anthocyanidins affects different cellular signaling cascades regulating cell proliferation. *Molecular Nutrition & Food Research*, *48*(4), 318–325.

Mattioli, R., Francioso, A., Mosca, L., & Silva, P. (2020). Anthocyanins: A comprehensive review of their chemical properties and health effects on cardiovascular and neurodegenerative diseases. *Molecules*, *25*(17), 3809.

Mauray, A., Milenkovic, D., Besson, C., Caccia, N., Morand, C., Michel, F., ... & Felgines, C. (2009). Atheroprotective effects of bilberry extracts in apo E-deficient mice. *Journal of Agricultural and Food Chemistry*, *57*(23), 11106–11111.

McGhie, T. K., & Walton, M. C. (2007). The bioavailability and absorption of anthocyanins: Towards a better understanding. *Molecular Nutrition & Food Research*, *51*(6), 702–713.

Mita, T., Goto, H., Azuma, K., Jin, W. L., Nomiyama, T., Fujitani, Y., ... & Watada, H. (2010). Impact of insulin resistance on enhanced monocyte adhesion to endothelial cells and atherosclerogenesis independent of LDL cholesterol level. *Biochemical and Biophysical Research Communications*, *395*(4), 477–483.

Miyazawa, T., Nakagawa, K., Kudo, M., Muraishi, K., & Someya, K. (1999). Direct intestinal absorption of red fruit anthocyanins, cyanidin-3-glucoside and cyanidin-3, 5-diglucoside, into rats and humans. *Journal of Agricultural and Food Chemistry*, *47*(3), 1083–1091.

Moseley, A. E., Williams, M. T., Schaefer, T. L., Bohanan, C. S., Neumann, J. C., Behbehani, M. M., ... & Lingrel, J. B. (2007). Deficiency in Na, K-ATPase α isoform genes alters spatial learning, motor activity, and anxiety in mice. *Journal of Neuroscience*, *27*(3), 616–626.

Muir, S. R., Collins, G. J., Robinson, S., Hughes, S., Bovy, A., Ric De Vos, C. H. V., ... & Verhoeyen, M. E. (2001). Overexpression of petunia chalcone isomerase in tomato results in fruit containing increased levels of flavonols. *Nature Biotechnology*, *19*(5), 470–474.

Muñoz-Espada, A. C., & Watkins, B. A. (2006). Cyanidin attenuates PGE_2 production and cyclooxygenase-2 expression in LNCaP human prostate cancer cells. *The Journal of Nutritional Biochemistry*, *17*(9), 589–596.

Nielsen, I. L. F., Dragsted, L. O., Ravn-Haren, G., Freese, R., & Rasmussen, S. E. (2003). Absorption and excretion of black currant anthocyanins in humans and Watanabe heritable hyperlipidemic rabbits. *Journal of Agricultural and Food Chemistry*, *51*(9), 2813–2820.

Özgen, M., Serçe, S., & Kaya, C. (2009). Phytochemical and antioxidant properties of anthocyanin-rich *Morus nigra* and *Morus rubra* fruits. *Scientia Horticulturae*, *119*(3), 275–279.

Pantan, R., Tocharus, J., Suksamrarn, A., & Tocharus, C. (2016). Synergistic effect of atorvastatin and cyanidin-3-glucoside on angiotensin II-induced inflammation in vascular smooth muscle cells. *Experimental Cell Research*, *342*(2), 104–112.

Papandreou, M. A., Dimakopoulou, A., Linardaki, Z. I., Cordopatis, P., Klimis-Zacas, D., Margarity, M., & Lamari, F. N. (2009). Effect of a polyphenol-rich wild blueberry extract on cognitive performance of mice, brain antioxidant markers and acetylcholinesterase activity. *Behavioural Brain Research*, *198*(2), 352–358.

Passamonti, S., Vrhovsek, U., Vanzo, A., & Mattivi, F. (2003). The stomach as a site for anthocyanins absorption from food. *FEBS Letters*, *544*(1–3), 210–213.

Pereira-Netto, A. B. (2018). Tropical fruits as natural, exceptionally rich, sources of bioactive compounds. *International Journal of Fruit Science*, *18*(3), 231–242.

Poorniammal, R., Prabhu, S., Dufossé, L., & Kannan, J. (2021). Safety evaluation of fungal pigments for food applications. *Journal of Fungi*, *7*(9), 692.

Pranprawit, A., Heyes, J. A., Molan, A. L., & Kruger, M. C. (2015). Antioxidant activity and inhibitory potential of blueberry extracts against key enzymes relevant for hyperglycemia. *Journal of Food Biochemistry*, *39*(1), 109–118.

Raman, S. T., Ganeshan, A. K. P. G., Chen, C., Jin, C., Li, S. H., Chen, H. J., & Gui, Z. (2016). *In vitro* and *in vivo* antioxidant activity of flavonoid extracted from mulberry fruit (*Morus alba* L.). *Pharmacognosy Magazine*, *12*(46), 128.

Ramirez, M. R., Izquierdo, I., Raseira, M. D. C. B., Zuanazzi, J. Â., Barros, D., & Henriques, A. T. (2005). Effect of lyophilised *Vaccinium* berries on memory, anxiety and locomotion in adult rats. *Pharmacological Research*, *52*(6), 457–462.

Rashwan, A. K., Karim, N., Xu, Y., Xie, J., Cui, H., Mozafari, M. R., & Chen, W. (2021). Potential micro-/nano-encapsulation systems for improving stability and bioavailability of anthocyanins: An updated review. *Critical Reviews in Food Science and Nutrition*, 1–24.

Rauf, A., Imran, M., Abu-Izneid, T., Patel, S., Pan, X., Naz, S., ... & Suleria, H. A. R. (2019). Proanthocyanidins: A comprehensive review. *Biomedicine & Pharmacotherapy*, *116*, 108999.

Reis, J. F., Monteiro, V. V. S., de Souza Gomes, R., do Carmo, M. M., da Costa, G. V., Ribera, P. C., & Monteiro, M. C. (2016). Action mechanism and cardiovascular effect of anthocyanins: A systematic review of animal and human studies. *Journal of Translational Medicine*, *14*(1), 1–16.

Rinaldo, D. (2020). Carbohydrate and bioactive compounds composition of starchy tropical fruits and tubers, in relation to pre and postharvest conditions: A review. *Journal of Food Science*, *85*(2), 249–259.

Rochette, L., Lorin, J., Zeller, M., Guilland, J. C., Lorgis, L., Cottin, Y., & Vergely, C. (2013). Nitric oxide synthase inhibition and oxidative stress in cardiovascular diseases: possible therapeutic targets? *Pharmacology & Therapeutics*, *140*(3), 239–257.

Roth, S., Spalinger, M. R., Gottier, C., Biedermann, L., Zeitz, J., Lang, S., ... & Scharl, M. (2016). Bilberry-derived anthocyanins modulate cytokine expression in the intestine of patients with ulcerative colitis. *PLoS ONE*, *11*(5), e0154817.

Salah, M., Mansour, M., Zogona, D., & Xu, X. (2020). Nanoencapsulation of anthocyanins-loaded β-lactoglobulin nanoparticles: Characterization, stability, and bioavailability *in vitro*. *Food Research International*, *137*, 109635.

Sasaki, R., Nishimura, N., Hoshino, H., Isa, Y., Kadowaki, M., Ichi, T., ... & Tsuda, T. (2007). Cyanidin 3-glucoside ameliorates hyperglycemia and insulin sensitivity due to downregulation of retinol binding protein 4 expression in diabetic mice. *Biochemical Pharmacology*, *74*(11), 1619–1627.

Seeram, N. P., Zhang, Y., & Nair, M. G. (2003). Inhibition of proliferation of human cancer cells and cyclooxygenase enzymes by anthocyanidins and catechins. *Nutrition and Cancer*, *46*(1), 101–106.

Sengul, H., Surek, E., & Nilufer-Erdil, D. (2014). Investigating the effects of food matrix and food components on bioaccessibility of pomegranate (*Punica granatum*) phenolics and anthocyanins using an *in-vitro* gastrointestinal digestion model. *Food Research International*, *62*, 1069–1079.

Septembre-Malaterre, A., Stanislas, G., Douraguia, E., & Gonthier, M. P. (2016). Evaluation of nutritional and antioxidant properties of the tropical fruits banana, litchi, mango, papaya, passion fruit and pineapple cultivated in Réunion French Island. *Food Chemistry*, *212*, 225–233.

Seymour, E. M., Tanone, I. I., Urcuyo-Llanes, D. E., Lewis, S. K., Kirakosyan, A., Kondoleon, M. G., ... & Bolling, S. F. (2011). Blueberry intake alters skeletal muscle and adipose tissue peroxisome proliferator-activated receptor activity and reduces insulin resistance in obese rats. *Journal of Medicinal Food*, *14*(12), 1511–1518.

Shi, N., Riedl, K. M., Schwartz, S. J., Zhang, X., Clinton, S. K., & Chen, T. (2016). Efficacy comparison of lyophilised black raspberries and combination of celecoxib and PBIT in prevention of carcinogen-induced oesophageal cancer in rats. *Journal of Functional Foods*, *27*, 84–94.

Shi, J., Simal-Gandara, J., Mei, J., Ma, W., Peng, Q., Shi, Y., ... & Lv, H. (2021). Insight into the pigmented anthocyanins and the major potential co-pigmented flavonoids in purple-coloured leaf teas. *Food Chemistry*, *363*, 130278.

Shishtar, E., Rogers, G. T., Blumberg, J. B., Au, R., & Jacques, P. F. (2020). Long-term dietary flavonoid intake and change in cognitive function in the Framingham Offspring cohort. *Public Health Nutrition*, *23*(9), 1576–1588.

Sivasinprasasn, S., Pantan, R., Thummayot, S., Tocharus, J., Suksamrarn, A., & Tocharus, C. (2016). Cyanidin-3-glucoside attenuates angiotensin II-induced oxidative stress and inflammation in vascular endothelial cells. *Chemico-Biological Interactions*, *260*, 67–74.

Solverson, P. M., Rumpler, W. V., Leger, J. L., Redan, B. W., Ferruzzi, M. G., Baer, D. J., ... & Novotny, J. A. (2018). Blackberry feeding increases fat oxidation and improves insulin sensitivity in overweight and obese males. *Nutrients*, *10*(8), 1048.

Staudacher, H. M., & Loughman, A. (2021). Gut health: Definitions and determinants. *The Lancet Gastroenterology & Hepatology*, *6*(4), 269.

Suganuma, T., Pattenden, S. G., & Workman, J. L. (2008). Diverse functions of WD40 repeat proteins in histone recognition. *Genes & Development*, *22*(10), 1265–1268.

Suh, H. J., Noh, D. O., Kang, C. S., Kim, J. M., & Lee, S. W. (2003). Thermal kinetics of color degradation of mulberry fruit extract. *Nahrung Food*, *47*(2), 132–135.

Sui, X., Zhang, Y., & Zhou, W. (2016). *In vitro* and in silico studies of the inhibition activity of anthocyanins against porcine pancreatic α-amylase. *Journal of Functional Foods*, *21*, 50–57.

Sukprasansap, M., Chanvorachote, P., & Tencomnao, T. (2017). *Cleistocalyx nervosum* var. *paniala* berry fruit protects neurotoxicity against endoplasmic reticulum stress-induced apoptosis. *Food and Chemical Toxicology*, *103*, 279–288.

Sun, C., Zhang, B., Zhang, J. K., Xu, C. J., Wu, Y. L., Li, X., & Chen, K. S. (2012). Cyanidin-3-glucoside-rich extract from Chinese bayberry fruit protects pancreatic β cells and ameliorates hyperglycemia in streptozotocin-induced diabetic mice. *Journal of Medicinal Food*, *15*(3), 288–298.

Sun, C., Huang, H., Xu, C., Li, X., & Chen, K. (2013). Biological activities of extracts from Chinese bayberry (*Myrica rubra* Sieb. et Zucc.): A review. *Plant Foods for Human Nutrition*, *68*(2), 97–106.

Talavera, S., Felgines, C., Texier, O., Besson, C., Lamaison, J. L., & Rémésy, C. (2003). Anthocyanins are efficiently absorbed from the stomach in anesthetized rats. *The Journal of Nutrition*, *133*(12), 4178–4182.

Tena, N., Martín, J., & Asuero, A. G. (2020). State of the art of anthocyanins: Antioxidant activity, sources, bioavailability, and therapeutic effect in human health. *Antioxidants*, *9*(5), 451.

Teng, H., Fang, T., Lin, Q., Song, H., Liu, B., & Chen, L. (2017). Red raspberry and its anthocyanins: Bioactivity beyond antioxidant capacity. *Trends in Food Science & Technology*, *66*, 153–165.

Thoppil, R. J., Bhatia, D., F Barnes, K., Haznagy-Radnai, E., Hohmann, J. S , Darvesh, A., & Bishayee, A. (2012). Black currant anthocyanins abrogate oxidative stress through Nrf2-mediated antioxidant mechanisms in a rat model of hepatocellular carcinoma. *Current Cancer Drug Targets*, *12*(9), 1244–1257.

Thummayot, S., Tocharus, C., Pinkaew, D., Viwatpinyo, K., Sringarm, K., & Tocharus, J. (2014). Neuroprotective effect of purple rice extract and its constituent against amyloid beta-induced neuronal cell death in SK-N-SH cells. *Neurotoxicology*, *45*, 149–158.

Tian, L., Tan, Y., Chen, G., Wang, G., Sun, J., Ou, S., ... & Bai, W. (2019). Metabolism of anthocyanins and consequent effects on the gut microbiota. *Critical Reviews in Food Science and Nutrition*, *59*(6), 982–991.

Tsuda, T., Horio, F., & Osawa, T. (2000). The role of anthocyanins as an antioxidant under oxidative stress in rats. *Biofactors*, *13*(1-4), 133–139.

Ullah, I., Park, H. Y., & Kim, M. O. (2014). Anthocyanins protect against kainic acid-induced excitotoxicity and apoptosis via ROS-activated AMPK pathway in hippocampal neurons. *CNS Neuroscience & Therapeutics*, *20*(4), 327–338.

Unusan, N. (2020). Proanthocyanidins in grape seeds: An updated review of their health benefits and potential uses in the food industry. *Journal of Functional Foods*, *67*, 103861.

Venancio, V. P., Cipriano, P. A., Kim, H., Antunes, L. M., Talcott, S. T., & Mertens-Talcott, S. U. (2017). Cocoplum (*Chrysobalanus icaco* L.) anthocyanins exert anti-inflammatory activity in human colon cancer and non-malignant colon cells. *Food & Function*, *8*(1), 307–314.

Wallace, T. C., Slavin, M., & Frankenfeld, C. L. (2016). Systematic review of anthocyanins and markers of cardiovascular disease. *Nutrients*, *8*(1), 32.

Wang, H., Wang, W., Zhang, P., Pan, Q., Zhan, J., & Huang, W. (2010a). Gene transcript accumulation, tissue and subcellular localization of anthocyanidin synthase (ANS) in developing grape berries. *Plant Science*, *179*(1–2), 103–113.

Wang, X., Tong, H., Chen, F., & Gangemi, J. D. (2010b). Chemical characterization and antioxidant evaluation of muscadine grape pomace extract. *Food Chemistry*, *123*(4), 1156–1162.

Wang, Y., Zhang, Y., Wang, X., Liu, Y., & Xia, M. (2012a). Cyanidin-3-O-β-glucoside induces oxysterol efflux from endothelial cells: Role of liver X receptor alpha. *Atherosclerosis*, *223*(2), 299–305.

Wang, Y., Zhang, Y., Wang, X., Liu, Y., & Xia, M. (2012b). Supplementation with cyanidin-3-O-β-glucoside protects against hypercholesterolemia-mediated endothelial dysfunction and attenuates atherosclerosis in apolipoprotein E–deficient mice. *The Journal of Nutrition*, *142*(6), 1033–1037.

Wei, Y. Z., Hu, F. C., Hu, G. B., Li, X. J., Huang, X. M., & Wang, H. C. (2011). Differential expression of anthocyanin biosynthetic genes in relation to anthocyanin accumulation in the pericarp of *Litchi chinensis* Sonn. *PloS ONE*, *6*(4), e19455.

Wei, T., Ji, X., Xue, J., Gao, Y., Zhu, X., & Xiao, G. (2021). Cyanidin-3-O-glucoside represses tumor growth and invasion *in vivo* by suppressing autophagy via inhibition of the JNK signaling pathways. *Food & Function*, *12*(1), 387–396.

Weisel, T., Baum, M., Eisenbrand, G., Dietrich, H., Will, F., Stockis, J. P., ... & Janzowski, C. (2006). An anthocyanin/polyphenolic-rich fruit juice reduces oxidative DNA damage and increases glutathione level in healthy probands. *Biotechnology Journal: Healthcare Nutrition Technology*, *1*(4), 388–397.

Winter, A. N., Ross, E. K., Wilkins, H. M., Stankiewicz, T. R., Wallace, T., Miller, K., & Linseman, D. A. (2018). An anthocyanin-enriched extract from strawberries delays disease onset and extends survival in the hSOD1G93A mouse model of amyotrophic lateral sclerosis. *Nutritional Neuroscience*, *21*(6), 414–426.

Wu, X., Liang, L., Zou, Y., Zhao, T., Zhao, J., Li, F., & Yang, L. (2011). Aqueous two-phase extraction, identification and antioxidant activity of anthocyanins from mulberry (*Morus atropurpurea* Roxb.). *Food Chemistry*, *129*(2), 443–453.

Xia, Y., Tian, L. M., Liu, Y., Guo, K. S., Lv, M., Li, Q. T., ... & Qiu, C. H. (2019). Low aose of cyanidin-3-O-glucoside alleviated dextran sulfate sodium-induced colitis, mediated by CD169+ macrophage pathway. *Inflammatory Bowel Diseases*, *25*(9), 1510–1521.

Xiao, D., Zhu, L., Edirisinghe, I., Fareed, J., Brailovsky, Y., &Burton-Freeman, B. (2019). Attenuation of postmeal metabolic indices with red raspberries in individuals at risk for diabetes: A randomized controlled trial. *Obesity*, *27*, 542–550.

Xie, X., Zhao, R., & Shen, G. X. (2012). Impact of cyanidin-3-glucoside on glycated LDL-induced NADPH oxidase activation, mitochondrial dysfunction and cell viability in cultured vascular endothelial cells. *International Journal of Molecular Sciences*, *13*(12), 15867–15880.

Xu, W., Dubos, C., & Lepiniec, L. (2015). Transcriptional control of flavonoid biosynthesis by MYB–bHLH–WDR complexes. *Trends in Plant Science*, *20*(3), 176–185.

Xu, L., Zhang, Y., & Wang, L. (2016). Structure characteristics of a water-soluble polysaccharide purified from dragon fruit (*Hylocereus undatus*) pulp. *Carbohydrate Polymers*, *146*, 224–230.

Xu, W., Zhou, Q., Yao, Y., Li, X., Zhang, J. I., Su, G. H., & Deng, A. P. (2016). Inhibitory effect of Gardanblue bluberry anthocyanin extracts on liposaccharide-stimulated inflammation response in RAW 264 7 cell. *Journal of Zhejiang University-Science B (Biomedicine & Biotechnology)*, *17*, 425–436.

Xu, Y., Xie, L., Xie, J., Liu, Y., & Chen, W. (2019). Pelargonidin-3-O-rutinoside as a novel α-glucosidase inhibitor for improving postprandial hyperglycemia. *Chemical Communications*, *55*(1), 39–42.

Xu, Y., Li, Y., Xie, J., Xie, L., Mo, J., & Chen, W. (2021). Bioavailability, absorption, and metabolism of pelargonidin-based anthocyanins using Sprague–Dawley rats and Caco-2 cell monolayers. *Journal of Agricultural and Food Chemistry*, *69*(28), 7841–7850.

Yan, F., Dai, G., & Zheng, X. (2016). Mulberry anthocyanin extract ameliorates insulin resistance by regulating PI3K/AKT pathway in HepG2 cells and db/db mice. *The Journal of Nutritional Biochemistry*, *36*, 68–80.

Yang, M., Koo, S., Song, W., & Chun, O. (2011). Food matrix affecting anthocyanin bioavailability. *Current Medicinal Chemistry*, *18*(2), 291–300.

Yoshimoto, M., Okuno, S., Yamaguchi, M., & Yamakawa, O. (2001). Antimutagenicity of deacylated anthocyanins in purple-fleshed sweet potato. *Bioscience, Biotechnology, and Biochemistry*, *65*(7), 1652–1655.

You, Q., Wang, B., Chen, F., Huang, Z., Wang, X., & Luo, P. G. (2011). Comparison of anthocyanins and phenolics in organically and conventionally grown blueberries in selected cultivars. *Food Chemistry*, *125*(1), 201–208.

Zhang, Z., Xuequn, P., Yang, C., Ji, Z., & Jiang, Y. (2004). Purification and structural analysis of anthocyanins from litchi pericarp. *Food Chemistry*, *84*(4), 601–604.

Zhang, W., Li, X., Zheng, J., Wang, G., Sun, C., Ferguson, I. B., & Chen, K. (2008). Bioactive components and antioxidant capacity of Chinese bayberry (*Myrica rubra* Sieb. and Zucc.) fruit in relation to fruit maturity and postharvest storage. *European Food Research and Technology*, *227*(4), 1091–1097.

Zhang, C., Lu, X., Tan, Y., Li, B., Miao, X., Jin, L., ... & Cai, L. (2012). Diabetes-induced hepatic pathogenic damage, inflammation, oxidative stress, and insulin resistance was exacerbated in zinc deficient mouse model. *PLoS ONE*, *7*(12), e49257.

Zhang, X., Yang, Y., Wu, Z., & Weng, P. (2016a). The modulatory effect of anthocyanins from purple sweet potato on human intestinal microbiota *in vitro*. *Journal of Agricultural and Food Chemistry*, *64*(12), 2582–2590.

Zhang, Y., Chen, S., Wei, C., Gong, H., Li, L., & Ye, X. (2016b). Chemical and cellular assays combined with *in vitro* digestion to determine the antioxidant activity of flavonoids from Chinese bayberry (*Myrica rubra* Sieb. et Zucc.) leaves. *PLoS ONE*, *11*(12), e0167484.

Zhang, G., Chen, S., Zhou, W. U., Meng, J., Deng, K., Zhou, H., ... & Suo, Y. (2019a). Anthocyanin composition of fruit extracts from *Lycium ruthenicum* and their protective effect for gouty arthritis. *Industrial Crops and Products*, *129*, 414–423.

Zhang, J., Wu, J., Liu, F., Tong, L., Chen, Z., Chen, J., ... & Huang, C. (2019b). Neuroprotective effects of anthocyanins and its major component cyanidin-3-O-glucoside (C3G) in the central nervous system: An outlined review. *European Journal of Pharmacology*, *858*, 172500.

Zhao, R., Xie, X., Le, K., Li, W., Moghadasian, M. H., Beta, T., & Shen, G. X. (2015). Endoplasmic reticulum stress in diabetic mouse or glycated LDL-treated endothelial cells: Protective effect of Saskatoon berry powder and cyanidin glycans. *The Journal of Nutritional Biochemistry*, *26*(11), 1248–1253.

Zhao, C. L., Yu, Y. Q., Chen, Z. J., Wen, G. S., Wei, F. G., Zheng, Q., ... & Xiao, X. L. (2017). Stability-increasing effects of anthocyanin glycosyl acylation. *Food Chemistry*, *214*, 119–128.

Zhou, L., Xie, M., Yang, F., & Liu, J. (2020). Antioxidant activity of high purity blueberry anthocyanins and the effects on human intestinal microbiota. *LWT*, *117*, 108621.

Zhu, Y., Sun, H., He, S., Lou, Q., Yu, M., Tang, M., & Tu, L. (2018). Metabolism and prebiotics activity of anthocyanins from black rice (*Oryza sativa* L.) *in vitro*. *PLoS ONE*, *13*(4), e0195754.

Zuo, Y., Peng, C., Liang, Y., Ma, K. Y., Yu, H., Chan, H. Y. E., & Chen, Z. Y. (2012). Black rice extract extends the lifespan of fruit flies. *Food & Function*, *3*(12), 1271–1279.

2 Chemical Properties of Anthocyanins Sourced from Different Subtropical Fruits

Sabbu Sangeeta, Sweta Rai, Shivani Bisht, and Mohd. Nazim

CONTENTS

DOI: 10.1201/9781003242598-2

2.1   INTRODUCTION

An attractive factor of different fruits and flowers is their pigmentation. Anthocyanins are one of these important natural pigments present in vascular plants, contributing the orange, red, pink, violet to blue colors in fruits and flowers. Anthocyanins are harmless water-soluble compounds that are also easily dissolved in aqueous media. Due to the toxicity of artificial colors, anthocyanins are gaining demand as natural food colors in both national and international food and pharmaceutical industries (Wills et al., 2009). Belonging to the class of flavonoids, anthocyanin synthesis follows the phenylpropanoid pathway. Anthocyanidins along with sugar moiety form the basic structure of anthocyanins. They are neither aromatic nor flavorful, although they have a moderately astringent taste. The rich sources of anthocyanins include grapes, red cabbage, berries, apples, tulips, radishes, olive oil, teas, honey, cocoa, asparagus, banana, pea, fennel, and pear. Several studies have exploited the antioxidant activity of anthocyanins and their preventive action against chronic diseases such as hypertension, liver disorders, inflammation, memory loss, eyesight damage, microbial activities, tumors, and mutations. Other lifestyle-related disorders like hyperglycemia, cancer, and neurological and cardiovascular diseases can also be controlled by consuming anthocyanin-based foods (Shaik et al., 2018). This chapter aims to compile information about the chemical properties of anthocyanins sourced from different subtropical fruits like red grape, blood orange, pomegranate, jamun, red guava, karonda, phalsa, tamarillo, fig, and acerola.

2.2   CHEMICAL STRUCTURE OF ANTHOCYANINS

Instead of only a natural pigment, anthocyanins have the potential to play other roles like antioxidant activity due to their chemical structure. The glycosylation of anthocyanidins (aglycones) forms anthocyanin molecules. The skeleton of this compound is the flavylium cation (Figure 2.1), which is hydroxylated at various positions (mostly at carbon numbers 3, 5, 6, 7, and 3', 4', 5'). Variation in hydroxylation patterns leads to the formation of different anthocyanin structures.

Flavylium is the central molecule of anthocyanidin that consists of one aromatic ring, named the A ring, one heterocyclic benzopyran ring, named the C ring, and one phenyl component, named the B ring. Anthocyanidins carry a positive charge in their structure due to the presence of double bonds in the C ring (Figure 2.2). Regardless of the fact that the structure of these compounds carries an oxonium group, the flavonoid backbone supports a charged oxygen atom with its ring nomenclature on the C ring (Castañeda-Ovando et al., 2009). Constant glycosylation at carbon position 3 with hydroxyl substituent provides thermal constancy. On the other hand, glycosylation at position 5 occasionally takes place. Also, a hydroxy or methoxy substituent is present at 2-phenyl- or B-ring (Bridle and Timberlake, 1997). The most commonly attached sugar moieties are glucose, galactose, rhamnose, or arabinose. The moiety might be a mono or disaccharide sugar and either phenolic or aliphatic acid is involved in its acylation (Hocine et al., 2018).

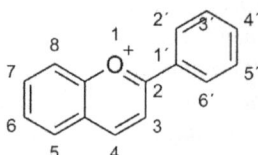

FIGURE 2.1   Chemical structure of the flavylium cation.

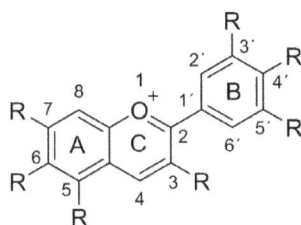

FIGURE 2.2   Chemical structure of anthocyanidin backbone.

TABLE 2.1
Substitution Pattern of Common Anthocyanidins

Common Anthocyanidins	Substitution Pattern						
	3	5	6	7	3′	4′	5′
Pelargonidin (Pg)	OH	OH	H	OH	H	OH	H
Cyanidin (Cy)	OH	OH	H	OH	OH	OH	H
Delphinidin (Dp)	OH	OH	H	OH	OH	OH	OH
Peonidin (Pn)	OH	OH	H	OH	OMe	OH	H
Petunidin (Pt)	OH	OH	H	OH	OMe	OH	OH
Malvidin (Mv)	OH	OH	H	OH	OMe	OH	OMe

Many structural forms of anthocyanins have been identified, which differ from each other princi-pally in the structure of B ring, number and type of the sugar attached, limited pH, and glycosylated segment. A number of anthocyanins occur naturally in fruits and vegetables; the most common are cyanidin, delphinidin, pelargonidin, peonidin, malvidin, and petunidin. The properties of these anthocyanidins depend on different substitution patterns on different positions (Table 2.1).

2.3   CLASSIFICATION OF ANTHOCYANINS

One of the classes of phenolic phytonutrients is anthocyanin. Anthocyanin exists as a glyco-side, whereas anthocyanidin exists as an aglycone. Anthocyanidins are divided into three cat-egories: 3-hydroxyanthocyanidins, 3-deoxyanthocyanidins, and O-methylated anthocyanidins, whereas anthocyanins are divided into two categories: anthocyanidin glycosides and acylated anthocyanins. Aside from common anthocyanins, acylated anthocyanins have been discovered in plants, namely acrylated anthocyanin, caffeoylated anthocyanin, coumaroylated anthocyanin, and malonylated anthocyanins (Khoo et al., 2017).

Cyanidin, delphinidin, pelargonidin, peonidin, petunidin, and malvidin are the most prevalent anthocyanidins in the human diet (Figure 2.3). The contribution of these anthocyanidins in fruits and vegetables is 50%, 12%, 12%, 12%, 7%, and 7%, respectively (Castañeda-Ovando et al., 2009).

2.3.1   CYANIDIN

Cyanidin is a magenta (reddish-purple) pigment found in nature. The pigment is responsible for the color of berries and other red-colored foods like red sweet potato and purple maize. Its characteristic color varies depending on pH: red at pH 3, violet at pH 7–8, and blue at pH > 11 (Khoo et al., 2017). Apples, apricots, bilberries, blackberries, mulberries, plums, pomegranates, blueberries, cranberries,

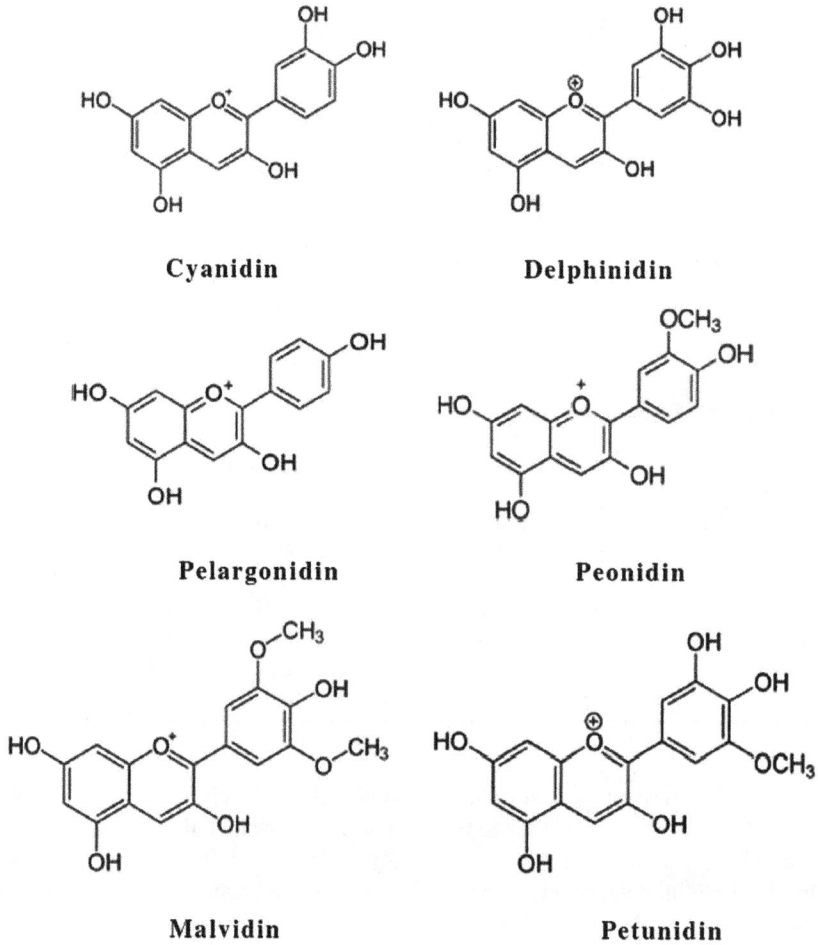

FIGURE 2.3   Major anthocyanidins in the human diet.

and red cabbage are all good sources of cyanidin, with the highest concentrations found in the fruit's skin (Zhang et al., 2019). Cyanidin possess anticarcinogenic, anti-inflammatory, anti-obesity, vasoprotective, antioxidant, anti-diabetes, and radical-scavenging characteristics. The glycosides of cyanidin are readily absorbed into the blood stream.

2.3.2   DELPHINIDIN

Delphinidin carries chemical properties similar to the majority of anthocyanidins. In the plant, it gives a blue-reddish or purple pigment. It is responsible for the blue hue of the flowers (Katsumoto et al., 2007). Delphinidin is pH-sensitive, changing color from blue to red in basic to an acidic solution. Bilberry, blueberry, grape, mulberry, saskatoon berry, pomegranate, pepper, soybean, rye, and cranberries contain a good amount of delphinidin (Zhang et al., 2019). Delphinidin acts against oxidation, mutation, inflammation, cancer, and angiogenesis. These activities are mediated through suppression of vascular endothelial growth factor receptor-2 phosphorylation, platelet-derived growth factor ligand/receptor signaling, cancer cell proliferation, and Met receptor phosphorylation regulation (Patel et al., 2013).

2.3.3 Pelargonidin

Pelargonidin is unique among anthocyanidins. It exists as a red-colored pigment in nature (Bakowska, 2005). Pelargonidin produces an orange color in flowers and reddish color in some fruits and berries (Robinson and Robinson, 1932, Jaakola, 2013). Pelargonidin is present in a variety of berries, including ripe raspberries, strawberries, blueberries, cranberries, and blackberries as well as in saskatoon and chokeberries. It is also present in pomegranates and plums. Kidney beans also contain high concentrations of this pigment (Lin et al., 2008). Pelargonidin has good antioxidant properties and also has a significant inhibitory impact on cancer cell proliferation in colon, stomach, lung, breast, and central nervous system (Parisi et al., 2014).

2.3.4 Peonidin

Another kind of anthocyanidin prevalent in plants is methylated anthocyanidin, such as peonidin. Its apparent color is magenta (Bakowska, 2005). It is pH-sensitive, like other anthocyanidins, and changes from red to blue as pH rises. Peonidin is cherry red at pH 2.0, bright yellowish pink at pH 3.0, grape red-purple at pH 5.0, and deep blue at pH 8.0; unlike many anthocyanidins, it is stable at higher pH. Peonidin imparts purplish-red colors to flowers like peonies and roses, from which it gets its name. Peonidin may be found in large quantities in berries (bilberry, blueberry, cranberry, haskap), grapes, and red wines (Zhang et al., 2019). Peonidin has significant apoptotic and inhibitory effects on cancer cells, most notably in metastatic cancer cells of the human breast (Jung et al., 2007).

2.3.5 Malvidin

O-methylated anthocyanidin is commonly known as malvidin. It has a purple hue and is present in ample quantities in blue blooms. Malvidin is the primary red pigment found in red wine (Mazza and Francis, 1995). In aged red wines, it occurs as a deeper dusky red (Barnard et al., 2011). In nature, malvidin exists as malvidin-3-glucoside and malvidin-3-galactoside, which are glycosylated forms with the sugar molecule bonded at position 3 on the C ring. Malvidin solutions that are somewhat acidic or neutral have a red color, whereas basic malvidin solutions have a blue tint (Huang et al., 2016). Many fruits and vegetables, including bilberries, grapes, blueberries, beans, potatoes, and black rice, contain malvidin and its glycosides, which give them a red to blue pigmentation. The blue color of the primula polyanthus plant is also due to malvidin (Zhang et al., 2019). Malvidin has a high antioxidant capacity and significant free radical-scavenging abilities. It also exhibits antihypertensive and anti-inflammatory properties via inhibiting the angiotensin I-converting enzyme and blocking the NF-κB pathway (Huang et al., 2016).

2.3.6 Petunidin

Petunidin is a water-soluble dark red or purple pigment (Bakowska, 2005). It is the B-ring O-methylated derivative of delphinidin. The molecule name comes from the word petunia. In neutral and alkaline pH, derivatives of petunidin may chelate with metal ions, Al^{3+} and Fe^{3+}, leading to an improvement in color and stability. These metal-chelated pigments may be responsible for the stable pigmentation in food with a range of vibrant violet, blue, and green colors. Petunidin is present in bilberry, black bean, and eggplant in the form of petunidin-3-galactoside, petunidin-3-glucoside, and petunidin-3-(p-coumaroylrutinoside)-5-glucoside respectively (Zhang et al., 2019). Peonidin 3-glucoside and cyanidin-3-glucoside reduced the invasion of SCC-4 cells and HeLa cells in oral and cervical cancer (Tena et al., 2020).

2.4 STABILITY AND DEGRADATION OF ANTHOCYANINS

The isolated anthocyanins are very unstable and easily degraded. Several factors influence their stability, including pH, storage temperature, chemical structure, concentration, light, oxygen, solvents, and the presence of enzymes, flavonoids, proteins, and metallic ions. The chemical stabilization of anthocyanins has been the subject of current research due to their numerous potential uses, positive effects, and use as an alternative to artificial colorants (Castañeda-Ovando et al., 2009). The use of anthocyanins in food is limited due to their participation in a variety of processes that result in decolorization (Lee and Khng, 2002). Because of their vibrant colors ranging from orange-red to blue, water solubility, and non-toxicity, anthocyanins would be suitable alternatives for synthetic colorants. Nonetheless, low stability has limited the application of these pigments in the food industry (Troise and Fogliano, 2013). Many factors influence anthocyanin stability, and this has been extensively addressed in the literature. The influence of anthocyanin-specific structures (glycosylation, acylation with aliphatic or aromatic acids) as well as the effects of sulfur dioxide (SO_2), pH, light, temperature, oxygen, presence of metal ions, and sugar has been discussed and somewhat elucidated (Brat et al., 2008).

2.4.1 pH

Anthocyanins are very sensitive to pH alteration. When anthocyanin is dissolved in water, the flavylium cation undergoes to series of secondary structure formations including acid-base hydration and tautomeric reactions (Chandrapala et al., 2012).

In highly acidic conditions, the flavylium cation is dominated in equilibrium form. At various pH levels, tautomeric quinoidal bases are produced by deprotonation of flavylium cation and hemiacetal and chalcone forms linked to the flavylium via nucleophilic interaction with water, as shown in Figure 2.4 (Andersen et al., 2008). The flavylium cation is the dominating species at pH 1 and contributes to the purple and red colors (Figure 2.4A). The quinoidal blue species predominate at pH levels ranging from 2 to 4 (Figure 2.4B–D). Only two colorless species can be seen at pH values between 5 and 6 – a carbinol pseudobase (Figure 2.4E) and a chalcone (Figure 2.4F), respectively. Anthocyanins are destroyed depending on their substituent groups at pH levels greater than 7. At pH levels ranging from 4 to 6, the coexistence of four structural forms of anthocyanin – flavylium cation, anhydrous quinoidal base, colorless carbinol base, and pale yellow chalcone – is observed (Castañeda-Ovando et al., 2009).

Under the simulated gastric environment, the majority of anthocyanins present in wine stay stable. The breakdown of anthocyanins produces syringic acid, protocatechuic acid, and vanillic acid, with anthocyanin chalcone serving as key intermediate degradative products (Yang et al., 2018).

The half-life of anthocyanins extracted from black carrot in citrate phosphate buffer solutions at 90°C with different values of pH 2.5, 3.0, 4.0, 5.0, 6.0, and 7.0 suggested as 5.6, 6.3, 6.3, 5.6, 5.6, and 5.0, respectively. The reduction in half-life values is associated with poor stability of anthocyanins over pH 5.0 (Kirca et al., 2007). In the case of purple sweet potato extract, the half-lives of anthocyanin degradation occurring at pH values of 3.0, 5.0, and 7.0 are 10.27, 12.42, and 4.66 h, respectively at 90°C, which also indicates that anthocyanin degradation is accelerated in a neutral environment and pH has a significant impact on color changes due to degradation of anthocyanin (Jiang et al., 2019).

2.4.2 TEMPERATURE

Temperature greatly influences the stability of anthocyanin. An increase in temperature has reduced the stability of anthocyanins and all pigments present in foods due to structure alteration (Lee and Khng, 2002; Troise and Fogliano, 2013). The intensity of monomer anthocyanins and the associated colorant diminished with the time-temperature combination, but the polymer fraction (brown

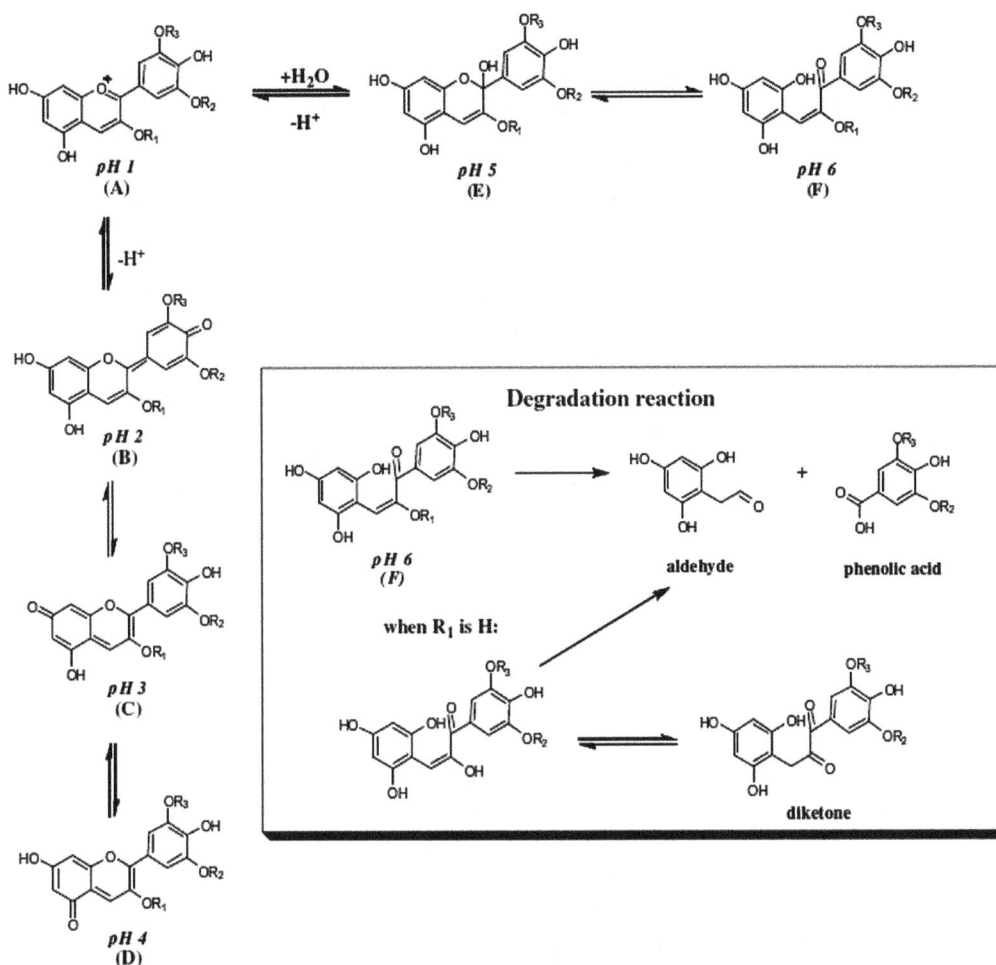

FIGURE 2.4 Structural transformations of anthocyanins at different pH.

pigments) did not (Brat et al., 2008). Foods containing anthocyanins should be processed at high temperatures for a short time to maximize the stability of anthocyanins (Skrede and Wrolstad, 2002).

Degradation kinetics of purified anthocyanins extracted from purple-fleshed potato (cv. Purple Majesty) follow the first-order reaction with reaction rate constants (k values) of 0.0262–0.2855/min, activation energy of 72.89 kJ/mol, thermal death times (D values) of 8.06–8789 min, and z value of 47.84°C. The activation enthalpy and entropy were 59.97 kJ/mol and –116.46 J/mol K, respectively (Nayak et al., 2011).

In dark-colored juice such as blood orange juice and its concentrates the anthocyanin content is reduced as the °Brix value and temperature increase (Troise and Fogliano, 2013). The thermostability of anthocyanins and antioxidant activity up to 120°C is quite stable but beyond this temperature both are reduced through intramolecular co-pigmentation and intermolecular self-association interactions in purple maize extracts, which are used in a variety of processing requiring high-temperature, short-time parameters (Aprodu et al., 2020).

Temperature and pH combination also greatly influence the stability of anthocyanin. The heat treatment time can be depleted remarkably from 167 to 72 h at neutral pH (7) as compared to low pH (3) from 167 to 12 h in the temperature range of 2–75°C during processing of aqueous extract of European cranberry bush (*Viburnum opulus* L.) fruit (Moldovan et al., 2012).

2.4.3 OXYGEN

Oxygen is detrimental to anthocyanins, and it is well known that anthocyanins kept in a vacuum, argon, or nitrogen environment are more stable than anthocyanins exposed to molecular oxygen-based environment (Skrede and Wrolstad, 2002). Anthocyanin concentration declined in all studied atmospheres; however, a greater reduction is reported in the presence of high oxygen concentration (Gonzalez-Aguilar, 2010). This impact is explained through the ability of anthocyanins to block radical activity at high oxygen concentrations, resulting in antioxidant pigment depletion (Gonzalez-Aguilar, 2010). The oxygen may directly or indirectly destroy anthocyanins by oxidizing molecules, which then further degrade the anthocyanins; this degrading action is most evident with the presence of high ascorbic acid concentration along with oxygen (Skrede and Wrolstad, 2002).

Strawberry purée was processed by Howard et al. (2014) under carbon dioxide, nitrogen, and air, followed by pasteurization and stored at 25°C. The researchers observed that purées processed with carbon dioxide or nitrogen preserved more anthocyanins (75 and 82%, respectively) than those made with air (60%). Purées processed under carbon dioxide or nitrogen exhibited better color stability ($L*$, $a*$, and $b*$ values) as compared to purees stored in air. Due to oxidative activities, the color of strawberry purée becomes unstable during processing and storage.

2.4.4 LIGHT

Being good absorbers of visible light, they appear as colored substances, responsible for the distinctive orange/red/bluish colors of grapes and berries. The color is mainly influenced by the substitution pattern of the aglycon's B ring, as opposed to the pattern of glycosylation of the flavan structure, which impacts color formation to a lesser degree (Skrede and Wrolstad, 2002). Light-induced degradation is affected by the amount of molecular oxygen present. When anthocyanin pigments are exposed to fluorescent light, they lose most of their stability. This can be prevented by using packaging materials that have excellent light barrier properties, especially ultraviolet ranges of the spectrum (Giusti and Wallace, 2009).

The stability and half-life of anthocyanin extracts from mangosteen peel under the influence of different light sources (fluorescent, incandescent, infrared, and ultraviolet) with controlled oxygen concentration revealed that fluorescent light had the longest half-life (597 h) followed by incandescent (306 h), ultraviolet (177 h), and infrared (100 h), which means that fluorescence luminosity is responsible for the least amount of anthocyanin loss (Chiste et al., 2010). The monomeric anthocyanin content of blackberry (*Rubus fructicosus*) extract is found to be reduced to 76% when it was exposed to 3968.30 lux of light source for 1 week, whereas only 29% reduction was observed in dark storage conditions (Contreras-Lopez et al., 2014).

2.4.5 ENZYMES

Many writers have associated the browning of the pericarp with anthocyanin breakdown by polyphenol oxidase. Polyphenol oxidase and peroxidase (POD) are membrane-bound enzymes, whereas anthocyanins are vacuolar. Under certain conditions, enzymes may hydrolyze glycoside substituents to destroy anthocyanins and other pigments found in fruit. Hydrolyzed anthocyanins in their pure aglycon form are very unstable and degrade rapidly, resulting in the loss of color (Giusti and Wallace, 2009). Anthocyanin is also lost owing to the activity of degrading enzymes (e.g., polyphenol oxidase) found in fruits. Certain crude fungal enzymes, especially liberated from *Aspergillus* spp., are responsible for decolorizing the extract of pigments. Enzymatic hydrolysis of the anthocyanin to anthocyanidin and sugar as well as a sudden change of aglucone into colorless derivatives are involved in the overall decolorization process (Huang, 1955).

The skin of sweet cherries has an enzyme system capable of degrading cyanidin-3-glucoside in the absence of phenols; however, the pulp homogenate destroys the anthocyanin only in the presence of phenols. The anthocyanin structure at different pH levels, as well as the type of quinone produced by enzymatic oxidation, affected the decolorization. The anhydrobase form of anthocyanin appears to be the most prone to oxidation (Pifferi and Cultrera, 1974).

2.4.6 SULFUR DIOXIDE

SO_2 has been widely utilized in the fruit and vegetable sector, primarily as a microbial growth inhibitor and an inhibitor of enzymatic and non-enzymatic browning (Giusti and Wallace, 2009). The ability of flavonoid to bind with anthocyanins is enhanced by sulfonation, which inhibits the hydration process and shifts the equilibrium toward the colored quinonoidal base (Troise and Fogliano, 2013). The reaction of anthocyanins with SO_2 produces colorless components. The process can be reversed through heating, which liberates some of the SO_2, and renewing the color of anthocyanin to some extent; the addition of acid also helps to regenerate anthocyanin (Skrede and Wrolstad, 2002). SO_2 can also cause a nucleophilic action of negatively charged bisulfate ion on the oxonium ion's (flavylium cation) C-4 position, leading to loss of anthocyanin color. This can be reversed through a bleaching process, which typically happens when fruits are exposed to 500–3000 ppm SO_2 (Giusti and Wallace, 2009).

In the wine-making industry, the presence of SO_2 impairs the organoleptic pattern of red wine, owing to the reaction products of SO_2 with anthocyanins. However, in white wine the antioxidant activity of SO_2 is mainly focused with the clear statement that without SO_2, white wines would lose their well-known qualities (Usseglio-Tomasset, 1992).

2.4.7 IMPACT OF WATER ACTIVITY (a_w)

Several studies have demonstrated that anthocyanin stability rises when a_w decreases. In freeze-dried strawberry purée, Erlandson and Wrolstad (1972) found that anthocyanin degradation was proportional to rise in water activity. Even though Bronnum-Hansen and FIink (1985) also discovered that a_w only affects anthocyanin losses considerably above a_w of 0.51 in stored freeze-dried elderberry, they also showed that lower pH has a favorable impact on stabilizing anthocyanins in dry systems. Dry anthocyanin powders ($a_w \leq 0.3$) can be kept in hermetically sealed containers for several years and have been found to be more stable in dry crystalline form or on dry paper chromatograms (Markakis et al., 1957; Skrede and Wrolstad, 2002). The loss of anthocyanin stability is greater at higher a_w and higher pH (Thakur and Arya, 1989). The stability of different anthocyanins such as pelargonidin 3-glucoside, pelargonidin 3-sophoroside, and pelargonidin 3-sophoroside 5-glucoside acylated was examined by Garzon and Wrolstad (2001) with malonic and cinnamic acids for various a_w levels (0.37, 0.63, 0.86, 0.90, and 1.0). They observed that the degree of degradation for all types of anthocyanin pigment rises as a_w increases.

2.4.8 INTERACTION WITH ASCORBIC ACID

Ascorbic acid generally acts as antioxidant and chelating agent in most horticultural produce. The effect of ascorbic acid on the stability of anthocyanin has been demonstrated by many researchers. Generally ascorbic acid accelerates the degradation of anthocyanin and in the presence of oxygen and copper ions the process of deterioration is speeded up (Skrede and Wrolstad, 2002; Brat et al., 2008). Ascorbic acid may condense with anthocyanin in the absence of oxygen, forming unstable products that deteriorate into colorless molecules (Figure 2.5). Condensation of anthocyanin with flavonols is thought to limit the formation of complexes between anthocyanin and ascorbic acid, most likely due to competitiveness with anthocyanin in the selectivity for condensation reactions reducing the degradative impact of ascorbic acid (Skrede and Wrolstad, 2002; Troise and Fogliano, 2013).

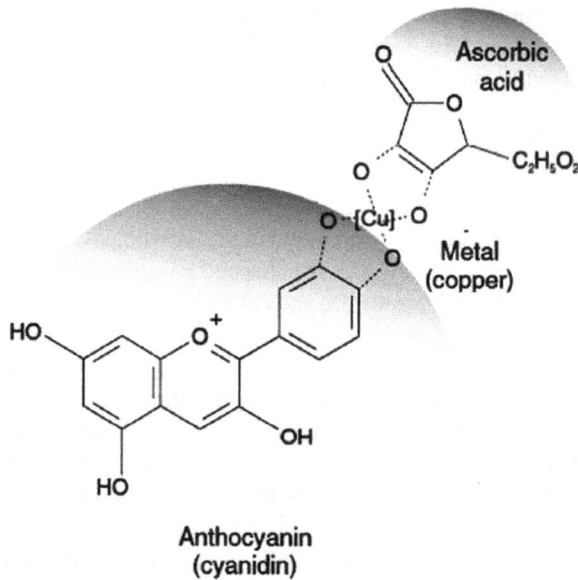

FIGURE 2.5 Anthocyanin–metal–ascorbic acid complex.

In blood orange juice anthocyanin and ascorbic acid losses occurred simultaneously during refrigerated storage, although the rate of reaction varied depending on the anthocyanin structure. According to CIELAB values in refrigerated conditions, all the ascorbic acid in blood orange juice is degraded in the absence of anthocyanin, while 15 and 23% of ascorbic acid remained in the presence of malvidin 3-glucoside and malvidin 3,5-diglucoside, respectively, because this effect is attributed to the antioxidant properties of anthocyanin (Troise and Fogliano, 2013). The ascorbic acid in the presence of peroxides such as hydrogen peroxide enhanced the degradation process of anthocyanin (Ozkan, 2002).

2.4.9 EFFECT OF SUGARS

Sugar solution helps to block enzymatic activities, which preserve anthocyanins during frozen storage (Giusti and Wallace, 2009). The effect of extra sugar on anthocyanin stability varies depending on the structure of the anthocyanin, its concentration, and the type of sugar used. The effect of sugars on anthocyanin stability is still a matter of debate. While anthocyanidin degradation in the presence of sugar is commonly reported in the literature (Brat et al., 2008), some writers do not describe any impact using a model solution of commercial anthocyanidin-based pigments with or without sugar. Glycosylation of anthocyanin or a greater number of sugar residue improves the stability and water solubility of anthocyanin. The thermostability of anthocyanin pigments, on the other hand, decreased linearly with increasing fructose content, most likely owing to the production of furaldehydes but in the case of sucrose it dropped up to 20% concentration. As the concentration of sucrose increased to 40% there was a favorable effect on pigment stability because sucrose exhibited a substantial protective role on anthocyanin pigment concentration, as well as retarding browning and polymeric color development (Nikkhah et al. 2007; Troise and Fogliano, 2013). Fructose, lactose, arabinose, and sorbose are more harmful than glucose, maltose, and sucrose, although oxygen accelerates the destruction of anthocyanin by sugar (Skrede and Wrolstad, 2002).

Individually quick-frozen strawberries were packaged with the addition of 10, 20, and 40% sugar by weight. Nikkhah et al. (2007) measured total monomeric anthocyanin pigment, polymeric color, browning, and color density after 3 years of storage at −15°C. Sucrose addition exhibited a

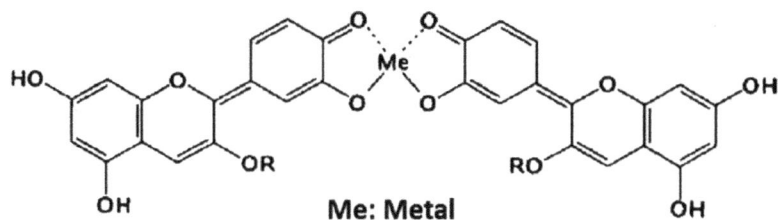

Me: Metal

FIGURE 2.6   Anthocyanin–metal complex.

substantial protective role on anthocyanin pigment concentration, as well as retarding browning and polymeric color development, according to one- and two-way analysis of variance.

2.4.10   METALLIC ION INTERACTIONS

The generation of chelates between metals and flavylium salts is originally thought to explain the diversity of colors in flowers (Clifford, 2000). Despite the lack of interest in anthocyanin–metal complexations in the food business, this interaction is a realistic solution for color stabilization, especially if the metals involved do not pose a health concern or are part of the necessary minerals in the diet. The properties of anthocyanins and anthocyanidins with o-dihydroxyl groups in the B ring (Cy, Dp, Pt) to form a metal–anthocyanin complex is one of their primary features (Figure 2.6) (Boulton, 2001). In some plants the blue colors are caused by the complexation of anthocyanins with metals such as Al, Fe, Cu, and Sn (Starr and Francis, 1973) or Mg and Mo (Hale et al., 2001). The complex of Al (III)–anthocyanin is carried out with Cy and other flavonoid derivatives and stabilizes the blue quinoidal base by preventing its oxidation (Moncada et al., 2003). Research also found that complexation of o-di-hydroxyl anthocyanins with Fe (III) or Mg (II) ions at pH 5 is required for the production of blue color in plants, especially if the anthocyanin: Fe (III) stoichiometrical ratio is 1:6, or greater for the Mg (II) ions (Yoshida et al., 2006). The complex of anthocyanin and molybdenum also stabilizes the blue color pigment in cabbage tissues (Hale et al., 2001).

2.4.11   CO-PIGMENTATION EFFECT

Co-pigmentation is a process in which colorants and other colorless organic molecules, such as metallic ions, establish the molecular or complex interactions that result in a change or increase in color intensity (Boulton, 2001). According to several studies, the major method of color stabilization in plants is anthocyanin co-pigmentation, with other chemicals commonly called as co-pigments (Mazza and Brouillard, 1990; Davies and Mazza, 1993). Pigments are electron-rich systems that can interact with electron-poor flavylium ions. In the second position of the flavylium ion, this association protects against water nucleophilic attack, and in the fourth position, it protects against additional species such as peroxides and SO_2 (Mazza and Brouillard, 1987; Garcia-Viguera and Bridle, 1999). When co-pigments are combined with an anthocyanin solution, they undergo an interaction that results in a hyperchromic response and a bathochromic change in the absorption spectra (UV-Vis region). Flavonoids, alkaloids, amino acids, organic acids, nucleotides, polysaccharides, metals, or other anthocyanins can all be used as co-pigments. Depending on the participating species, the anthocyanin–co-pigment interaction can be carried out in a variety of ways (Figure 2.7). When the co-pigment is another anthocyanin, a self-association or intramolecular co-pigmentation occurs (Figure 2.7 A and B); when the interaction is with a metal, a complexation occurs (Figure 2.7C); when co-pigments with free electron pairs interact, an intermolecular co-pigmentation occurs (Figure 2.7D); and in the most complex case, the co-pigmentation can be carried out by aglycon, sugar, co-pigment, and proton.

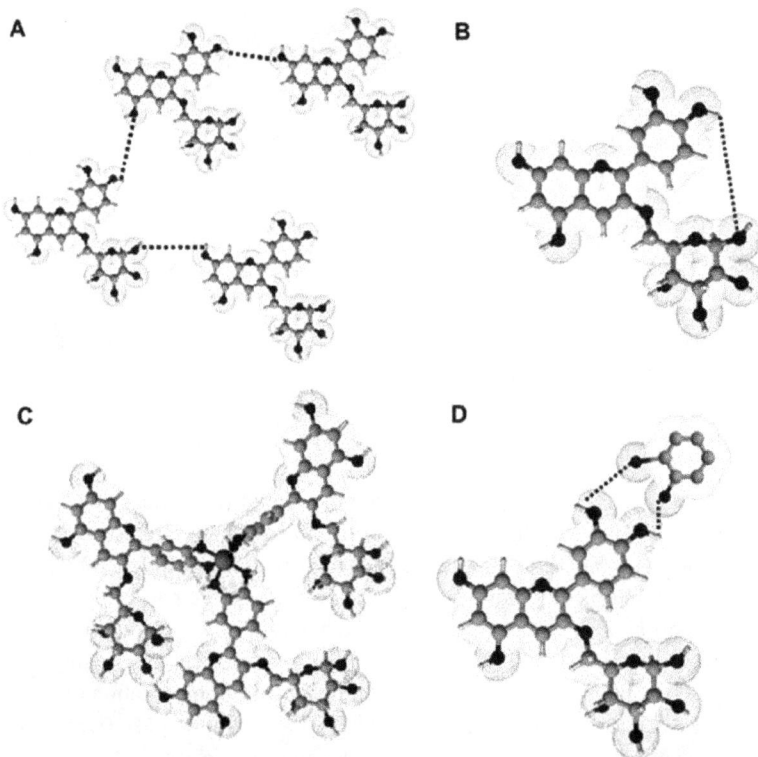

FIGURE 2.7   Anthocyanin interactions. (A) Self-association; (B) intramolecular co-pigmentation; (C) metal complexation; and (D) intermolecular co-pigmentation (Castaneda-Ovando et al., 2009).

Spray drying is generally used as an effective method of encapsulation and stabilization of antho-cyanin extracts. The additional co-pigments such as rutin and ferulic acid in dried anthocyanin also help to reduce their losses. The antioxidative activities of the co-pigments, as well as the inhibition of anthocyanin hydration, are the factors responsible for this stabilizing impact (Weber et al., 2017). The co-pigmented anthocyanin-containing nanoparticles also have great anthocyanin stability in black soybean (Ko et al., 2017).

2.4.12   High-Pressure Processing

High pressure with the combination of heat has remarkably affected anthocyanin degradation. The covalent interaction (Figure 2.3) of anthocyanins with other flavonols or organic acids leads to the creation of a new pyran ring via cycloaddition, which is promoted by high pressure and/or temperatures. When compared to thermal processing, HPP preservation (at low, room, and moderate temperatures) has a negligible influence on anthocyanin degradation.

The anthocyanins present in raspberries and strawberries, such as pelargonidin-3-glucoside and pelargonidin-3-rutinoside, respectively, are found to be stable during processing under high pressure of 800 MPa at 18–22°C for 15 min (Garcia-Palazon et al., 2004), whereas in some fruits, espe-cially red-colored fruits, minor changes in anthocyanins are observed under HPP at mild and high temperatures (Ferrari et al., 2011; Verbeyst et al., 2012; Terefe et al., 2013; Marszałek et al., 2016). Some researchers have suggested that no significant changes have been found in anthocyanins when treated by HPP up to 600 MPa at ambient temperature but as the temperature rises to 70°C, the loss of anthocyanin pigments occurs up to 25%. As compared to typical thermal treatment at the same temperature, this degradation is 20% greater, showing that HPP combined with higher temperatures promotes anthocyanin breakdown (Corrales et al., 2008; Barba et al., 2013). Condensation of anthocyanins causes color changes, such as in red wine during storage, by creating complexes of brown pigments (Corrales et al., 2008). It means that damaging influence of the synergistic effect of pressure and temperature should be taken into account while applying high-pressure processing.

A first-order kinetic model well characterizes anthocyanin degradation during thermal processing and HPP, according to different authors. The kinetic rate constant for anthocyanin degradation ranges from 0.7 to 7.1 10^{-2}/min and increases with increasing pressure and temperature. The most important benefits of HPP are the small variations in anthocyanin concentrations at cold and room temperatures under pressures up to 600 MPa. Unfortunately, moderate heating is necessary for significant tissue enzyme deactivation. The thermal breakdown and condensation process of HPP pigments are accelerated when the temperature is raised over 50°C (Marszałek et al. 2017). Studies of the stability of anthocyanins under HPP by various researchers are detailed in Table 2.2.

2.5 ANTHOCYANIN PROFILE OF SUBTROPICAL FRUITS

2.5.1 RED GRAPE (*VITIS VINIFERA*)

Anthocyanins are very important in grapes (Figure 2.8A) and wines for two reasons: first, they are an important part of the sensory attributes, as their levels, various forms, and derivatives directly affect the pigmentation of the final product; and second, they are thought to have a variety of biological characteristics, so they are classified as secondary metabolites (Stintzing and Carle, 2004). Six anthocyanins – delphinine, cyanidin, petunidin, peonidin, malvidin, and malvidin *p*-coumarate 3-O-glucosides – are identified in the red grape at 520 nm. Malvidin 3-O-glucoside is the most abundant pigment (82.53 mg/100 g) followed by its coumarate derivative (29.26 mg/100 g), peonidin 3-O-glucoside (10.84 mg/100 g), petunidin 3-O-glucoside (7.80 mg/100 g), cyanidin 3-O-glucoside (5.67 mg/100 g), and delphinidin 3-O-glucoside (1.28 mg/100 g) in freshly harvested grapes (Kallithraka et al., 2009).

Anthocyanins in grapes are made up of five aglycones with different substitution patterns in the B ring. In *Vitis vinifera* cultivars, they are only found as 3-glucosides. The absence of pelargonidin in the Vitaceae family indicates that the first flavanone undergoes the first hydroxylation in ring B of the anthocyanin molecule, with cyanidin serving as the precursor ingredient for the other anthocyanidins (Hrazdina et al., 1982). The total anthocyanin concentration in cultivars grown at 35°C is less than half that of berries grown at 25°C (Mori et al., 2007). The absence or low concentration of ascorbic acid and co-pigmentation generally does not affect anthocyanin degradation, although it helps in its preservation. Anthocyanin concentration is enhanced during ripening because of enzymatic activities. Cyanidin is a precursor of other anthocyanidins. During the biosynthesis pathway of anthocyanin, cyanidin is transformed into peonidin by 3-O-methyltransferase or delphinidin by 3-hydroxylase and 3′5′-O-methyltransferase converts delphinidin to malvidin via petunidin (Boss et al., 1996; Pomar et al., 2005; Fournand et al., 2006).

2.5.2 PHALSA (*GREWIA ASIATICA*)

Phalsa (Figure 2.8B) is a fruit-bearing shrub native to the South and East. It is a valuable indigenous fruit to be utilized as a coloring agent and has nutraceutical value (Zia-Ul-Haq et al., 2013; Sinha et al., 2015). The peel, pulp, and seed of phalsa fruit have been identified for their antioxidative and antiradical properties (Khattab et al., 2015).

Non-acylated (peonidin-3-O-glucoside, delphinidin-3-O-glucoside, pelargonidin-3-O-malonyl glucoside), (cyaniding-O-6″-acetylglucoside, peonidin-O-6″-acetylglucosideand pelargonidin-3-O-6″-acetylglucoside), and pyranoanthocyanin (malvidin-3-O-glucoside pyruvic acid) types of anthocyanin classes are found in phalsa. Cyanidin-3-O-6″-acetylglucoside is the most abundant

TABLE 2.2
Effect of High-Pressure Processing on Anthocyanin Stability

Food Item	Type of Anthocyanin	Processing Parameters			Effect on Anthocyanin	Reference
		Pressure (MPa)	Temp. (°C)	Time (Min)		
Orange juice	Cy-3-glc	600	20	15	Retention of 99 % of Cy-3-glc	Torres et al. (2011)
Blueberries	Total anthocyanin content	200–600	25	5–15	Negligible changes	Barba et al. (2013)
Pomegranate juice	Total anthocyanin content	400–600	25–50	5–10	Small decrease with increase in temperature and pressure	Ferrari et al. (2010)
Strawberry pulp	Cy-3-glc Pg-3-glc Pg-3-rut	400–600	Room temperature	5–25	Negligible changes in Cy-3-glc and Pg-3-glc, 6% loss of Pg-3-rut at 400 MPa for 10 min	Cao et al. (2012)
Strawberry and raspberry pastes and juices	Cy-3-glc Pg-3-glc Pg-3-ara Cy-3-soph Cy-3-rut	400–700	20–110	20	23% changes at temperature less than 80°C and 80% losses at temperature more than 80°C	Verbeyst et al. (2012)
Strawberry paste	Cy-3-rut	200–700	80–110	50	Accelerated degradation	Verbeyst et al. (2010)
Strawberries	Total anthocyanin content	300–600	20–60	2–10	Negligible changes	Terefe et al. (2009)
Strawberry purée	Cy-3-glc Pg-3-glc Pg-3-rut	300–500	0–50	5–15	Increase in losses with increasing temperature; no effect of pressure and time	Marszalek et al. (2015)
Strawberry purée	Cy-3-glc Pg-3-glc Pg-3-rut	600	50	15	20% loss	Marszalek et al. (2016)

FIGURE 2.8 (A) Grape (*Vitis vinifera*); (B) phalsa (*Grewia asiatica*); (C) pomegranate (*Punica granatum*); (D) jamun (*Eugenia jambolana*); (E) blood orange (*Citrus sinensis*); (F) red guava (*Psidium cattleyanum*); (G) tamarillo (*Solanum betaceum*); (H) acerola (*Malphigia emarginata*); (I) karonda (*Carissa carandas*).

acylatedanthocyanin present in phalsa; it accounts for 44–63% (695 g/g) of total anthocyanins, followed by 3–30% of peonidin-3-O-glucoside (163.6 g/g) and 8–14% of pelargonidin-3-O-6″-acetylglucoside (140.4 g/g). The optimum solvent for extracting anthocyanins from Phalsa is acidic methanol (Talpur et al., 2017). Furthermore, anthocyanins extracted in dark conditions are more stable than those subjected to ultraviolet light at temperature ranges of 10–40 °C.

2.5.3 POMEGRANATE (*PUNICA GRANATUM*)

Pomegranate (*Punica granatum* L.) has various health benefits and its production has exploded worldwide (Seeram et al., 2006). The red color of pomegranate (Figure 2.8C) is imparted by anthocyanins derived mostly from the aril (Hernandez et al., 1999). The pomegranate fruit contains six anthocyanin pigments: 3-mono- and 3,5-diglucosides of cyanidin, delphinidin, and pelargonidin

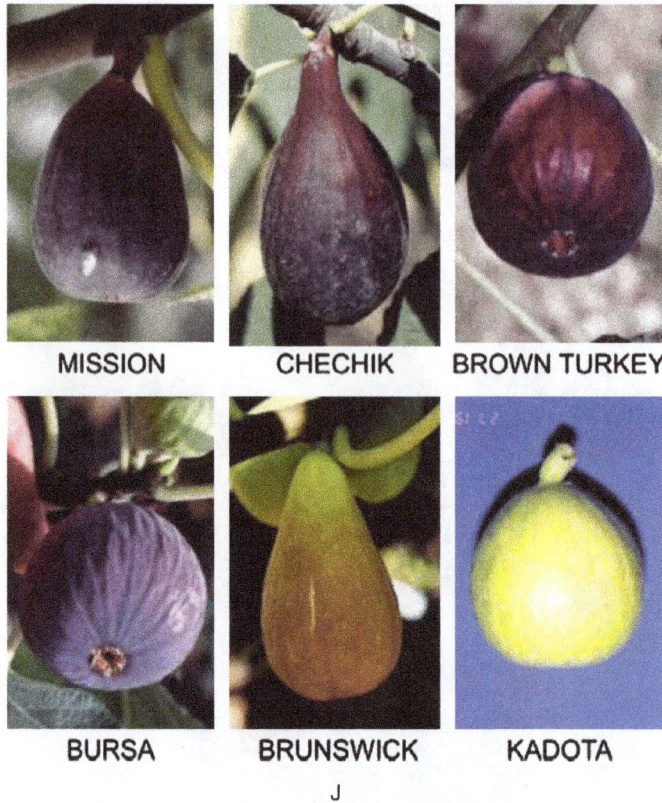

FIGURE 2.8   (Continued) (J) fig (*Ficus carica*).

(Du et al., 1975; Noda et al., 2002). The development of anthocyanin in plants is influenced by environmental factors. Low temperatures promote the formation of anthocyanins, but high temperatures
decrease pigment accumulation (Borochov-Neori et al., 2011). The amount of anthocyanin production changes inversely with the temperature of the season in all plants. Pomegranate is abundant in cyanidins but delphinidin also accumulates in higher quantities in colder months. At lower
temperatures, monoglucosylated anthocyanins predominate, but as the season progresses, the
amount of diglucosides increases (Welch et al., 2008).

The UV-C light treatment of pomegranate juice may promote photo-degradation of anthocyanins
owing to pigment discoloration. Higher doses of gamma irradiation (3.5–10 kGy) also have a substantial negative impact on the overall anthocyanin content of pomegranate juice (Alighourchi et al.,
2008). Heat treatment also leads to loss of anthocyanin pigment in pomegranate juice by 23.64%.
Punicalin, a hydrolyzable tannin frequently found in high amounts in pomegranate juice, participates
in co-pigmentation of anthocyanins (Kerimi et al., 2017). Furthermore, anthocyanin content is directly linked to the antioxidant and radical-scavenging activity of pomegranate (Kostka et al., 2020).

2.5.4 JAMUN (*EUGENIA JAMBOLANA*)

Eugenia jambolana Lam. (syn. *Syzygium cumini* Skeels) produces delicious fruits that acquire a
deep purple color (Figure 2.8D) when ripe and are widely utilized in ethnomedicine for several
purposes in India (Sagrawat et al., 2006; Helmstadter 2008). Water-soluble anthocyanins, a family
of secondary metabolites, are responsible for the purple color of the ripe Jamun fruits. Delphinidin,
cyanidin, petunidin, peonidin, and malvidin are the five kinds of anthocyanins present in jamun. Its

anthocyanin profile consists of two glucosyl units linked to the 3 and 5 locations of the anthocyanidin units separately (Li et al., 2009).

As cyanidin-3-glucoside equivalents, jamun pulp powder contains 0.54% anthocyanins. But after acid hydrolysis the anthocyanin content is reduced to 0.23% as cyanidin-3-glucoside. The loss of sugar during the transformation of anthocyanins to anthocyanidins is responsible for the reduction in anthocyanin concentration (hydrolysis). The seed extracts of jamun contain no anthocyanin (Aqil et al., 2012).

2.5.5 Blood Orange (*Citrus sinensis*)

Anthocyanin is responsible for the red color of sweet orange (Figure 2.8E). Cyanidin-3-(6"-malonylglucoside) is the most abundant anthocyanin in Budd blood oranges, contributing 44.8% of total anthocyanins, followed by cyanidin-3-glucoside with 33.6%. Malonated anthocyanins make up the majority of the anthocyanins in Budd blood orange, i.e., greater than 51% (Lee, 2002). Because current nutritional and epidemiological research has documented numerous favorable health benefits of water-soluble anthocyanins for protection against diseases like cardiovascular disease, cancer, and aging, this red pigment of sweet orange is of great interest.

Stability of anthocyanin in orange juice can be improved by pasteurization using microwaves (Maccarone et al., 1983), and the inclusion of moderately acidic and antioxidant agents like tartaric acid and gluthathione, respectively, enhanced anthocyanin retention. The anthocyanin–co-pigment complexes utilizing phenolic chemicals, such as rutin and caffeic acid as co-pigments, provide the best stability (Maccarone et al., 1985). The retention of color in the juices is attributed to the anthocyanins' restricted availability to participate in nucleophilic interactions with water or other agents, resulting in the development of stable interparticle complexes (Sweeny et al., 1981). The use of pectolytic enzyme and bentonite for clarification purposes and high temperature during storage reduces anthocyanin stability.

2.5.6 Red Guava (*Psidium cattleyanum*)

In red guava (Figure 2.8F), the anthocyanin content is approximately 685.8 µg/g on a dry-weight basis (Table 2.3). Cyanidin chloride, malvidin 3-glycoside, and cyanidin 3-glycoside are the three anthocyanins present in large amounts in red guava. The most common anthocyanin is cyanidin 3-glycoside, which accounted for 51.7% of the overall anthocyanin content of the fruit, followed by malvidin 3-glycoside (35.5%) and cyanidin chloride (12.7%) (Nora et al., 2014).

Fresh red guava has more anthocyanin content, owing to the presence of cyanidin 3-glucoside. This fruit has fragile skin and contains various endogenous enzymes such as glycosidases, peroxidases, and polyphenol oxidases, which facilitate the breakdown and oxidation of the anthocyanins during the freezing and thawing process (Francis, 1989). The chemical make-up of red guava may also contribute to its decomposition. During processing, cyanidin-3-glucoside becomes very reactive. After hot-air drying, it is reduced by more than 98%. The anthocyanins are rapidly degraded by oxidation in the hot-air drying process because of high temperatures and high atmospheric oxygen

TABLE 2.3
Anthocyanin Composition of Red Guava

Anthocyanin Detected	Concentration (µg g^{-1} dry fruit)
Cy 3-glu	354.66 ± 45.09
Mv 3-glu	243.58 ± 2.09
Cy	87.57 ± 5.77

level. Maintaining anthocyanin level in red guava by freeze drying is similarly ineffective. Total anthocyanin content in freeze-dried samples declines by 54.9% after 30 days in contrast to fresh samples, and by 7.3% in frozen samples for 30 days (Nora et al., 2014).

2.5.7 TAMARILLO (*SOLANUM BETACEUM*)

Tamarillo (Figure 2.8G) belongs to the family Solanaceae and is also known as tree tomato due to its flesh's resemblance to that of a tomato. Depending on the cultivar, the mature fruit develops different colors (yellow, orange, red, or purple) and has a somewhat bitter, sour, and astringent flavor with a distinct odor (Diep et al., 2020).

The anthocyanin proportion is influenced by several factors, including cultivation sites, fruit colors (e.g., yellow vs. purple), fruit parts (e.g., flesh vs. peel), extraction methods, and storage conditions (Wang and Zhu, 2020). Anthocyanins found in tamarillo fruits extend the shelf-life of food by reducing lipid oxidation rate and thus act as a potential antioxidant (Hurtado et al., 2009; Castro-Vargas et al., 2013). The flesh of tamarillo fruits contains more total anthocyanin, i.e., 4.15 mg/100 g, than the peel, i.e. 1.36 mg/100 g on a dry-weight basis (Hassan and Bakar, 2013). Pelargonidin 3-O-rutinoside (77.0–78.0 mg/100 g on a dry-weight basis), delphinidin 3-O-rutinoside (21.8–87.42 mg/100 g on a dry-weight basis), and cyanidin 3-O-rutinoside (2.5–4.5 mg/100 g on a dry-weight basis) are the most abundant anthocyanin pigments available in the tamarillo fruit pulp (Espin et al., 2016). The most prevalent anthocyanin in peel and flesh are cyanidin-3-rutinoside and delphinidin-3-rutinoside, respectively (Wrolstad and Heatherbell, 1974).

Seeds and the surrounding jelly of fresh Colombian ripe tamarillo fruits (20.03 mg delphinidin 3-glucoside/L sample) had a better total anthocyanin content than peel crude extract (0.20 mg delphinidin 3-glucoside/L sample). The anthocyanin content of seed and surrounding jelly extract is more sensitive to pH variation than the peel extract regarding the presence of phenolic acids, flavonols, flavones, flavanones, and organic acids, which help to stabilize the color of the peel extract (Eiro and Heinonen, 2002; Hurtado et al., 2009). Hydroxylated and glycosylated anthocyanin structures affect the color and stability of rutinosides isolated from Colombian tamarillo at varying pH levels (2.0–6.2) (Hurtado et al., 2009). The monomeric form of anthocyanin is unstable during refrigeration storage. According to Villegas-Ruiz et al. (2013), total monomeric anthocyanin content of tamarillo fruits decreased from 0.91 to 0.61 mg cyanidin-3-glucoside/g after 60 days of refrigeration storage.

2.5.8 ACEROLA (*MALPHIGIA EMARGINATA*)

Acerola (*Malphigia emarginata*) are small round fruit, with intense red or orange peel and orange-red flesh, and are also known as Barbados cherries (De Brito et al., 2007) (Figure 2.8H). They may be red, purple, or yellow when fully mature. The weight of the fruit ranges between 3 and 16 g, depending on the plant's potential and the farming environment (De Assis et al., 2008). Acerola is considered a functional food because of its high vitamin C (957–1074 mg/100 g) (Vendramini and Trugo, 2000; De Assis et al., 2001), carotenoid (371–1881 mg/100 g) (De Rosso and Mercadante, 2005), and anthocyanin content (Santini and Huyke, 1993; Hanamura et al., 2005).

The total anthocyanin content of acerola was found to be 6.5–8.4 mg/100 g (De Rosso et al., 2008). Initially, Hanamura et al. (2005) discovered cyanidin 3-rhamnoside and pelargonidin 3-rhamnoside as anthocyanin present in acerola, whereas their free aglycones cyanidin (De Rosso et al., 2008) and pelargonidin (Vendramini and Trugo, 2004) were identified later. The predominant anthocyanin detected is cyanidin 3-rhamnoside, which accounts for 76–78% of total anthocyanin content, followed by 13–16% pelargonidin 3-rhamnoside, 6–8% cyanidin, and 2–3% pelargonidin (De Rosso et al., 2008).

The loss of red color in frozen acerola pulp and processed juice is caused by anthocyanin degradation, the primary concern during commercial storage of these foods. Freitas et al. (2006) measured the anthocyanin concentration of non-sweetened glass-bottled fruit juice prepared from acerola using the hot-fill method before and after storage at 28 ± 2°C for 350 days. After 350 days of storage, a drop of 47.79% was observed when compared to the initial concentration. The presence of high levels of ascorbic acid has been found to have a deleterious effect on the stability of acerola anthocyanins, resulting in mutual degradation of these molecules (De Rosso and Mercadante, 2007). The process postulated by Jurd (1972) for the degradation of anthocyanins in the presence of ascorbic acid, which was later confirmed by Poei-Langston and Wrolstad (1981), involves direct condensation of ascorbic acid on the carbon 4 of the anthocyanin molecule, resulting in the loss of both components.

Degradation kinetics of acerola anthocyanin was studied by Silva et al. (2017), where it was pasteurized in a tubular system operated at a temperature of 70°C and a residence period of around 20 s. By increasing the residence time at the same temperature, anthocyanin losses rise from 0.1 to 0.2%, indicating that residence time plays an essential role in anthocyanin degradation. De Rosso and Mercadante (2007) investigated the effects of sugar, salt, and light on the stability of anthocyanins in an acerola isotonic soft-drink system, finding that adding sugar and salt decreased anthocyanin stability. The degradation of anthocyanin was 1.2 times faster in the presence of light than in its absence.

2.5.9 Karonda (*Carissa carandas*)

Karonda (Figure 2.8I) is a shrub that grows on the Indian subcontinent and belongs to the Apocynaceae family. The fruits of the karonda tree are said to be antiscorbutic, astringent, and a treatment for constipation and stomach discomfort. High-performance liquid chromatography analysis revealed that the total anthocyanin concentration in karonda is 108.6 mg/kg and therefore, this fruit has great nutraceutical value (Sarkar et al., 2018).

Cyanidin-3-glucoside is the most prominent anthocyanin in karonda. Cyanidin-3-O-galactoside, delphinidin-3-O-glucoside, delphinidin-3-O-galactoside, 5-carboxypyranopelargonidin-3-O-beta-glucopyranoside, delphinidin, delphinidin-3-O-rutinoside, and delphinidin-3-O-arabinoside are minor anthocyanins found in karonda. The form of anthocyanin present is important for antioxidant activity. Anthocyanin is a fully conjugated structure which only exists in the lower pH range, contributing the most to its antioxidant activities (van Acker et al., 1996). In the case of conjugated structures, greater electron delocalization also leads to antioxidant action. The bioaccessibility of karonda crude anthocyanin extract (12.15%) was lower than that of purified anthocyanin (18.94%); because of low pH in the gut, breakdown of anthocyanin occurs, which reduces its bioaccessibility.

The change in anthocyanin concentration and degree of polymerization of anthocyanin in Karonda are all influenced by temperature variation during extraction. When extraction temperature increased from 30 to 50°C, the anthocyanin concentration is higher, at 6.67 mg/g. The content of anthocyanin in the extract drops considerably when the temperature rises beyond 50°C. *Carissa carandas* fruits have a high anthocyanin content (max. 13.65 mg/g), which rises as they mature. As a result, for anthocyanin extraction, ripe fruits should be utilized. The citric acid (3.0–5.0 g/L) in aqueous extract solutions stabilizes the anthocyanin color and decreases its degradation rate. Furthermore, to prevent anthocyanin loss in the fruit, the anthocyanin extract of karonda should be stored in the freezer at 18°C (Le et al., 2019).

2.5.10 Fig (*Ficus carica*)

Fig is one of the oldest cultivated fruit trees and belongs to the family Moraceae. The color of the fig can range from dark purple to green. It can be consumed whole and raw; however, most people peel figs and eat the flesh while discarding the skin (Solomon et al., 2006). The presence of

TABLE 2.4
Anthocyanin Content of Different Varieties of Mature Fig

Fig Type	Total Anthocyanins (mg of cyn-3-glu/100 g)		
	Fig Fruit	Fig Skin	Fig Pulp
Mission	10.9 ± 1.3	27.3 ± 2.3	0.3 ± 1.3
Chechick	1.8 ± 0.2	7.7 ± 0.5	0.1 ± 0.2
Brown-Turkey	1.3 ± 0.1	6.5 ± 0.7	0.1 ± 0.1
Bursa	0.3 ± 0.03	4.1 ± 0.3	0.1 ± 0.03
Brunswick	ND	0.7 ± 0.1	ND
Kadota	ND	ND	ND

Note: ND = not detectable.

anthocyanins is responsible for color variation in fig peel, with dark purple types having the highest expression of genes controlling the anthocyanin pathway (Dueñas et al., 2008). The anthocyanidin skeleton of fig is made only from cyanidin. Mission, Chechick, Bursa, Brown-Turkey, Brunswick, and Kadota are major commercial varieties of fig with differences in color (black, red, yellow, and green) (Figure 2.8J). Most of the anthocyanins accumulate in the skin of the unripe fruit.

The Mission variety of immature fig has greater amount of total anthocyanin, i.e., 3.0 mg/100 g, whereas the Bursa variety contains 0.8 mg/100 g total anthocyanin. Brunswick and Kadota varieties of immature fig have low total anthocyanin content. The anthocyanin concentration of fig increases as the immature fruit begins to ripen.

The quantity of anthocyanins in mature fruits varies from 11.0 mg/100 g for Mission to 0.3 mg/100 g for Bursa, with lower anthocyanin levels of 0.1 mg/100 g for Brunswick and Kadota types. When compared to the other dark type, Chechick, the Mission cultivar has five times the anthocyanin content. When compared to the other purple variant Bursa, Brown-Turkey has a much greater anthocyanin concentration. The lighter variants Brunswick and Kadota have the lowest levels of anthocyanins, with no noticeable differences between them.

Cyanidin-3-rhamnoglucoside, cyanidin 3,5-diglucoside, and pelargonidin-3-rhamnoglucoside are the main anthocyanin groups found in the skin of the Mission variety of fig fruits (Solomon et al., 2006). The stability of anthocyanin extract of the Mission variety of figs is up to 14 days under suitable storage conditions at pH 3.0 and temperature 4°C in darkness (Aguilera-Ortiz et al., 2009).

The skin of fig fruits has more anthocyanin content (almost 100 times as high in the Mission variant) than the fleshy part in both ripe and unripe conditions. The dark color of the fruit skin is rich in more anthocyanins than in purple or lighter variants (Table 2.4).

Brown-Turkey and Mission varieties of fig have 28 and 36% relative antioxidant capacity for the anthocyanin component, respectively. In fig skin, 92% of overall antioxidant capacity is mainly contributed by cyanidin-3-rutinoside. The antioxidant capacity and anthocyanin content have a strong relationship. Because anthocyanin is found mostly in the skin of ripe fruits, many researchers suggest that the whole fruit should be eaten (Solomon et al., 2006).

REFERENCES

Aguilera-Ortíz, M., Alanis-Guzmán, M. G., García-Díaz, C. L., and Hernández-Brenes, C. M. (2009). Characterisation and stability of mission variety fig anthocyanins. *Universidad y Ciencia*, 25(2), 151–158.

Alighourchi, H., Barzegar, M., and Abbasi, S. (2008). Effect of gamma irradiation on the stability of anthocyanins and shelf-life of various pomegranate juices. *Food Chemistry, 110*, 1036–1040.

Andersen, Ø. M., Daayf, F., and Lattanzio, V. (2008). Recent advances in the field of anthocyanins – Main focus on structures. In: Daayf, F., and Lattanzio, V. (eds.) *Recent Advances in Polyphenol Research.* Singapore: Blackwell; pp. 167–201.

Aprodu, I., Milea, Ş. A., Enachi, E., Râpeanu, G., Bahrim, G. E., and Stănciuc, N. (2020). Thermal degradation kinetics of anthocyanins extracted from purple maize flour extract and the effect of heating on selected biological functionality. *Foods, 9*(11), 1593.

Aqil, F., Gupta, A., Munagala, R., Jeyabalan, J., Kausar, H., Sharma, R. J., ... and Gupta, R. C. (2012). Antioxidant and antiproliferative activities of anthocyanin/ellagitannin-enriched extracts from *Syzygium cumini* L. (Jamun, the Indian blackberry). *Nutrition and Cancer, 64*(3), 428–438.

Bąkowska-Barczak, A. (2005). Acylated anthocyanins as stable, natural food colorants – A review. *Polish Journal of Food Nutrition Science, 14*(2), 107–116.

Barba, F. J., Esteve, M. J., and Frigola, A. (2013). Physicochemical and nutritional characteristic of blueberry juice after high pressure processing. *Food Research International, 50*(2), 545–549.

Barba, F. J., Terefe, N. S., Buckow, R., Knorr, D., and Orlien, V. (2015). New opportunities and perspectives of high pressure treatment improve health and safety attributes of foods. *Food Research International,* 77(4), 725–742.

Barnard, H., Dooley, A. N., Areshian, G., et al. (2011). Chemical evidence for wine production around 4000 BCE in the Late Chalcolithic Near Eastern highlands. *Journal of Archaeological Science, 38*(5), 977–984.

Borochov-Neori, H., Judeinstein, S., Harari, M., Bar-Ya'akov, I., Patil, B. S., Lurie, S., and Holland, D. (2011). Climate effects on anthocyanin accumulation and composition in the pomegranate (*Punica granatum* L.) fruit arils. *Journal of Agricultural and Food Chemistry, 59*(10), 5325–5334.

Boss, P. K., Davies, C. and Robinson, S. P. (1996). Analysis of the expression of anthocyanin pathway genes in developing *Vitis vinifera* L. cv Shiraz grape berries and the implications for pathway regulation. *Plant Physiology, 111*(4), 1059–1066.

Boulton, R. (2001). The copigmentation of anthocyanins and its role in the color of red wine: A critical review. *American Journal of Enology and Viticulture, 52*(2), 67–87.

Brat, P., Tourniaire, F., and Amiot-Carlin, M. J. (2008). Stability and analysis of phenolic pigments. In: Socaciu, C. (ed.) *Food Colorants, Chemical and Functional Properties.* Boca Raton, FL: CRC Press; pp. 71–86

Bridle, P., and Timberlake, C. (1997). Anthocyanins as natural food colours – selected aspects. *Food Chemistry,* 58, 103–109.

Bronnum-Hansen, K., and Flink, J. M. (1985). Anthocyanin colourants from elderberry (*Samhucus nigra* L.). Storage stability of the freeze-dried product. *Journal of Food Technology, 20*(6), 725–733.

Cao, X., Bi, X., Huang, W., Wu, J., Hu, X., and Liao, X. (2012). Changes of quality of high hydrostatic pressure processed cloudy and clear strawberry juices during storage. *Innovative Food Science and Emerging Technologies, 16,* 181–190.

Castañeda-Ovando, A., de Lourdes Pacheco-Hernández, M., Páez-Hernández, M. E., Rodríguez, J. A., and Galán-Vidal, C. A. (2009). Chemical studies of anthocyanins: A review. *Food Chemistry, 113*(4), 859–871.

Castro-Vargas, H. I., Benelli, P., Ferreira, S. R. S., and Parada-Alfonso, F. (2013). Supercritical fluid extracts from tamarillo (*Solanum betaceum* Sendtn) epicarp and its application as protectors against lipid oxidation of cooked beef meat. *Journal of Supercritical Fluids, 76,* 17–23.

Chandrapala, J., Oliver, C., Kentish, S., and Ashok Kumar, M. (2012). Ultrasonics in food processing – Food quality assurance and food safety. *Trends in Food Science and Technology, 26*(2), 88–98.

Chisté, R. C., Lopes, A. S., and De Faria, L. J. (2010). Thermal and light degradation kinetics of anthocyanin extracts from mangosteen peel (*Garcinia mangostana* L.). *International Journal of Food Science and Technology, 45*(9), 1902–1908.

Clifford, M. N. (2000). Anthocyanins – nature, occurrence and dietary burden. *Journal of the Science of Food and Agriculture, 80*(7), 1063–1072.

Contreras-Lopez, E., Casta, A., González-Olivares, L. G., and Jaimez-Ordaz, J. (2014). Effect of light on stability of anthocyanins in ethanolic extracts of *Rubus fruticosus. Food and Nutrition Sciences, 2014.*

Corrales, M., Butz, P., and Tausher, B. (2008). Anthocyanin condensation reactions under high hydrostatic pressure. *Food Chemistry, 110*(3), 627–635.

Davies, A. J., and Mazza, G. (1993). Copigmentation of simple and acylated anthocyanins with colorless phenolic compounds. *Journal of Agricultural and Food Chemistry, 41*(5), 716–720.

De Assis S. A., Fernandes F. P., Martins A. B. G., and Oliveira O. M. M. (2008). Aerola: importance, culture conditions, production and biochemical aspects. *Fruits*, *63*(2), 93–101.

De Assis, S. A., Lima, D. C., and de Faria-Oliveira, O. M. M. (2001). Activity of pectinmethyl-esterase, pectin content and vitamin C in acerola fruit at various stages of development. *Food Chemistry*, *74*(2), 133–137.

De Brito, E. S., De Araujo, M. C. P., Alves, R. E., Carkeet, C., Clevidence, B. A., and Novotny, J. A. (2007). Anthocyanins present in selected tropical fruits: acerola, jambolão, jussara, and guajiru. *Journal of Agricultural and Food Chemistry*, *55*(23), 9389–9394.

De Rosso, V. V. and Mercadante, A. Z. (2007). The high ascorbic acid content is the main cause of the low stability of anthocyanin extracts from acerola. *Food Chemistry*, *103*(3), 935–943.

De Rosso, V. V., and Mercadante, A. Z. (2005). Carotenoid composition of two Brazilian genotypes of acerola (*Malpighia punicifolia* L.) from two harvests. *Food Research International*, *38*, 1073–1077.

De Rosso, V. V., Hillebrand, S., Montilla, E. C., Bobbio, F. O., Winterhalter, P., and Mercadante, A. Z. (2008). Determination of anthocyanins from acerola (Malpighia emarginata DC.) and açai (Euterpe oleracea Mart.) by HPLC–PDA–MS/MS. *Journal of Food Composition and Analysis*, *21*(4), 291–299.

Diep, T., Pook, C., and Yoo, M. (2020). Phenolic and anthocyanin compounds and antioxidant activity of tamarillo (*Solanum betaceum* Cav.). *Antioxidants*, *9*(2), 169.

Du, C. T., Wang, P. L., and Francis, F. J. (1975). Anthocyanins of pomegranate, *Punica granatum*. *Journal of Food Science*, *40*(2), 417–418.

Dueñas, M., Pérez-Alonso, J. J., Santos-Buelga, C., and Escribano-Bailón, T. (2008). Anthocyanin composition in fig (*Ficus carica* L.). *Journal of Food Composition and Analysis*, *21*(2), 107–115.

Eiro, M. J., and Heinonen, M. (2002). Anthocyanin color behavior and stability during storage: Effect on intermolecular copigmentation. *Journal of Agricultural and Food Chemistry*, *50*(25), 7461–7466.

Erlandson, J. A., and Wrolstad, R. E. (1972). Degradation of anthocyanins in limited water concentration. *Journal of Food Science*, *37*(4), 592–595.

Espin, S., Gonzalez-Manzano, S., Taco, V., Poveda, C., Ayuda-Duran, B., Gonzalez-Paramas, A. M., et al. (2016). Phenolic composition and antioxidant capacity of yellow and purple-red Ecuadorian cultivars of tree tomato (*Solanum betaceum* Cav). *Food Chemistry*, *194*(1), 1073–1080.

Ferrari, G., Maresca, P., and Ciccarone, R. (2010). The application of high hydrostatic pressure for the stabilization of functional foods: Pomegranate juice. *Journal of Food Engineering*, *100*(2), 245–253.

Ferrari, G., Maresca, P., and Ciccarone, R. (2011). The effects of high hydrostatic pressure on the polyphenols and anthocyanins in red fruit products. *Procedia Food Science*, *1*, 847–853.

Fournand, D., Vicens, A., Sidhoum, L., Souquet, J. M., Moutounet, M., and Cheynier, V. (2006). Accumulation and extractability of grape skin tannins and anthocyanins at different advanced physiological stages. *Journal of Agricultural and Food Chemistry*, *54*(19), 7331–7338.

Francis, F.J. (1989). Food colorants: Anthocyanins. *Critical Review of Food Science and Nutrition*, *28*(4), 273–314.

Freitas, C. A. S. D., Maia, G. A., Sousa, P. H. M. D., Brasil, I. M., and Pinheiro, A. M. (2006). Storage stability of acerola tropical fruit juice obtained by hot fill method. International *Journal of Food Science & Technology*, *41*(10), 1216–1221.

Garcia-Palazon, A., Suthanthangjai, W., Kajda, P., and Zabetakis, I. (2004). The effects of high pressure on β-glucosidase, peroxidase and polyphenoloxidase in red raspberry (*Rubus idaeus*) and strawberry (*Fragaria × ananassa*). *Food Chemistry*, *88*(1), 7–10.

Garzón, G. A., and Wrolstad, R. E. (2001). The stability of pelargonidin-based anthocyanins at varying water activity. *Food Chemistry*, *75*(2), 185–196.

Giusti, M. M., and Wallace, T. C. (2009). Flavonoids as natural pigments. In: Bechtold, T., and Mussak, R. (eds.) *Handbook of Natural Colorants*. Chichester, West Sussex, UK: John Wiley; pp. 255–275.

Gonzalez-Aguilar, G., Villa-Rodriguez, J., Ayala-Zavala, J., and Yahia, M. E. (2010). Improvement of the antioxidant status of tropical fruits as a secondary response to some postharvest treatments. *Trends in Food Science and Technology*, *21*(10), 475–482.

Hale, K. L., McGrath, S. P., Lombi, E., Stack, S. M., Terry, N., Pickering, I. J., et al. (2001). Molybdenum sequestration in *Brassica* species: A role for anthocyanins? *Plant Physiology*, *126*(4), 1391–1402.

Hanamura, T., Hagiwara, T., and Kawagishi, H. (2005). Structural and functional characterization of polyphenols isolated from acerola (Malpighia emarginata DC.) fruit. *Bioscience Biotechnology and Biochemistry*, *69*(2), 280–286.

Hassan, A. S. H., and Bakar, A. M. F. (2013). Antioxidative and anticholinesterase activity of *Cyphomandra betacea* fruit. *Science World Journal, 2013*, 1–7.

Helmstädter, A. (2008). *Syzygium cumini* (L.) Skeels (Myrtaceae) against diabetes – 125 years of research. *Pharmazie, 63*(2), 91–101;

Hernandez, F., Melgarejo, P., Tomas-Barberan, F. A., and Artes, F. (1999). Evolution of juice anthocyanins during ripening of new selected pomegranate (*Punica granatum*) clones. *European Food Research Technology, 210*, 39–42.

Hocine, R., Farid, D., Yasmine, S., Khodir, M., Kapranov, V. N., and Kiselev, E. F. (2018). Recent advances on stability of anthocyanins. *RUDN Journal of Agronomy and Animal Industries, 13*(4), 257–286.

Howard, L. R., Brownmiller, C., and Prior, R. L. (2014). Improved color and anthocyanin retention in strawberry puree by oxygen exclusion. *Journal of Berry Research, 4*(2), 107–116.

Hrazdina, G. (1982). Anthocyanins. In: Harborne, J. B. and Marbray, T. J. (eds.) *The Flavonoids, Advances in Research*. New York: Chapman and Hall; pp. 135–188.

Huang, H. T. (1955). Fruit color destruction, decolorization of anthocyanins by fungal enzymes. *Journal of Agricultural and Food Chemistry, 3*(2), 141–146.

Huang, W., Zhu, Y., Li, C., Sui, Z., and Min, W. (2016). Effect of blueberry anthocyanins malvidin and glycosides on the antioxidant properties in endothelial cells. *Oxidative Medicine and Cellular Longevity, 2*, 1–10.

Hurtado, N. H., Morales, A. L., Gonzalez-Miret, M. L., Escudero-Gilete, M. L., and Heredia, F. J. (2009). Colour, pH stability and antioxidant activity of anthocyanin rutinosides isolated from tamarillo fruit (*Solanum betaceum* Cav.). *Food Chemistry, 117*(1), 88–93.

Jaakola, L. (2013). New insights into the regulation of anthocyanin biosynthesis in fruits. *Trends in Plant Science, 18*(9), 477–483.

Jiang, T., Mao, Y., Sui, L., Yang, N., Li, S., Zhu, Z., ... and He, Y. (2019). Degradation of anthocyanins and polymeric color formation during heat treatment of purple sweet potato extract at different pH. *Food Chemistry, 274*, 460–470.

Jurd. L. (1972). Some advances in the chemistry of anthocyanin type plant pigments. In: Chichester, C. O. (eds.) *The Chemistry of Plant Pigments*. New York: Academic Press; pp. 123–142.

Kallithraka, S., Aliaj, L., Makris, D. P., and Kefalas, P. (2009). Anthocyanin profiles of major red grape (*Vitis vinifera* L.) varieties cultivated in Greece and their relationship with in vitro antioxidant characteristics. *International Journal of Food Science and Technology, 44*(12), 2385–2393.

Katsumoto, Y., Fukuchi-Mizutani, M., Fukui, Y., et al. (2007). Engineering of the rose flavonoid biosynthetic pathway successfully generated blue-hued flowers accumulating delphinidin. *Plant Cell Physiology, 48*(11), 1589–1600.

Kerimi, A., Nyambe-Silavwe, H., Gauer, J.S., Tomás-Barberán, F. A., and Williamson, G. (2017). Pomegranate juice, but not an extract, confers a lower glycemic response on a high-glycemic index food: Randomized, crossover, controlled trials in healthy subjects. *American Journal of Clinical Nutrition, 106*(6), 1384–1393.

Khattab, H. A., El-Shitany, N. A., Abdallah, I. Z., Yousef, F. M., and Alkreathy, H. M. (2015). Antihyperglycemic potential of *Grewia asiatica* fruit extract against streptozotocin-induced hyperglycemia in rats: Anti-inflammatory and antioxidant mechanisms. *Oxidative Medicine and Cellular Longevity, 2015*, 1–7.

Khoo, H. E., Azlan, A., Tang, S. T., and Lim, S. M. (2017). Anthocyanidins and anthocyanins: Colored pigments as food, pharmaceutical ingredients, and the potential health benefits. *Food and Nutrition Research, 61*(1), 1–21.

Ko, A., Lee, J. S., Sop Nam, H., and Gyu Lee, H. (2017). Stabilization of black soybean anthocyanin by chitosan nanoencapsulation and copigmentation. *Journal of Food Biochemistry, 41*(2), 1–8.

Kostka, T., Ostberg-Potthoff, J. J., Briviba, K., Matsugo, S., Winterhalter, P., and Esatbeyoglu, T. (2020). Pomegranate (*Punica granatum* L.) extract and its anthocyanin and copigment fractions – Free radical scavenging activity and influence on cellular oxidative stress. *Foods, 9*(11), 1617.

Kırca, A., Özkan, M., and Cemeroğlu, B. (2007). Effects of temperature, solid content and pH on the stability of black carrot anthocyanins. *Food Chemistry, 101*(1), 212–218.

Kwon, J. Y., Lee, K. W., Hur, H. J., and Lee, H. J. (2007). Peonidin inhibits phorbol-ester-induced COX-2 expression and transformation in JB6 P+ cells by blocking phosphorylation of ERK-1 and -2. *Annals of the New York Academy of Sciences, 1095*(1), 513–520.

Le, X. T., Huynh, M. T., Pham, T. N., Than, V. T., Toan, T. Q., Bach, L. G., and Trung, N. Q. (2019). Optimization of total anthocyanin content, stability and antioxidant evaluation of the anthocyanin extract from Vietnamese *Carissa carandas* L. fruits. *Processes, 7*(7), 468.

Lee, H. S. (2002). Characterization of major anthocyanins and the color of red-fleshed Budd blood orange (*Citrus sinensis*). *Journal of Agricultural and Food Chemistry, 50*(5), 1243–1246.

Lee, Y. K., and Khng, H. P. (2002). Natural color additives. In: Branen, A. L., Davidson, P. M., Salminen, S., and Thorngate, III J. H. (eds.) *Food Science and Technology*. New York: Marcel Dekker; pp. 501–522.

Li, L., Zhang, Y., and Seeram, N. P. (2009). Structure of anthocyanins from *Eugenia jambolana* fruit. *Natural Product Communications, 4*(2), 217–219.

Lin, L.-Z., Harnly, J. M., Pastor-Corrales, M. S., and Luthria, D. L. (2008). The polyphenolic profiles of common bean (*Phaseolus vulgaris* L.). *Food Chemistry, 107*(1), 399–410.

Maccarone, E., Maccarrone, A., Perrini, G. E. A., and Rapisarda, P. (1983). Anthocyanins of the Moro orange juice. *Annali di Chimica, 73*(9–10), 533–539.

Maccarone, E., Maccarrone, A., and Rapisarda, P. (1985). Stabilization of anthocyanins of blood orange fruit juice. *Journal of Food Science, 50*(4), 901–904.

Markaris, P., Livingston, G. E., and Fellers, C. R. (1957). Quantitative aspects of strawberry pigment degradation a, b. *Journal of Food Science, 22*(2), 117–130.

Marszałek, K., Mitek, M., and Skapska, S. (2015). The effect of thermal pasteurization and high pressure processing at cold and mild temperatures on the chemical compositions, microbial and enzyme activity in strawberry puree. *Innovative Food Science and Emerging Technologies, 27*, 48–56.

Marszałek, K., Woz´niak, Ł., and Ska. pska, S. (2016). Application of high pressure mild temperature processing for prolonging shelf-life of strawberry purée. *High Pressure Research, 36*, 220–234.

Marszałek, K., Woźniak, Ł., Kruszewski, B., and Skąpska, S. (2017). The effect of high pressure techniques on the stability of anthocyanins in fruit and vegetables. *International Journal of Molecular Sciences, 18*(2), 277.

Mazza, G., and Brouillard, R. (1990). The mechanism of co-pigmentation of anthocyanins in aqueous solutions. *Phytochemistry, 29*(4), 1097–1102.

Mazza, G., and Francis, F. J. (1995). Anthocyanins in grapes and grape products. *Critical Reviews in Food Science and Nutrition, 35*(4), 341–371.

Moldovan, B., David, L., Chişbora, C., and Cimpoiu, C. (2012). Degradation kinetics of anthocyanins from European cranberrybush (*Viburnum opulus* L.) fruit extracts. Effects of temperature, pH and storage solvent. *Molecules, 17*(10), 11655–11666.

Moncada, M. C., Moura, S., Melo, M. J., Roque, A., Lodeiro, C., and Pina, F. (2003). Complexation of aluminum (III) by anthocyanins and synthetic flavylium salts: A source for blue and purple color. *Inorganica Chimica Acta, 356*, 51–61.

Mori, K., Goto-Yamamoto, N., Kitayama, M., and Hashizume, K. (2007). Loss of anthocyanins in red-wine grape under high temperature. *Journal of Experimental Botany, 58*(8), 1935–1945.

Nayak, B., Berrios, J. D. J., Powers, J. R., and Tang, J. (2011). Thermal degradation of anthocyanins from purple potato (Cv. Purple Majesty) and impact on antioxidant capacity. *Journal of Agricultural and Food Chemistry, 59*(20), 11040–11049.

Nikkhah, E., Khayamy, M., Heidari, R., and Jamee, R. (2007). Effect of sugar treatment on stability of anthocyanin pigments in berries. *Journal of Biological Sciences, 7*(8), 1412–1417.

Noda, Y., Kaneyuki, T., Mori, A., and Packer, L. (2002). Antioxidant activities of pomegranate fruit extract and its anthocyanidins: Delphinidin, cyanidin and pelargonidin. *Journal of Agricultural and Food Chemistry, 50*(1), 166–171.

Nora, C. D., Jablonski, A., Rios, A. D. O., Hertz, P. F., de Jong, E. V., and Flôres, S. H. (2014). The characterisation and profile of the bioactive compounds in red guava (*Psidium cattleyanum* Sabine) and guabiju (*Myrcianthes pungens* (O. Berg) D. Legrand). *International Journal of Food Science and Technology, 49*(8), 1842–1849.

Özkan, M. (2002). Degradation of anthocyanins in sour cherry and pomegranate juices by hydrogen peroxide in the presence of added ascorbic acid. *Food Chemistry, 78*(4), 499–504.

Parisi, O. I., Casaburi, I., Sinicropi, M. S., Avena, P., Caruso, A., Givigliano, F., ... and Puoci, F. (2014). Most relevant polyphenols present in the Mediterranean diet and their incidence in cancer diseases. In: Watson, R., Preedy, V., and Zibady, S. (eds.) *Polyphenols in Human Health and Disease*. New York: Academic Press; pp. 1341–1351.

Patel, K., Jain, A., and Patel, D. K. (2013). Medicinal significance, pharmacological activities, and analytical aspects of anthocyanidins 'delphinidin': A concise report. *Journal of Acute Disease*, *2*(3), 169–178.

Pifferi, P. G., and Cultrera, R. (1974). Enzymatic degradation of anthocyanins: The role of sweet cherry polyphenol oxidase. *Journal of Food Science.* *39*(4), 786–791.

Poei-Langston, M. S., and Wrolstad, R. E. (1981). Color degradation in an ascorbic acid-anthocyanin-flavanol model system. *Journal of Food Science*, *46*(4), 1218–1236.

Pomar, F., Novo, M., and Masa, A. (2005). Varietal differences among the anthocyanin profiles of 50 red table grape cultivars studied by high performance liquid chromatography. *Journal of Chromatography A*, *1094*, 34–41.

Robinson, G. M., and Robinson, R. (1932). A survey of anthocyanins. *Biochemical Journal*, *6*(5), 1647.

Sagrawat, H., Mann, A. S., and Kharya, M. D. (2006). Pharmacological potential of *Eugenia jambolana*: A review. *Pharmacognosy Magazine*, *2*(6), 96–105.

Santini, R., and Huyke, A. J. (1993). Identification of the anthocyanin present in the acerola which produces color changes in the juice on pasteurization and canning. In: Mazza, G., and Miniati, E. (eds.) *Anthocyanins in Fruits, Vegetables and Grains*. London: CRC Press; pp. 171–176.

Sarkar, R., Kundu, A., Banerjee, K., and Saha, S. (2018). Anthocyanin composition and potential bioactivity of karonda (*Carissa carandas* L.) fruit: An Indian source of biocolorant. *LWT*, *93*, 673–678.

Seeram, N. P., Zhang, Y., and Heber, D. (2006). Commercialization of pomegranates: Fresh fruit, beverages, and botanical extracts. In: Seeram, N. P., Schulman, R. N., and Heber, D. (eds.) *Pomegranates:Ancient Roots to Modern Medicine.* Boca Raton, FL: CRC Press; pp. 187–198.

Shaik, A., Killari, K. N., and Panda, J. (2018). A review on anthocyanins: A promising role on phytochemistry and pharmacology. *International Research Journal of Pharmacy*, *9*(1), 1–9.

Sinha, J., Purwar, S., Chuhan, S. K., and Rai, G. (2015). Nutritional and medicinal potential of *Grewia subinaequalis* DC. (syn. *G. asiatica*) (phalsa). *Journal of Medicinal Plants Research*, *9*(19), 594–612.

Skrede, C., and Wrolstad, R. E. (2002). Flavonoids from berries and grapes. In: Shi, J., Mazza, G., Le Maguer, M., and Boca, R. (eds.) *Functional Foods: Biochemical and Processing Aspects.* Boca Raton, FL: CRC Press; pp. 71–133.

Solomon, A., Golubowicz, S., Yablowicz, Z., Grossman, S., Bergman, M., Gottlieb, H. E., ... and Flaishman, M. A. (2006). Antioxidant activities and anthocyanin content of fresh fruits of common fig (*Ficus carica* L.). *Journal of Agricultural and Food Chemistry*, *54*(20), 7717–7723.

Starr, M. S., and Francis, F. J. (1973). Effect of metallic ions on color and pigment content of cranberry juice cocktail. *Journal of Food Science*, *38*(6), 1043–1046.

Stintzing, F. C., and Carle, R. (2004). Functional properties of anthocyanins and betalains in plants, food, and in human nutrition. *Trends in Food Science and Technology*, *15*(1), 19–38.

Sweeny, J. G., Wilkinson, M. M., and Iacobucci, G. A. (1981). Effect of flavonoid sulfonates on the photobleaching of anthocyanins in acid solutions. *Journal of Agricultural and Food Chemistry*, *29*, 563–567.

Talpur, M. K., Talpur, F. N., Balouch, A., Nizamani, S. M., Surhio, M. A., Shah, M. R., ... and Afridi, H. I. (2017). Analysis and characterization of anthocyanin from phalsa (*Grewia asiatica*). *MOJ Food Process Technology*, *5*(3), 299–305.

Tena, N., Martín, J., and Asuero, A. G. (2020). State of the art of anthocyanins: Antioxidant activity, sources, bioavailability, and therapeutic effect in human health. *Antioxidants*, *9*(5), 11–27.

Terefe, N. S., Matthies, K., Simons, L., and Versteeg, C. (2009). Combined high-pressure-mild temperature processing for optimal retention of physical and nutrition quality of strawberries (*Fragaria ananassa*). *Innovative Food Science and Emerging Technologies*, *10*(3), 297–307.

Terefe, N. S., Kleintschek, T., Gamage, T., Fanning, K. J., Netzel, G., Versteeg, C., and Netzel, M. (2013). Comparative effect of thermal and high-pressure processing on phenolic phytochemicals in different strawberry cultivars. *Innovative Food Science and Emerging Technology*, *19*, 57–65.

Thakur, B. R., and Arya, S. S. (1989). Studies on stability of blue grape anthocyanins. *International Journal of Food Science and Technology*, *24*(3), 321–326.

Torres, B., Tiwari, B. K., Partas, A., Cullen, P. J., Brunton, N., and O'Donnel, C. P. (2011). Stability of anthocyanins and ascorbic acid of high pressure processes blood orange juice during storage. *Innovative Food Science and Emerging Technologies*, *12*(2), 93–97.

Troise, A. D., and Fogliano, V. (2013). Reactants encapsulation and Maillard reaction. *Trends in Food Science and Technology.* 33(1), 63–74.

Usseglio-Tomasset, L. (1992). Properties and use of sulphur dioxide. *Food Additives and Contaminants*, *9*(5), 399–404.

van Acker, S. A. B. E., van den Berg, D. J., Tromp, M. N. J. L., Griffioen, D. H., van Bennekom, W. P., et al. (1996). Structural aspects of antioxidant activity of flavonoids. *Free Radical Biology and Medicine*, *20*, 331–342.

Vendramini, A. L., and Trugo, L. C. (2000). Chemical composition of acerola fruit (*Malpighia punicifolia* L.) at three stages of maturity. *Food Chemistry*, *71*(2), 195–198.

Vendramini, A. L., and Trugo, L. C. (2004). Phenolic compounds in acerola fruit (*Malpighia punicifolia* L.). *Journal of the Brazilian Chemical Society*, *15*(5), 664–668.

Verbeyst, L., Oey, I., Plancken, V. I., Hendrickx, M., and Loey, A. (2010). Kinetic study on the thermal and pressure degradation of anthocyanins in strawberries. *Food Chemistry*, *123*(2), 269–274.

Verbeyst, L., Hendrickx, M., and Loey, A. (2012). Characterisation and screening of the process stability of bioactive compounds in red fruit paste and red fruit juice. *European Food Research Technology*, *234*, 593–605.

Villegas-Ruiz, X., Rodriguez-Armas, D. N., Guerrero-Beltran, J.A., and Barcenas-Pozos, M. E. (2013). Stability of a sweet product of tamarillo (*Cyphomandra betacea*) preserved by combined methods. *Scientia Agropecuaria, 4*, 89–100.

Wang, S., and Zhu, F. (2020). Tamarillo (*Solanum betaceum*): Chemical composition, biological properties, and product innovation. *Trends in Food Science and Technology*, *95*, 45–58.

Weber, F., Boch, K., and Schieber, A. (2017). Influence of copigmentation on the stability of spray dried anthocyanins from blackberry. *LWT*, *75*, 72–77.

Welch, C. R., Wu, Q., and Simon, J. E. (2008). Recent advances in anthocyanin analysis and characterization. *Current Analytical Chemistry, 4*, 75–101.

Wills, J. M., Schmidt, D. B., Pillo-Blocka, F., and Cairns, G. (2009). Exploring global consumer attitudes toward nutrition information on food labels. *Nutrition Reviews, 67*(1), 102–106.

Wrolstad, R. E., Skrede, G., Lea, P. E. R., and Enersen, G. (1990). Influence of sugar on anthocyanin pigment stability in frozen strawberries. *Journal of Food Science*, *55*(4), 1064–1065.

Yang, P., Yuan, C., Wang, H., Han, F., Liu, Y., Wang, L., and Liu, Y. (2018). Stability of anthocyanins and their degradation products from Cabernet Sauvignon red wine under gastrointestinal pH and temperature conditions. *Molecules*, *23*(2), 1–20.

Yoshida, K., Kitahara, S., Ito, D., and Kondo, T. (2006). Ferric ions involved in the flower color development of the Himalayan blue poppy, *Meconopsis grandis. Phytochemistry*, *67*(10), 992–998.

Zhang, J., Celli, G. B., and Brooks, M. S. (2019). *Natural Sources of Anthocyanins*. In: Brooks, M. S., and Celli, G. B. (eds.) *Anthocyanins from natural sources: Exploiting targeted delivery for improved health*. London: Royal Society of Chemistry; pp. 1–33.

Zia-Ul-Haq, M., Stanković, M. S., Rizwan, K., and Feo, V. D. (2013). *Grewia asiatica* L., a food plant with multiple uses. *Molecules*, *18*(3), 2663–2682.

3 Extraction of Anthocyanins from Subtropical Fruits Using Thermal Processing Methods

M. Selvamuthukumaran

CONTENTS

3.1 INTRODUCTION

The color of fruit, which may range from red to purple, may be ascribed to the presence of anthocyanins. Anthocyanin can be used as a food additive, i.e. colorant, and is safe to use in food. There is a lot of scope for the extraction of anthocyanins from various subtropical fruits and also from their processed products, including the waste generated during processing, by adopting conventional heat extraction techniques. The major advantage of using a conventional heat-assisted extraction process is that it does not require any sophisticated infrastructure facilities like instruments and therefore it can be implemented at mass level in the food industries (Table 3.1).

3.2 EXTRACTION OF ANTHOCYANINS BY CONVENTIONAL THERMAL PROCESSING METHODS

Various conventional thermal processing methods are available to extract anthocyanins efficiently from subtropical fruits, including distillation process, solvent extraction method, pressing and sublimation (Figure 3.1). The mode of selecting the extraction process, which is solely dependent on the stability of the anthocyanin, has a direct correlation with its beneficial properties. Compounds such as proteins, lipids and other substances need to be removed before anthocyanin extraction; otherwise they may hinder the extraction process.

There is another method, known as the Soxhlet method, of extraction of anthocyanin. This includes mixing powdered solid particles with solvents in the Soxhlet apparatus, helping to extract anthocyanin by continuous extraction under controlled temperature and time period. The solvents and acids that are generally used to extract anthocyanins are shown in Figures 3.2 and 3.3.

DOI: 10.1201/9781003242598-3

TABLE 3.1
Advantages of Using Thermal Processing Methods for the Extraction of Anthocyanins from Subtropical Fruits

Method	Advantage
Thermal	It does not require modernized instrumentation facilities
	It can easily be implemented on a small scale
	Mode of operation is simple and easy
	It does not require huge investment
	It does not require an expert to handle the extraction process

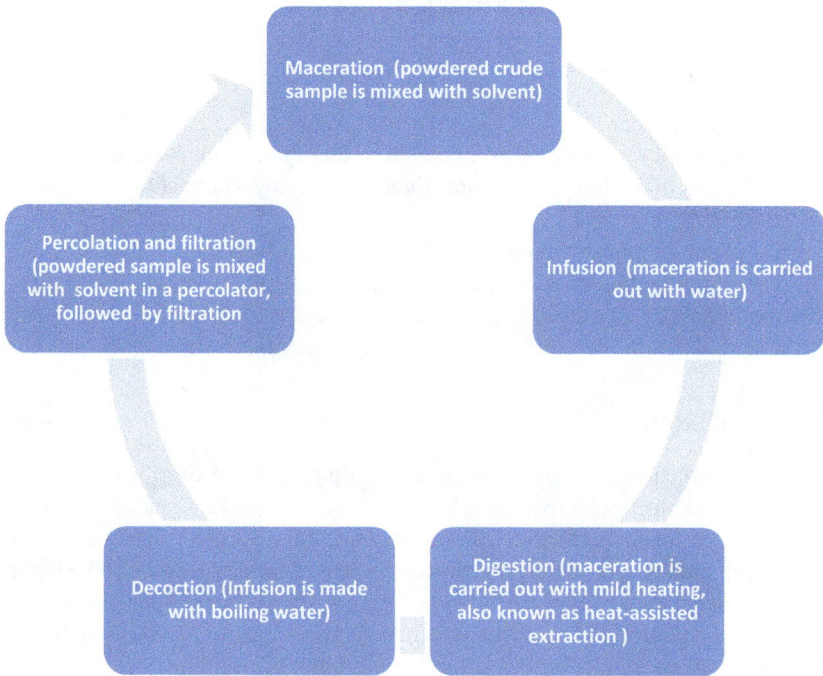

FIGURE 3.1 Conventional thermal methods of extraction of anthocyanins from various subtropical fruits.

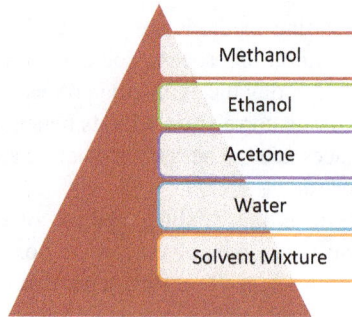

FIGURE 3.2 The solvents used for the extraction of anthocyanins.

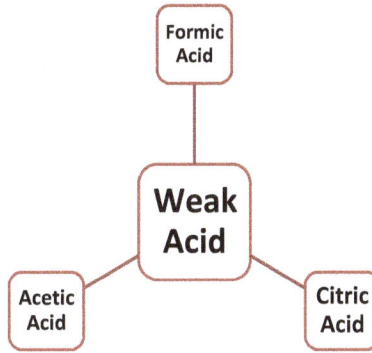

FIGURE 3.3   The acids used for the extraction of anthocyanins.

The use of weak acid is highly recommended because strong acid has a destabilizing effect on anthocyanin molecules. The extraction or yield can be further enhanced by incorporating water in the mixture of solvents due to the polar nature of the anthocyanin component. Various parameters need to be optimized before extracting anthocyanin; these include particle size, temperature and timing of extraction, solvent and solid ratio, so that maximum recovery is possible.

3.2.1   EXTRACTION OF ANTHOCYANINS FROM BLUEBERRIES USING THERMAL PROCESSING METHODS

Blueberries are a rich source of anthocyanins and have many health benefits, such as anticancer, antitumor, anti-inflammatory and antidiabetic activity (Luo et al., 2017).

The anthocyanin content of blueberries is around 35% and this may vary depending on the varieties available of the fruit (Cásedas et al., 2017). Anthocyanin exhibits potent antioxidant activity, thereby scavenging the formation of free radicals, as well as exhibiting antimutagenic activity cum bacteriostasis (Bakuradze et al., 2019; Zhou et al., 2018).

Anthocyanins can be efficiently extracted using heat extraction techniques and used as a functional ingredient in the preparation of several functional and drug-based products. Li et al. (2021) extracted anthocyanins from blueberries using a conventional solvent extraction process. The common extracting agents used to extract anthocyanins from blueberries were ethanol, acetone, methanol and alkanes (Costa et al., 2018), as these anthocyanins were highly soluble or miscible in these organic solvents. The extent of solubility of such anthocyanins is higher in ethanol compared to other solvents. During the anthocyanin extraction process, certain chemicals like acid hydrazone are incorporated into the alcohol, mainly to adjust the pH ratio or else a denaturation and structural transformation process may occur because the alkaline environment needs to be adjusted. The conventional method of anthocyanin extraction does not need higher investment and sophisticated equipment facilities. The mode of operation is not complex but the method does use huge quantities of organic solvents during the extraction process.

Li et al. (2021) used response surface methodology to obtain anthocyanins from blueberry fruit. They used acidified ethanol as an extraction solvent. They optimized the extraction parameters as a material-to-liquid ratio of 1:20 (g/mL), ethanol extract volume fraction volume of 60%, extraction temperature and time of 50°C and 122 min under pH conditions of 3.0. The extractants achieved an anthocyanin content of around 326 mg/100 g of the extract, which projected a higher degree of extraction, which is possible using solvents and heat.

3.2.2 EXTRACTION OF ANTHOCYANINS FROM GRAPES USING THERMAL PROCESSING METHODS

Ohmic heating, also known as moderate electric fields, is a popular heat-processing technique that can be used toextract anthocyanin from grapes. In ohmic heat processing the food product is being placed in direct contact with ohmic heater electrodes, in which an electric field is applied in the form of a bipolar or sinusoidal pulsed square wave with frequency ranging from 50 to 100 kHz (Sarkis et al., 2013). Pereira et al. (2016) reported that uniform and rapid heat is achieved as a result of using ohmic heating and this can be applied to particulate foods with the possibility of achieving higher efficient energy transfer. As a result, the processing time is highly reduced, resulting in enhanced flavor, better nutrition retention and finally, enhanced product quality. Several researchers (El Darra et al., 2013; Loypimai et al., 2015; Pereira et al. (2016) successfully extracted polyphenols and also polysaccharides like pectin from foods including food waste. The extraction of anthocyanin from grape skin, i.e. wine-making residue, was carried out using ohmic heating by Pereira et al. (2020). They used flash heating with a temperature ranging from 40 to 100°C for less than 20 s and also carried out electric pretreatment with temperature of 40°C and and time of 20 min. These authors found that ohmic heating highly enhanced the soluble solids and total phenolic compound content, thereby increasing the red color intensity. The use of higher temperature resulted in an anthocyanin yield of around maximum 78%.

Junqua et al. (2021) applied ohmic heating for the extraction of anthocyanin from grape must before the alcoholic fermentation in the wine-making process. The application of ohmic heating led to rupture of grape skin cell membrane and significantly improved phenolic compound release into the must compared with conventional heat treatment processes. The phenolic compound concentration in the final wine obtained from musts which are heated by ohmic heating showed similar results as wine produced from musts heated by traditional or conventional heat treatment process. The electroporation effect led to the rapid release of phenolic compounds as a result of the application of ohmic heat treatment. This process can produce prominent results when compared to wine produced by traditional methods. The researchers further concluded that ohmic heating can be significantly applied on a commercial scale in the food industry for extraction of anthocyanins from grape must.

The wine industry provide surplus amounts of byproducts, which are currently discarded and pose a significant pollution problem in the environment with respect to waste disposal. Pereira et al. (2021) extracted anthocyanins from winery waste using ohmic heating methods and compared this process with conventional extraction methods using solvents of food grade, i.e. methanol and water. The phenolic content recorded in conventional extraction methods was found to be 2.84 mg/g whereas for ohmic heating it was around 3.28 mg/g in terms of gallic acid equivalent. The authors quantified major anthocyanins in the extracts as cyanidin-3-O-glucoside, delphinidin-3-O-glucoside, petunidine-3-O-glucoside, peonidine-3-O-glucoside and malvidin-3-O-acetylglucoside. They found that the great advantage of using ohmic heating for anthocyanin extraction is less treatment time with more or less similar yield when compared to conventional methods of anthocyanin extraction. They recommended that the combination of ohmic heating with citric acid and water helps to extract anthocyanins efficiently and safely compared to the use of chemical solvents like methanol, which may leave toxins in the extractants if the limit level is exceeded, and anthocyanin from ohmic heating using citric acid and water can be successfully utilized in the production of functional food and drug-based products.

3.3 CONCLUSION

Natural colorants, i.e. anthocyanin, from subtropical fruits can be successfully extracted by adopting ohmic heating or solvent extraction processes under controlled conditions. This conventional method of extracting anthocyanin does not require sophisticated instrument facilities, minimizing costs of the manufacturing process for the food industries.

REFERENCES

Bakuradze, T., Tausend, A., Galan, J., Groh, I. A. M., Berry, D., Tur, J. A., et al. (2019). Antioxidative activity and health benefits of anthocyanin rich fruit juice in healthy volunteers. *Free Radical Research*, *53*(Suppl 1), 1045–1055.

Cásedas, G., Les, F., Gómez-Serranillos, M. P., Smith, C., & López, V. (2017). Anthocyanin profile, antioxidant activity and enzyme inhibiting properties of blueberry and cranberry juices: A comparative study. *Food Function*, *8*(11), 4187–4193.

Costa, D. V., Almeida, D. P., & Pintado, M. (2018). Effect of postharvest application of ethylene on the profile of phenolic acids and anthocyanins in three blueberry cultivars (*Vaccinium corymbosum*). *Journal Science of Food Agriculture*, *98*(13), 5052–5061.

El Darra, N., Grimi, N., Vorobiev, E., Louka, N., & Maroun, R. (2013). Extraction of polyphenols from red grape pomace assisted by pulsed ohmic heating. *Food and Bioprocess Technology*, *6*(5), 1281–1289. https://doi.org/10.1007/s11947-012-0869-7.

Junqua, R., Carullo, D., Ferrari, G., Pataro, G., & Ghidossi, R. (2021). Ohmic heating for polyphenol extraction from grape berries: An innovative prefermentary process. *OENO One*, *55*(3). DOI: https://doi.org/10.20870/oeno-one.2021.55.3.4647

Li, X., Zhu, F., & Zeng, Z. (2021). Effects of different extraction methods on antioxidant properties of blueberry anthocyanins. *Open Chemistry*. https://doi.org/10.1515/chem-2020-0052

Loypimai, P., Moongngarm, A., Chottanom, P., & Moontree, T. (2015). Ohmic heating-assisted extraction of anthocyanins from black rice bran to prepare a natural food colourant. *Innovative Food Science & Emerging Technologies*, *27*, 102–110. https://doi.org/10.1016/j.ifset.2014.12.009

Luo, S. Z., Chen, S. S., Pan, L. H., Qin, X.-S., Zheng, Z., Zhao, Y.-Y., et al. (2017). Antioxidative capacity of crude camellia seed oil: Impact of lipophilization products of blueberry anthocyanin. *International Journal of Food Properties*, *20*(Suppl 2), 1627–1636.

Pereira, R. N., Rodrigues, R. M., Genisheva, Z., Oliveira, H., de Freitas, V., Teixeira, J. A., & Vicente, A. A. (2016). Effects of ohmic heating on extraction of food-grade phytochemicals from colored potato. *LWT*, *74*, 493–503. https://doi.org/10.1016/j.lwt.2016.07.074

Pereira, R. N., Coelho, M. I., Genisheva, Z., Fernandes, J. M., Vicente, A. A., & Pintado, M. E. (2020). Using ohmic heating effect on grape skins as a pretreatment for anthocyanins extraction. *Food and Bioproducts Processing*, *124*, 320–328. https://doi.org/10.1016/j.fbp.2020.09.009

Pereira, E., Rodrigues, A. S., Teixeira, J. A., & Pintado, M. (2021). Anthocyanin recovery from grape by-products by combining ohmic heating with food-grade solvents: phenolic composition, antioxidant, and antimicrobial properties. *Molecules*, *26*, 3838. https://doi.org/10.3390/molecules26133838

Raudsepp, P., Koskar, J., Anton, D., Meremäe, K., Kapp, K., Laurson, P., et al. (2019). Antibacterial and antioxidative properties of different parts of garden rhubarb, blackcurrant, chokeberry and blue honeysuckle. *Journal of Science and Food Agriculture*, *99*(5), 2311–2320.

Sarkis, J. R., Jaeschke, D. P., Tessaro, I. C., & Marczak, L. D. (2013). Effects of ohmic and conventional heating on anthocyanin degradation during the processing of blueberry pulp. *LWT-Food Science and Technology*, *51*(1), 79–85. https://doi.org/10.1016/j.lwt.2012.10.024

Zhou, F., Wang, T., Zhang, B. L., & Zhao, H. (2018). Addition of sucrose during the blueberry heating process is good or bad? Evaluating the changes of anthocyanins/anthocyanidins and the anticancer ability in HepG-2 cells. *Food Research International*, *107*, 509–517.

4 Extraction of Anthocyanins from Subtropical Fruits Using Non-Thermal Processing Methods

M. Selvamuthukumaran

CONTENTS

4.1 INTRODUCTION

Anthocyanins contribute to fruit color; they possess excellent antioxidant effect (Speer et al., 2020; Tena et al., 2020) and can enhance the activity of gut microbiota. Anthocyanins can be obtained from naturally available raw materials and used as natural colorants (Jackman et al., 1987; Mateus and de Freitas, 2008; Roy and Rhim, 2020). In order to achieve maximum recovery it is essential to adopt non-thermal processing methods that can retain the color to the greatest extent.

4.2 EXTRACTION OF ANTHOCYANINS FROM GRAPES BY NON-THERMAL PROCESSING METHODS

Anthocyanins are concentrated in the skin region of the grape, known as the exocarp region, although in a few varieties anthocyanins can be seen in the pulp (Morata et al., 2019). Pectin, a kind of polysaccharide, needs to be disintegrated in order to extract anthocyanin. A maceration processing step, temperature used in fermentation and mechanical treatment help depolymerize the cell wall and achieve effective polysaccharide fiber separation, especially during the wine-making process (Morata et al., 2019). Busse-Valverde et al. (2011) extracted anthocyanins using

DOI: 10.1201/9781003242598-4

FIGURE 4.1 Non-thermal methods for anthocyanin extraction from subtropical fruit, i.e. grapes.

dry ice as a substitute for fermentation. Rio Segade et al. (2015) use enzymatic treatment to extract anthocyanins, thereby degrading cell wall pectins, which can facilitate the efficient extraction of anthocyanins. Non-thermal processing methods are available to extract anthocyanins efficiently from grapes (Figure 4.1), and these are described below.

4.2.1 HIGH HYDROSTATIC PRESSURE

In this method, higher pressure is applied to the food using water. Water is sent to a steel vessel, which is loaded with food products; here the food is subjected to a pressure treatment of 400–600 MPa (Buzrul, 2012). The high hydrostatic pressure treatment effect on plant tissues can damage cell wall integrity, resulting in smaller pores that can help extract metabolites from the cell wall. It was reported that anthocyanin was effectively extracted from grape using high hydrostatic pressure techniques (Corrales et al., 2009; Morata et al., 2014). The researchers found that the efficiency of anthocyanin using this method increased in the range of 24–82%. Morata et al. (2014) concluded that it is an excellent technology for extracting anthocyanins from grape pomace, enhancing antioxidant activity to a greater extent at temperatures below 30°C with pressure of 550 MPa for 10 min. During this treatment the researchers observed that anthocyanins had migrated from skin to seed as well as pulp, which clearly demonstrated that pressure led to migrating anthocyanin components into the berry as a result of cell wall poration.

4.2.2 PULSE ELECTRIC FIELD

The pulse electric field is nothing but food subjected to a higher-intensity electric field ranging from 3 to 40 kV/cm for a short time. This treatment led to cell poration on a nano-scale level; these pores can enhance cell permeability, enhancing anthocyanin extraction. Luengo et al. (2013) reported that grape and its byproducts – especially pomace – can be successfully processed using this pulse electric field technique; grapes can be handled at a maximum flow rate of 1900 kg/h (Luengo et al., 2013). The grapes were crushed and pumped using a peristaltic pump followed by the application of either decay pulses or squared pulses with an electric field strength ranging from 2 to 10 kV/cm (Tedjo et al., 2002; Corrales et al., 2008; Lopez et al., 2008; Puertolas et al., 2010). Using this process the extraction of anthocyanin varies from 17 to 100%, with varying post-maceration time and processing conditions.

4.2.3 ULTRASOUND

Ultrasound is mechanic waves with a frequency higher than 20 kHz, which is not perceptible to the human ear; human hearing varies in the range of 20 Hz to 20 kHz (Kumar et al., 2021). Bioactive compounds like anthocyanins can be effectively extracted using this technique and it is deemed to be one of the green extraction techniques (Chemat et al., 2017), as it does not utilize solvents and is resistant to heat-sensitive compounds (Tiwari, 2015). A cavitation is formed, ascribed to product compression, which is produced by ultrasound waves and this can further reduce the size and lead to expanding bubbles. The collapse of the bubbles may lead to a release of greater energy with a temperature of 5000K and pressure and 200 MPa (Chemat et al., 2011). These mechanisms can depolymerize the biostructures, further facilitating the extraction of biomolecules from plant tissues. In grapes cell wall polysaccharide depolymerization can lead to release of anthocyanins from grape skin. The temperature rise due to the cavitation effect can lead to the extraction of anthocyanin in water within a fraction of a few minutes. The best extraction efficiency was achieved in the range from 20 to 25 kHz (Chemat et al., 2017).

On a commercial basis, ultrasound waves can be applied using a continuous tubular exchanger with the external surface in which the sonoplates were distributed, when mash or liquid flows through the exchanger. The section can be hexagonal for a fine sonoplate distribution on the exchange surface. This technology can extract anthocyanins efficiently from wine obtained from Tempranillo grapes with a continuous skin maceration period of 72 h. On a lab scale González et al. (2020) reported that ultrasound has increased the anthocyanin extraction rate by 50% when compared to controls. The frequency of ultrasound waves has been analyzed, ranging from 12.5 to 37.5 kHz; it was observed that anthocyanin extraction efficiency was increased to 18% in the grape pomace at a lower frequency range (Dranca and Oroian, 2019).

4.2.4 E-BEAM IRRADIATION

This technique uses accelerated electrons at a higher energy, i.e. around 10 MeV (Morata et al., 2015,2017; Elias et al., 2020). The doses of irradiation were measured in either Grays (Gy) or kGrays (kGy). It can be adopted commercially as a continuous process (Morata et al., 2017). It is a non-thermal processing technique using increments of temperature less than 5°C with doses of up to 10 kGy. The application of e-beam irradiation results in fibrillar polymer fragmentation, which aids in promoting and releasing the anthocyanin. Morata et al. (2015) reported enhanced extraction of anthocyanins from grapes at an irradiation dose of 10 kGy with increased color intensity (Table 4.1). The application of an irradiation dose at 1 kGy does not show increased anthocyanin recovery, but if the irradiation dose is increased to 10 kGy, the anthocyanins extracted in the juice were found to be 125 mg/L (Morata et al., 2015). Kong et al. (2014) observed that in blueberries, the application of irradiation dose less than 3 kGy did not affect the antioxidant activity and monomeric anthocyanins. Yoon et al. (2020) also reported that strawberries that received an irradiation dose 1 kGy significantly preserved anthocyanin and antioxidant activity.

4.3 EXTRACTION OF ANTHOCYANINS FROM MAGNOLIA BERRY BY NON-THERMAL PROCESSING METHODS

The magnolia berry, i.e. *Schisandra chinensis*, is largely cultivated in Asian countries, where it is used as a food additive (Hancke et al., 1999). The fruit is a rich source of bioactive constituents, especially anthocyanins.

TABLE 4.1

Effect of Application of E-Beam Irradiation on the Recovery of Anthocyanins from Subtropical Fruits

Dose of Irradiation	Type of Product Applied	Effect Achieved	Reference
10 kGy	Grapes	Increased anthocyanin extraction with more color intensity	Morata et al. (2015)
Near to 3 kGy	Blueberries	No degradation of monomeric anthocyanins	Kong et al. (2014)
1 kGy	Strawberries	No degradation of anthocyanins	Yoon et al. (2020)

4.3.1 ULTRASOUND-ASSISTED EXTRACTION OF ANTHOCYANINS FROM *S. CHINENSIS*

Vilkhu et al. (2008) reported that using ultrasound as a non-thermal processing method can enhance the extraction efficiency of bioactive molecules. The major advantage of using this method in juice is minimal or reduced degradation of the anthocyanin effect that is being achieved (Tiwari et al., 2008, 2009).

The effect of the ultrasound processing technique on *S. chinensis* fruit anthocyanin stability was studied by Ma et al. (2012) in the dark with an ultrasonic bath; at its bottom, transducers of around 50 kHz were annealed. The rating of the bath power was 250 W and the bath water temperature was kept controlled at 25°C. The ultrasound process degradation time was kept in the range of 0–90 min. The authors found that, when the ultrsasound treatment time was 60 min, the residual anthocyanin content recorded was 95%, but when the treatment time was increased to 90 min the residual anthocyanin content was 93%. The authors further concluded that to achieve effective ultrasound-assisted extraction, the treatment time should be 60 min.

4.3.2 MICROWAVE-ASSISTED EXTRACTION OF ANTHOCYANINS FROM *S. CHINENSIS*

Natural ingredients can be successfully extracted using this microwave-assisted extraction technique (Chan et al., 2011). Liazid et al. (2011) and Yang and Zhai (2010) reported that using this microwave-assisted extraction technique anthocyanins were successfully extracted. The effect of microwave treatment on the anthocyanin stability of *S. chinensis* fruit was studied by Ma et al. (2012) in a microwave oven with output power of around 700 W. The degradation time of anthocyanin was kept in the range of 0–45 min. The residual fruit anthocyanin content was 95% when the treatment time was 5 min but when the time was increased to 45 min, the residual anthocyanin content of the fruit seemed to be 90%, proving that the longer treatment time slightly reduced the anthocyanin content of the *S. chinensis* fruit.

4.4 EXTRACTION OF ANTHOCYANINS FROM BLACKTHORN BY NON-THERMAL PROCESSING METHODS

Blackthorn, scientifically known as *Prunus spinosa*, is a wild shrub cultivated widely in European countries, especially in Spain and Portugal. The fruit has medicinal value and is used in the preparation of jams and liqueurs (Morales et al., 2013). The presence of anthocyanin in this fruit significantly contributes to its coloration (Usenik et al., 2008; Ieri et al., 2012).

Generally, one has to consider critical factors like temperature, pH, structure and concentration level, and oxygen level, as well as solvents used to extract anthocyanin from fruits (Rodriguez-Amaya, 2016). In order to achieve maximum yield it is essential to optimize the extraction method choice with extraction-variable optimization (Jiménez et al., 2018). Montesano et al. (2008) reported that extraction efficiency is highly correlated with observation of variability with different matrices. Leichtweis et al. (2019) adopted response surface methodology to extract anthocyanin efficiently from epicarp of blackthorn fruit using ultrasound as a non-thermal processing technique. They had chosen time, ultrasound power and ethanol content to extract anthocyanin from blackthorn fruit. They found that this fruit contains two major anthocyanins: peonidin 3-rutinoside and cyanidin 3-rutinoside. They achieved an anthocyanin concentration of 18.17 mg/g in an extract with the processing conditions of time of 5 min, ethanol of 48% and ultrasound power of 400 W. The authors further concluded that using ultrasound-assisted extraction of anthocyanins from this blackthorn fruit can enable a greater production of anthocyanins without much degradation.

4.5　CONCLUSION

Therefore, using several non-thermal processing techniques, anthocyanins can be successfully extracted from various kinds of subtropical fruit. This technique can be commercially implemented in order to maximize the efficiency of anthocyanin extraction. The degradation effect can be greatly minimized using this technique.

REFERENCES

Busse-Valverde, N., Gómez-Plaza, E., López-Roca, J. M., Gil-Muñoz, R., & Bautista-Ortín, A. B. The extraction of anthocyanins and proanthocyanidins from grapes to wine during fermentative maceration is affected by the enological technique. *J. Agric. Food Chem.* 2011; *59*, 5450–5455.

Buzrul, S. High hydrostatic pressure treatment of beer and wine: A review. *Innov. Food Sci. Emerg. Technol.* 2012; *13*, 1–12.

Chan, C. H., Yusoff, R., Ngoh, G. C., & Kung, F. W. L. Microwave-assisted extractions of active ingredients from plants. *J. Chromatogr. A.* 2011; 1218, 6213–6225.

Chemat, F., Zill-E-Huma, & Khan, M. K. Applications of ultrasound in food technology: Processing, preservation and extraction. *Ultrason. Sonochem.* 2011; *18*, 813–835.

Chemat, F., Rombaut, N., Sicaire, A. G., Meullemiestre, A., Fabiano-Tixier, A. S., & Abert-Vian, M. Ultrasound assisted extraction of food and natural products. Mechanisms, techniques, combinations, protocols and applications. A review. *Ultrason. Sonochem.* 2017; *34*, 540–560.

Corrales, M., Toepfl, S., Butz, P., Knorr, D., & Tauscher, B. Extraction of anthocyanins from grape by-products assisted by ultrasonics, high hydrostatic pressure or pulsed electric fields: A comparison. *Innov. Food Sci. Emerg. Technol.* 2008; *9*, 85–91.

Corrales, M., García, A. F. Butz, P., & Tauscher, B. Extraction of anthocyanins from grape skins assisted by high hydrostatic pressure. *J. Food Eng.* 2009; *90*, 415–421.

Dranca, F., & Oroian, M. Kinetic improvement of bioactive compounds extraction from red grape (*Vitis vinifera* Moldova) pomace by ultrasonic treatment. *Foods* 2019; *8*, 353.

Elias, M. I., Madureira, J., Santos, P. M. P., Carolino, M. M., Margaça, F. M. A., & Cabo Verde, S. Preservation treatment of fresh raspberries by e-beam irradiation. *Innov. Food Sci. Emerg. Technol.* 2020; *66*, 102487.

González, M., Barrios, S., Budelli, E., Pérez, N., Lema, P., & Heinzen, H. Ultrasound assisted extraction of bioactive compounds in fresh and freeze-dried *Vitis vinifera* cv Tannat grape pomace. *Food Bioprod. Process.* 2020; *124*, 378–386.

Hancke, J. L., Burgos, R. A., & Ahumada, F. *Schisandra chinensis* (Turcz.) Baill. *Fitoterapia* 1999; 70, 451–471.

Ieri, F., Pinelli, P., & Romani, A. Simultaneous determination of anthocyanins, coumarins and phenolic acids in fruits, kernels and liqueur of *Prunus mahaleb* L. *Food Chem.* 2012; 135, 2157–2162.

Jackman, R. L., Yada, R. Y., & Tung, M. A. (1987). A review: separation and chemical properties of anthocyanins used for their qualitative and quantitative analysis. *J. Food Biochem.*, *11*(4), 279–308. https://doi.org/10.1111/j.1745-4514.1987.tb00128.x

Jiménez, L. C., Caleja, C., Prieto, M. A., Barreiro, M. F., Barros, L., & Ferreira, I. C. F. R. Optimization and comparison of heat and ultrasound assisted extraction techniques to obtain anthocyanin compounds from *Arbutus unedo* L. fruits. *Food Chem.* 2018; *264*, 81–91.

Kong, Q., Wu, A., Qi, W., Qi, R., Carter, J. M., Rasooly, R., & He, X. Effects of electron-beam irradiation on blueberries inoculated with *Escherichia coli* and their nutritional quality and shelf life. *Postharvest Biol. Technol.* 2014; *95*, 28–35.

Kumar, K., Srivastav, S., & Sharanagat, V. S. Ultrasound assisted extraction (UAE) of bioactive compounds from fruit and vegetable processing by-products: A review. *Ultrason. Sonochem.* 2021; *70*, 105325.

Leichtweis, M. G., Carla Pereira, M. A., Prieto, M,. F., Barreiro, I. J., Baraldi, L. B., & Ferreira, I. C. F. R. Ultrasound as a rapid and low-cost extraction procedure to obtain anthocyanin-based colorants from *Prunus spinosa* L. fruit epicarp: Comparative study with conventional heat-based extraction. *Molecules* 2019; 24, 573. doi:10.3390/molecules24030573.

Liazid, A., Guerrero, R. F., Cantos, E., Palma, M., & Barroso, C. G. Microwave assisted extraction of anthocyanins from grape skins. *Food Chem.* 2011; 124, 1238–1243.

López, N., Puértolas, E., Condón, S., Álvarez, I., & Raso, J. Application of pulsed electric fields for improving the maceration process during vinification of red wine: Influence of grape variety. *Eur. Food Res. Technol.* 2008; *227*, 1099–1107.

Luengo, E., Franco, E., Ballesteros, F., Álvarez, I., & Raso, J. Winery trial on application of pulsed electric fields for improving vinification of Garnacha grapes. *Food Bioprocess Technol.* 2013; *7*, 1457–1464.

Ma, C., Lang, Y., Yang, F., Wang, W., Zhao, C., & Zu, Y. Content and color stability of anthocyanins isolated from *Schisandra chinensis* fruit. *Int. J. Mol. Sci.* 2012; *13*(11), 14294–14310.

Mateus, N., & de Freitas, V. Anthocyanins as food colorants. In: Winefield, C., Davies, K., and Gould, K. (eds.) *Anthocyanins*. New York, NY: Springer, 2008; pp. 284–304.

Montesano, D., Fallarino, F., Cossignani, L., Simonetti, M. S., Puccetti, P., & Damiani, P. Innovative extraction procedure for obtaining high pure lycopene from tomato. *Eur. Food Res. Technol.* 2008; 226, 327–335.

Morales, P., Ferreira, I. C. F. R., Carvalho, A. M., Fernández-Ruiz, V., Sánchez-Mata, M. S. O. S. C. C., Cámara, M., Morales, R., & Tardío, J. Wild edible fruits as a potential source of phytochemicals with capacity to inhibit lipid peroxidation. *Eur. J. Lipid Sci. Technol.* 2013; 115, 176–185.

Morata, A., Loira, I., Vejarano, R., Bañuelos, M. A., Sanz, P. D., Otero, L., & Suárez-Lepe, J. A. Grape processing by high hydrostatic pressure: Effect on microbial populations, phenol extraction and wine quality. *Food Bioprocess Technol.* 2014; *8*, 277–286.

Morata, A., Bañuelos, M. A., Tesfaye, W., Loira, I., Palomero, F., Benito, S., Callejo, M. J., Villa, A., González, M. C., & Suárez-Lepe, J. A. Electron beam irradiation of wine grapes: Effect on microbial populations, phenol extraction and wine quality. *Food Bioprocess Technol.* 2015; *8*, 1845–1853.

Morata, A., Loira, I., Vejarano, R., González, C., Callejo, M. J., & Suárez-Lepe, J. A. Emerging preservation technologies in grapes for winemaking. *Trends Food Sci. Technol.* 2017; *67*, 36–43.

Morata, A., Escott, C., Loira, I., Manuel Del Fresno, J., González, C., & Suárez-Lepe, J. A. Influence of *Saccharomyces* and non-*Saccharomyces* yeasts in the formation of pyranoanthocyanins and polymeric pigments during red wine making. *Molecules* 2019; *24*, 4490.

Puértolas, E., López, N., Condón, S., Álvarez, I., & Raso, J. Potential applications of PEF to improve red wine quality. *Trends Food Sci. Technol.* 2010; *21*, 247–255.

Río Segade, S., Pace, C., Torchio, F., Giacosa, S., Gerbi, V., & Rolle, L. Impact of maceration enzymes on skin softening and relationship with anthocyanin extraction in wine grapes with different anthocyanin profiles. *Food Res. Int.* 2015; *71*, 50–57.

Rodriguez-Amaya, D. B. Natural food pigments and colorants. *Curr. Opin. Food Sci.* 2016; *7*, 20–26.

Roy, S., & Rhim, J.-W. Anthocyanin food colorant and its application in pH-responsive color change indicator films. *Crit. Rev. Food Sci. Nutr.* 2020; *61*, 2297–2325.

Speer, H., D'Cunha, N. M., Alexopoulos, N. I., McKune, A. J., & Naumovski, N. Anthocyanins and human health – A focus on oxidative stress, inflammation and disease. *Antioxidants* 2020; *9*, 366.

Tedjo, W., Eshtiaghi, M. N., & Knorr, D. Einsatz nicht-thermischer Verfahren zur Zellpermeabilisierung von Weintrauben und Gewinnung von Inhaltsstoffen. *Fluss. Obs.* 2002; *69*, 578–585.

Tena, N., Martín, J., & Asuero, A. G. State of the art of anthocyanins: Antioxidant activity, sources, bioavailability, and therapeutic effect in human health. *Antioxidants* 2020; *9*, 451.

Tiwari, B. K. Ultrasound: A clean, green extraction technology. *TrAC Trends Anal. Chem.* 2015; *71*, 100–109.

Tiwari, B. K., O'Donnell, C. P., Patras, A., & Cullen, P. J. Anthocyanin and ascorbic acid degradation in sonicated strawberry juice. *J. Agric. Food Chem.* 2008; *56*, 10071–10077.

Tiwari, B. K., O'Donnell, C. P., & Cullen, P. J. Effect of sonication on retention of anthocyanins in blackberry juice. *J. Food Eng.* 2009; *93*: 166–171.

Usenik, V., Fabčič, J., & Štampar, F. Sugars, organic acids, phenolic composition and antioxidant activity of sweet cherry (*Prunus avium* L.). *Food Chem.* 2008; *107*, 185–192.

Vilkhu, K., Mawson, R., Simons, L., & Bates D. Applications and opportunities for ultrasound assisted extraction in the food industry – A review. *Innov. Food Sci. Emerg. Technol.* 2008; *9*, 161–169.

Yang, Z. D., & Zhai, W. W. Optimization of microwave-assisted extraction of anthocyanins from purple corn (*Zea mays* L.) cob and identification with HPLC-MS. *Innov. Food Sci. Emerg. Technol.* 2010; *11*, 470–476.

Yoon, Y.-S., Kim, J.-K., Lee, K.-C., Eun, J.-B., & Park, J.-H. Effects of electron-beam irradiation on postharvest strawberry quality. *J. Food Process. Preserv.* 2020; *44*, e14665.

5 Commercial Production of Anthocyanins from Subtropical Fruits

Sachin Kumar, Swati Mitharwal, Sourabh Kumar,
Khalid Bashir, Kulsum Jan, and Aman Kaushik

CONTENTS

5.1 INTRODUCTION

Tropical fruit crops are typically grown in the geographical zone stretching from 30° south to 30° north latitudes, where climatic conditions are suitable, with average temperatures of roughly 27°C, low photoperiod differences, and high humidity levels. Summers are hotter and winters are cooler in subtropical zones, with greater photoperiod variations and lower humidity. The fruit crops grown under a climatic condition between temperate and tropical are known as subtropical fruit crops (Barea-Álvarez et al., 2016). The most common subtropical fruits are citrus fruits such as oranges, grapefruits, lemons, and limes. Other subtropical fruits include dates, figs, olives, grapes, kiwi or Chinese gooseberry, persimmon, stone fruits (apricot, plum), pomegranates, certain types of avocados, guava, litchi, cherry, finger lime, mock strawberry, pear, naartijie, and tamarillo. Anthocyanins (in Greek *anthos* means flower and *kianos* means blue) are the phenolic compound referred to as flavonoids, and are widely present throughout the plant system (Sharif et al., 2020). Owing to their antioxidant effects, anthocyanins have been found to be effective against several ailments such as cancer, inflammation, and other neurological and cardiovascular diseases. Depending on the pH of the environment, this broad and significant category of water-soluble phytopigments provides color to many fruits and flowers, e.g., purple, blue, red, orange, and pink hues (Wang et al., 2017). Anthocyanins are the plant secondary metabolites that are mostly present

DOI: 10.1201/9781003242598-5

in fruits and flowers. Major sources of anthocyanins are colored fruits, e.g., black and red berries, dark-colored vegetables, e.g., black bean, eggplant, red onion, red cabbage, red radish, purple corn, and sweet potato (He & Giusti, 2010). A plant-based diet is the main route through which anthocyanins become part of a human diet (Santos-Buelga & González-Paramás, 2018). As a result, it may be used as a replacement for synthetic dyes in the commercial food industry, owing to their appealing hues and high water solubility, making them suitable for incorporation in aqueous-based food media. Further, anthocyanins have been recognized as food colorants by different countries, with E-163 being the specific color code allocated (Arroyo-Maya & McClements, 2015).

During the last decades, a wide range of research has been performed to examine the diverse protective and promoting effects of anthocyanins found in a variety of fruits and vegetables on human health. It is because of their antioxidant qualities that plant-based foods that are high in anthocyanins have shown pharmacological and therapeutic application (Carvalho et al., 2016). Furthermore, recent studies have discovered a relationship between anthocyanin intake and the prevention of neurological diseases, as well as the control of age-related decline in brain and cognitive functions, together with their role in the reduction of heart disease, enhancement of vision and brain performance, and as an anti-inflammatory in chronic diseases such as obesity and diabetes (Morais et al., 2016; Riaz et al., 2016; Tsuda, 2012). The consumption of anthocyanin is extremely intriguing; yet its chemical instability and various harsh food-processing conditions have created difficulties in increasing their bioavailability (de Moura et al., 2018). In this sense, encapsulation has the ability to minimize nutritional losses and might increase the half-life of the bioactive core material. The encapsulation technique involves encasing bioactive chemicals in a coating material which helps deliver the core compound at the appropriate time and location. Further, particle size distribution can be classified into three main categories: (1) micro (1–5000 m); (2) macro (more than 5000 m); and (3) nano (less than 1 m) (Jafari et al., 2008). Encapsulation has numerous advantages for the bioactive chemical, including improved oxidative, thermal, and photo stability, bioavailability, flavor masking, sustained and controlled release, and ease of handling (Sharif et al., 2020).

In general, several strategies for creating micro- and nanocarriers for encapsulating anthocyanins have been reported. The first category of carriers is those that require specific equipment to manufacture, such as spray drying, freeze drying, and electro-spraying/spinning (da Rosa et al., 2019; Prietto et al., 2018). Furthermore, encapsulation techniques employing lipid-based (fats, oils) carriers, e.g. liposomes and emulsions, have successfully been utilized to encapsulate anthocyanins obtained from various plant sources (Chi et al., 2019). Finally, micro/nanocarriers are made via an ionic gelation method, which is a type of encapsulation that involves dripping, atomization (extrusion, coextrusion), or spray operations (Arriola et al., 2019; da Silva et al., 2019). Several different types of biopolymers have been investigated for encapsulation of anthocyanins, irrespective of the procedures used for developing the carriers. Anthocyanin protection has been studied extensively in carbohydrates such as gum Arabic, maize starch, and maltodextrin, as well as proteins such as whey protein concentrate, gelatin, and soy protein isolate (Sharif et al., 2020). Thus, the focus of this chapter is to cover various traditional and emerging encapsulation strategies for protecting anthocyanins from harmful circumstances (gastrointestinal, environmental etc.), and improving their chemical stability and bioavailability.

5.1.1 What Are Anthocyanins?

Anthocyanins are heterosides of anthocyanidins (aglycone unit) connected with glycosides. Flavylium cation (C6–C3–C6) is the fundamental structure of anthocyanins, which can be connected to different sugars, hydroxyls, and methoxyl groups. The structure of anthocyanins depends upon several factors: (1) amount and location of hydroxyl group; (2) degree of methoxyl groups in anthocyanin molecules; (3) quantity, type, and positions of connected sugar; and (4) number, nature, and extent of aromatic or aliphatic acid coupled to the sugar. The most common sugar in anthocyanins

TABLE 5.1
Chemical Structures and Other Characteristics of Natural Anthocyanidins

Anthocyanidin	Chemical Structure	R1	R2	R3	Main Color	Examples
Cyanidin		–OH	–OH	–H	Reddish-orange	Apples, blackberries, cabbages, peaches, elderberries, plums, nectarines
Delphinidin		–OH	–OH	–OH	Purple, blue	Beans, eggplants, grapes, oranges
Malvidin		–OCH3	–OH	–OCH3	Purple	Grapes
Pelargonidin		–H	–OH	-H	Orange	Strawberries, some beans, red radishes
Peonidin		–OCH3	–OH	-H	Purplish-red	blueberries, cranberries, cherries, corn, plums, grapes, purple
Petunidin		–OH	–OH	–OCH3	Dark-red or purple	Grapes, red berries

Source: Adapted and modified from Mohammadalinejhad and Kurek (2021) and Sharif et al. (2020).

is glucose; others include arabinose, galactose, rhamnose, rutinose and xylose. The molecular weight may range between 400 and 1200 (medium-size biomolecules). Depending on the quantity of sugars attached, anthocyanins are classified as mono-, di-, or tri-glycosides (Tarone et al., 2020). Anthocyanin aglycones are referred as anthocyanidins. There are roughly 23 different varieties of anthocyanidins; the most frequent ones are cyanidin, delphinidin, malvidin, pelargonidin, petunidin, and peonidin (Table 5.1).

5.2 BOTANICAL SOURCES OF ANTHOCYANINS

Anthocyanins are generally isolated from a variety of plants, including carrots, cabbage, corn, hibiscus, and sweet potato. However, berries have been documented as a chief source of anthocyanin, accounting for 29.17% of all sources, followed by black rice (8.33%), black carrot (8.33%), sour cherry (8.33%), and grape (8.33%) (Table 5.2). Other anthocyanin sources include purple maize, pomegranate fruit, saffron, banana, and mao fruit (Sharif et al., 2020). Depending on the predominance, berries are categorized as cyanidin-based or pelargonidin-based sources. Interestingly, most of the botanical sources of anthocyanins are cyanidin-based, with the exception of strawberries (rich in pelargonidin) (Prior et al., 2008). Numerous articles have been published on cyanidin-based berry sources. The bioavailability of raspberry anthocyanins was examined by McDougall et al. (2005), who found that intact glycosides are not much affected by *in vitro* digestion. Similar research on *in vitro* digestion of anthocyanins was also reported for chokeberries, apples, gooseberries, grapes, and mulberries (Bermúdez-Soto et al., 2007; Bouayed et al., 2011; Chiang et al., 2013; Liang et al., 2012; Tagliazucchi et al., 2010). Aiyer et al. (2011) documented that the phytochemical profiles of blueberries and black raspberries were complementary to each other. Hager et al. (2008) studied the effect of thermal processing on monomeric anthocyanins, antioxidant activity, and percent polymeric color. According to the study, losses in monomeric anthocyanins were shown to coincide with an increase in polymeric color. Blackcurrant anthocyanins encapsulated in glucan gels (Xiong et al., 2006) and pomegranate juice both showed similar storage degradation kinetics (Fischer et al., 2013). Srivastava et al. (2010) studied the anti-inflammatory and antioxidant properties of three different blackberry cultivars (Navaho, Kiowa, and Ouachita). The results from mouse ear myeloperoxidase activity showed that the blackberry extracts from all cultivars had similar activity towards mouse ear edema. Further, grape anthocyanins have been found to exhibit anti-inflammatory

TABLE 5.2
Anthocyanin Quantity Reported from Several Sources

Name of Source	Content/Range (mg/100 g)	Reference
Acai pulp	282.5	Rosso and Mercadante (2007)
Acerola pulp	7.21	Rosso and Mercadante (2007).
A hybrid of fresh strawberries	71.8	Fan-Chiang and Wrolstad (2005); Ngo et al. (2007)
Banana bracts	32–250	Pazmiño-Durán et al. (2001)
Black raspberries	145–607	Tian et al. (2006)
Berries	23.7	Longo and Vasapollo (2006)
Capulin	31.7	Ordaz-Galindo et al. (1999)
Corncobs	290–1323	Jing and Giusti (2005)
Fresh blackberries	75	Fan-Chiang and Wrolstad (2005); Ngo et al. (2007)
Grape peel powder	171.42	Li et al. (2012)
Kokum	1000–2400	Nayak et al. (2010)
Red wine grapes	30–750	Mazza and Miniati (1993)
Roselle	230	Tsai et al. (2002)
Strawberry	13–315	da Silva et al. (2007)

Source: Yousuf et al. (2016).

properties, α-glucosidase activity, and antioxidant activity (Hogan et al., 2010 a, b). Alcohol-induced stomach lesions in rats were inhibited by strawberry extracts containing anthocyanins and other polyphenols (Alvarez-Suarez et al., 2011). The addition of anthocyanin-rich tart cherries to animal diets improved plasma antioxidant activity while lowering plasma glucose, hyperlipidemia, fatty liver, and hyperinsulinemia (Seymour et al., 2008). Several *in vitro* studies have documented blueberry extracts to be effective in lowering the inflammation and plasma cholesterol and inhibiting proliferation of colon cancer (Bornsek et al., 2012; Esposito et al., 2014; Liang et al., 2013; Yi et al., 2005). According to research on animals, diets supplemented with blueberry anthocyanins increased vascular sensitivity (Del Bo et al., 2012). Bilberry and blueberry anthocyanins have been encapsulated for culinary and nutraceutical uses (Betz & Kulozik, 2011; Betz et al., 2012; Flores et al., 2014;Oidtmann et al., 2012).

5.3   ANTHOCYANIN DEGRADATION

Anthocyanin pigments are not particularly stable. They may decay when still in fresh tissue, or they may be destroyed during the commodity's processing and storage.

5.3.1   ANTHOCYANIN DEGRADATION MECHANISM

Anthocyanins are glycosylated anthocyanidins, with sugars associated with anthocyanidin's 3-hydroxyl group at the fifth or seventh position of flavynium ion. The number of hydroxyls present within the molecule, the degree of methylation of OH groups, number and nature of sugar moiety connected with phenolic molecule, and to a lesser extent the nature and number of aromatic and aliphatic acids connected to it all contribute to differences in chemical structure. Sugar moieties are attached as 3-monosides, 3-biosides, 3-triosides, 3, 5-diglycosides, and 3, 7-diglycosides, with the main sugars being glucose, rhamnose, galactose, and xylose (McGhie & Walton, 2007). The major causes of anthocyanin degradation are oxidation, dissociation of covalent bonds, and enhanced oxidation processes during thermal processing. Further, depending on the extent and form of heating,

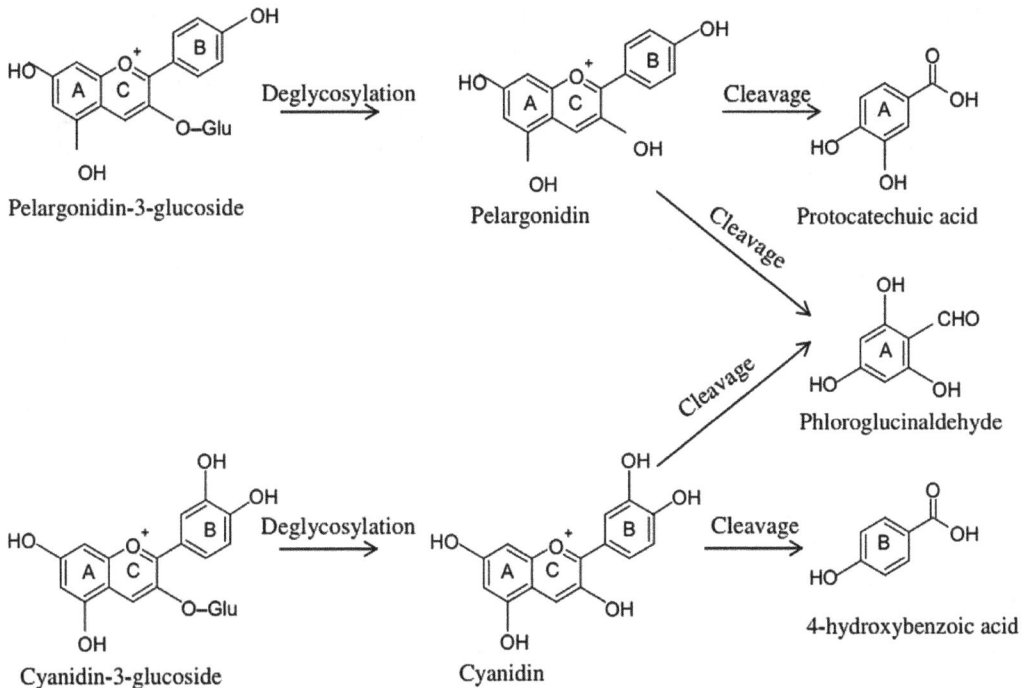

FIGURE 5.1 The mechanism of possible thermal degradation of two common anthocyanins (adapted from Patras, 2010).

anthocyanins can degrade into a number of species (Figure 5.1). This is vital to understand the pathways of anthocyanin degradation in order to maximize the nutritional and sensory properties of the food. The mechanism of deterioration is poorly understood while the chemical structure of anthocyanins and the presence of various other organic acids have a substantial influence on their breakdown mechanisms. The rate of deterioration of anthocyanins, for example, increases when the temperature rises during processing and storage (Palamidis & Markakis, 1975). Markaris et al. (1957) proposed that the opening of the pyrylium ring and generation of chalcones are preliminary steps in the degradation of anthocyanins. Adams (1973) claimed sugar moiety hydrolysis and creation of aglycones as a first degradation step, potentially due to the generation of cyclic adducts. The author also stated that heating anthocyanin results in the formation of a chalcone structure, which is then converted into a coumarin glucoside derivative with the loss of the B-ring. At pH 2–4, the aglycon–sugar bond is more labile, unlike other glycoside bonds, according to Adams (1973). However, at pH 1 all glycosidic linkages are accessible to hydrolysis because heating of cyanidin-3-rutinoside at pH 1 resulted in the synthesis of glucose and rhamnose, although merely traces of rutinose are formed. A study by Seeram et al. (2001) concluded that anthocyanins obtained from cherry when subjected to high temperature along with high pH can cause its degradation, resulting in the generation of three benzoic acid derivatives. However, another study documented that thermal breakdown of anthocyanin 3, 5-diglycosides includes coumarin 3, 5-diglycosides (Von Elbe & Schwartz, 1996).

5.3.2 Oxidative Degradation

Another important factor in the decomposition of anthocyanins is the presence of oxygen, which can accelerate the destruction of anthocyanins through the activity of oxidizing enzymes (Jackman et al.,

1987). Enzymes like polyphenol oxidase catalyze the oxidation of chlorogenic acid to the equivalent o-quinone (chlorogenoquinone) in the presence of oxygen. Brown condensation products are formed when this quinone interacts with anthocyanins (Kader et al., 1999). These findings support the idea that polyphenol oxidase is important for anthocyanin breakdown. In another study, the possible mechanism for degradation of anthocyanins (pelargonidin-3-glucoside) was reported to be the interaction between o-quinone and secondary oxidation products generated by the quinone and anthocyanin pigments (Kader et al., 2001). Sadilova et al. (2007) evaluated the thermal degradation of anthocyanins obtained from elderberry and strawberry and its degradation patterns using high-performance liquid chromatography-diode array detector mass spectrometry (HPLC DAD-MS). The results revealed that heating strawberry for 4 h formed a degradation product with maximum absorbance at 253 nm. Further, 3, 4-hydroxybenzoic acid (protocatechuic acid) was generated during the breakdown of pelargonidin B-ring while anthocyanin's A-ring was degraded to phloroglucinaldehyde Phloroglucinaldehyde and protocatechuic acid were found to be the major degradation products in elderberry concentrate heated for 4 h.

5.4   METHODS OF ENCAPSULATING ANTHOCYANIN

There are five types of anthocyanin-loaded carriers (Assadpour & Jafari, 2019). The five groups are: (1) spray-dried microparticles; (2) freeze-dried micro/nanoparticles; (3) emulsion technology; (4) ion gelation; and (5) nano/micro-gels.

Spray drying (33.33%) and electrospinning (4.17%) were the most and least prevalent encapsulation methods used for anthocyanins, respectively (Figure 5.2). The popularity of spray-drying encapsulation technology may be due to its simplicity as well as its high performance. One of the most promising ways to encapsulate sensitive chemicals is electrospinning. However, the applications for these techniques are confined to the pharma industry for drug delivery, with very few studies targeted towards encapsulation of anthocyanins for food application. The product's final

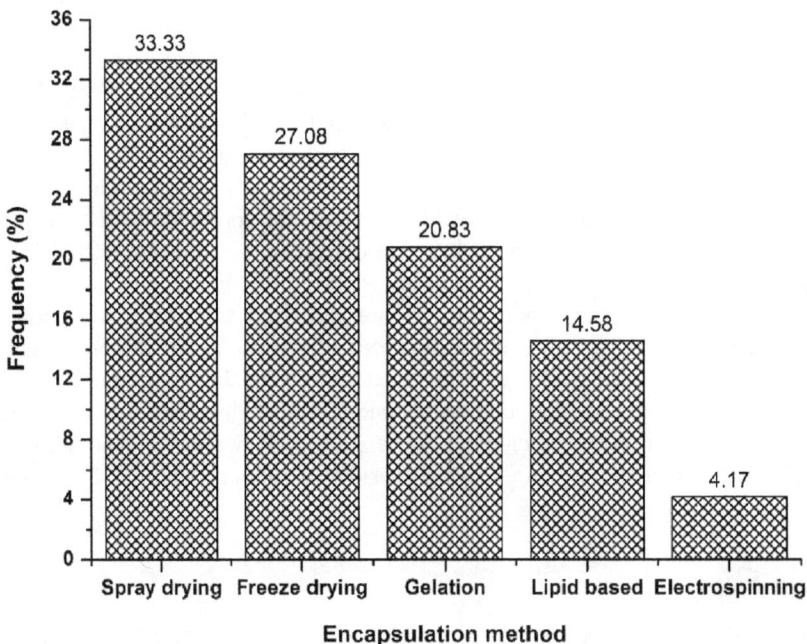

FIGURE 5.2   Frequency of anthocyanin encapsulation methods.

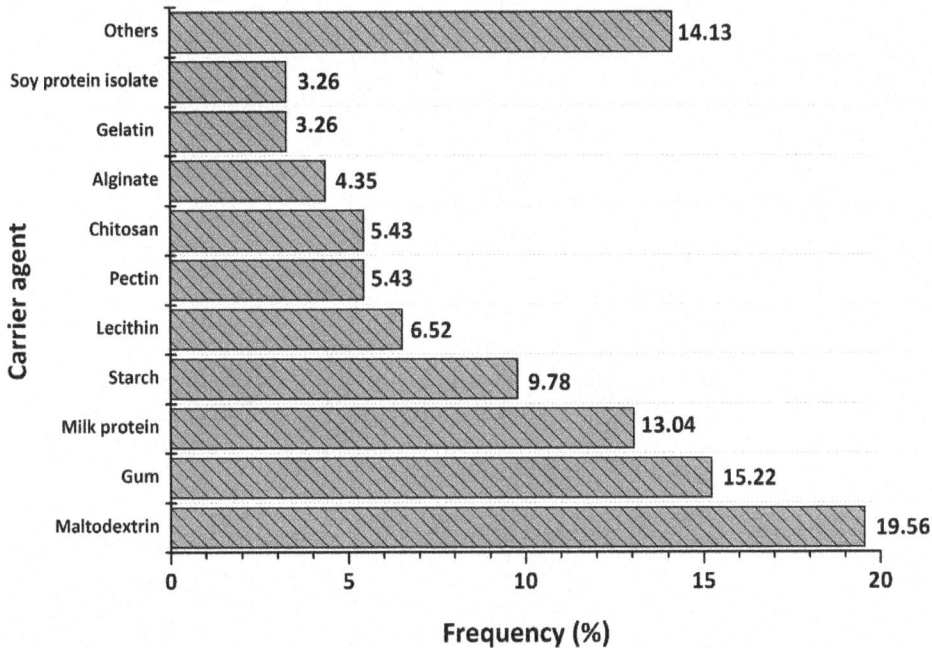

FIGURE 5.3 Encapsulation frequency of anthocyanins using various carrier agents.

cost is a key aspect in the commercial food industry, and the final production price for the electro-sprayed food ingredients is quite expensive, perhaps because of low yield and reproducibility capacity of food-grade capsules and fibers, huge investment costs, and the difficulty of dealing with biopolymers. As a result, additional study is needed to regulate the final cost of electrospun products (Sharif et al., 2020).

Maltodextrin, gums, starch, and milk proteins have been the most prominent polymers employed in anthocyanin encapsulation in recent years among the numerous carrier agents (Figure 5.3) (Sharif et al., 2020).

A variety of encapsulation techniques are available, some of which have been successfully used with anthocyanins (Table 5.3). Microencapsulation methods are chosen based on specific applications and selected parameters, e.g., physicochemical features of wall and core materials, adequate particle size, their release mechanisms, processing cost, and so on. Encapsulation has helped in the stabilization of anthocyanins in general, regardless of the encapsulation method employed. On the other hand, the degree of stabilization is apparently proportional to the parameters of the encapsulation technique used (Mahdavi et al., 2014).

5.4.1 Spray Drying

Spray drying is one of the oldest and most widely used microencapsulation techniques, with approximately 80–90% of extracted anthocyanins being successfully encapsulated via this technique (Mahdavi et al., 2014; Mohammadalinejhad et al., 2021). In comparison to alternative technologies already in use in the food business, spray drying provides anthocyanin powder with better storage stability, easy handling, and reduced transportation cost.

Additionally, it offers better control of particle size and requires less manufacturing time (Idham et al., 2012). In spray-drying technique, wall material surrounds the inner core substance, and this mixture is further atomized inside a heated chamber, resulting in expansion of dry particles and

TABLE 5.3
Brief Overview of Most Prevalent Microencapsulation Techniques Used for Anthocyanins and Its Characteristics

Encapsulation Process	Processing Steps	Particle Size (µm)	Advantages	Disadvantages
Spray drying	Innermost core material should be dissolved with aqueous wall material solutions Atomization Dehydration	10–400	Process costs are low A wide range of wall materials is available Encapsulation efficiency is high The finished product has a high level of stability Continuous large-scale production is a possibility	Can destroy substances that are extremely sensitive to temperature Proper control of particle size is difficult Small batch yields are moderate
Freeze drying	Water should be used to dissolve the core and wall materials Place the sample in the freezer Use low pressure to dry Grinding	20–5000	This approach can effectively encapsulate thermosensitive compounds that are unstable in aqueous solutions	Processing time is lengthy Process costs are high The capsules are expensive to store and deliver
Ionic gelation	In an ionic solution, core material along with wall material should be extruded as droplets Ionic contact creates the capsules		Extreme temperature and pH conditions, as well as use of organic solvents, are avoided	Only laboratory-scale application is employed The capsules possess a high porosity, which encourages intense bursts
Thermal gelation	The process is nearly identical to that of ionic gelation; however, it does not require the use of ionic solution to generate gelled drop; the gelation is solely attributable to thermal factors		The same as for ionic gelation	The same drawback as for ionic gelation
Emulsification	In a water or oil phase, dissolve the active and emulsifiers Under shear, combine the oil and water phases	0.2–5000	Non-polar, polar, or apolar; amphiphilic materials might be used	There are just a few emulsifiers that can be utilised Capsule development is difficult to manage
Emulsion method	An emulsion of protein solution containing the core material within an oil phase, followed by heating of the emulsion, resulted in the dispersed droplets to form gel	1–100	Micro-nanocapules having a narrow size distribution can be produced	When exposed to external environmental stresses, e.g. heating, drying, and so on, it becomes unstable There are just a few emulsifiers that can be utilized
Coacervation	In the oil phase, make oil-in-water emulsions with lipophilic active Mix under turbulent conditions Three immiscible phases should be induced Cool Crosslink	10–800	Because it is done at room temperature, it can be utilized to encapsulate heat-sensitive substances	Toxic chemicals are used The complex coacervates are extremely unstable On the capsules' surfaces, there exist residual solvents and coacervating agents The size range of spheres is small Costly and complicated

Source: Adapted from Mahdavi et al. (2014) and Mohammadalinejhad et al. (2021).

creation of voids (Jafari et al., 2008). Proper selection of wall material is the most crucial step in spray drying for better protection of core material. It depends on several factors, e.g., particle size, thermal or dissolution release, mechanical strength, and compatibility with food product (Akhavan Mahdavi et al., 2016). There are various types of encapsulating agents, including polysaccharides, lipids, and proteins, that can be utilised for this purpose (Gibbs et al., 1999). Maltodextrin was primarily employed as a wall material, either alone or in combination with other biopolymers. Maltodextrin has been discovered to be a useful drying agent for microencapsulated anthocyanins, assisting in their preservation. Maltodextrin has an economic advantage due to its low cost and widespread availability (Jafari et al., 2008). Thus, using costly carrier agents in combination with maltodextrin could give financial benefit (Nayak & Rastogi, 2010). Several studies have also demonstrated that combining biopolymers improves the effectiveness and stability of the encapsulation process when compared to using them alone (Norkaew et al., 2019; Pieczykolan & Kurek, 2019). On the contrary, Rocha et al. (2019) concluded that utilizing a combination of two or more different biopolymers can reduce encapsulation efficiency. Commonly used wall materials for the production of spray-dried microparticles include maltodextrin, gum Arabic, starch (maize starch, modified and phosphorylated starch), whey protein isolate/concentrate, β-glucan, other materials (black glutinous rice, β-cyclodextrin, guar gum, gelatin, inulin, soy protein isolate, pectin, polyvinylpyrrolidone). Polysaccharides are preferred over proteins for spray-drying encapsulation of anthocyanin owing to their good solubility, emulsifying capacity, high volatile component-holding capacity, and low viscosity (Shiga et al., 2004). Fredes et al. (2018) employed an spray-drying experiment for encapsulation of maqui juice using sodium alginate and insulin and observed encapsulation efficiencies of 47.3 and 68.6, respectively. The bioaccessibility of anthocyanins in the final product was 10% more than in maqui juice without any added encapsulating material. Despite being quick and economical, the spray-drying technique has a few limitations, such as high processing temperature, and uneven particle size and shape, which can damage the anthocyanin pigment and limit bioavailability and absorption. Furthermore, the frequent loss of tiny particulates in the exhaust air may result in significant material waste (Zhao et al., 2017). As a result, an adequate wall material should be selected with proper care. The encapsulation efficiency of diverse micro/nanocarriers can be influenced by their emulsifying capabilities, stiffness, porosity, and surface-active properties. As a result, it appears that good biopolymer selection and blending can have a significant impact on encapsulation efficiency (da Rosa et al., 2019). The performance and application of nano/microcarriers are heavily influenced by particle size. The particle size of spray-dried powder falls in the micro-scale range and thus this method is ineffective for producing nanoparticles. Furthermore, when producing particles with a spray drier, size management has been observed to be a difficult task. Atomizing pressure, liquid delivery rate, nozzle position, spray nozzle size, and solution concentration (such as viscosity) all have an impact on particle size (Mahdavi et al., 2014). Novel technology such as a "nano spray dryer" is a recent technique which is capable of creating encapsulated nanoparticles and is being commercialized (Arpagaus et al., 2018).

5.4.2 Freeze Drying

Freeze drying, also referred to as lyophilization, is one of the most effective ways to dry anthocyanins. Freeze-dried microcapsules have been found to exhibit a greater anthocyanin-stabile structure and antioxidant activity than spray-dried microcapsules (da Fonseca Machado et al., 2018; Idham et al., 2012). Freeze drying, operated under vacuum at low temperature, facilitates sublimation that dehydrates the frozen combinations of core and wall materials. As a result, items retain their chemical structure while also posing a low chance of undesired alterations (Ezhilarasi et al., 2013). This method is ideal for encapsulating heat-sensitive compounds like anthocyanins since it employs moderate temperatures (Souza et al., 2017). The powder prepared from a freeze-drying technique are of superior quality, and offers several advantages, such as no adverse effect on sensory properties, better retention of bio-functionality, greater reconstitution capacity, and a longer shelf-life. This approach is used to improve the thermal and color stability of anthocyanins in a variety of wall materials (Garavand et al., 2019). Similarly to the spray-drying technique, the type of carrier used influences the encapsulation efficiency and morphology of freeze-dried powder. The selected wall material should be stable, biodegradable, and non-reactive with core material. Additionally, it should preserve the structure and enhance the functionality of core material during various processing stages and varied storage environments (Šaponjac et al., 2017). Maltodextrin was used as the most prevalent wall material in encapsulation. Combining maltodextrin with different biopolymers can boost the performance of encapsulation while also increasing the process's economic justification. Gum Arabic and maltodextrin have been used as carrier material in the production of freeze-dried anthocyanin powders by several researchers (Jafari et al., 2016; Suravanichnirachorn et al., 2018). There is a lack of definitive data on the optimum materials for maximizing anthocyanin retention and encapsulation efficiency. Although the inconsistency of results discovered in numerous studies illustrates the challenge of anthocyanin encapsulation by freeze drying, the mixed use of these components is commonly cited as one of the better solutions (Stoll et al., 2016).

The viscosity of wall materials has been linked to encapsulation efficiency in several studies (Rosenberg et al., 1990). When the carrier viscosity is optimal, changing it will not appreciably improve encapsulation efficiency. The addition of xanthan gum to carboxymethyl starch (in 1:30 ratio) increased the viscosity of mixture but no significant improvement in the encapsulation efficiency was observed. However, integration of xanthan gum resulted in a better-compacted surface which inhibited the degradation caused by temperature and oxygen, thus boosting the antioxidant stability of the core material (Cai et al., 2019).

The major disadvantage with freeze-dried food powders is the non-uniformity of the particle size and there is no option for particle size control as a result of mixing the material after drying (Ozkan et al., 2019). Furthermore, sublimation of ice crystals created during the freezing step can result in porosity in all samples. Freeze-dried particles consist of a thin sheet-like structure with a porous and rough surface (da Fonseca Machado et al. 2018)

However, because of the vacuum technology used, this process is quite expensive and requires a longer dehydration time compared to other encapsulation techniques (Simões et al., 2017). Furthermore, the high porosity of the freeze-dried powder substantially impacts the encapsulation and retention efficiency as well as the stability of an encapsulated substance (Ezhilarasi et al., 2013).

5.4.3 Ionic Gelation Method

The ionic gelation method is also utilized in the production of nano/microparticles (Rajabi et al., 2019). Ionic gelation is a crucial technique among the various ways for encapsulating active compounds since it is technically simple and does not require sophisticated equipment, high temperatures, or organic solvents. Ionic gelation is a kind of microencapsulation accomplished by atomization, dripping (coextrusion, extrusion), or electrostatic spraying. Because high temperatures, intensive stirring, or organic solvents are not used in this technique, it allows for the encapsulation of compounds that would otherwise deteriorate under several conditions. The encapsulation of hydrophilic or low-molecular-weight molecules has shown concerns with simple diffusion and rapid release, regardless of pH, via the ionic gel matrix (De Moura et al., 2019). Several approaches (emulsion process, coating material) must be used to maintain the hydrophilic active compounds because only hydrophobic or low-solubility active compounds are directly applicable to ionic gelation. The ionic gelation technique for anthocyanin encapsulation has been effectively used by several authors. The authors noted that the larger particle size and low particle consistency are disadvantages of this method (mainly for hydrophilic active compounds). The desirable characteristics are minimal polydispersity and high encapsulation quality. Ionic gelation is gradually increasing in importance as it is simple to use and practical while avoiding high temperature requirements and chemical solvents. Particulate gels can be used as strengthening, structuring, and texturizing agents in a food system as well as to improve the sensory attributes of the product. They can also alter shape and scale, making it possible to release active substances in agricultural, pharmaceutical, and food goods with precision. The use of ionic gelation for the microencapsulation of anthocyanins has been studied in a number of ways (Mohammadalinejhad & Kurek, 2021).

The porosity of ionic gel particles reduces encapsulation efficacy. This is a significant disadvantage of the encapsulation technique for hydrophilic compounds. To lower the diffusion process and porosity of the beads, a combination of ionic gelation and complexation with cationic polyelectrolytes has been proposed. The electrostatic interaction between the cationic charges of proteins and anionic charges of polysaccharides causes complexation (Ghasemi et al., 2018). Ionic gelation is a relatively simple method for the encapsulation of anthocyanins. Micro/nanogels have a lower encapsulation efficiency and stability than spray-dried or freeze-dried particles. Ionic gels have a porous structure which accelerates the penetration of oxygen through the matrix, thereby facilitating the easy release of bioactive compounds. As a result, different pores may limit the functional properties of gels in the form of particles, lowering the efficiency of encapsulation (George & Abraham, 2006). A combination of ionic gelation with complexation techniques, for example, could enhance the functionality of enclosed particles (Hébrard et al., 2013). Hydrogels also have another significant issue in the creating high moisture content, which makes storage more difficult. As a result, in some circumstances, refrigeration is required for storage. Silica gels may be used to encapsulate bioactive substances like anthocyanins, albeit their use in the food sector is restricted (Sharif et al., 2020).

5.4.4 EMULSIONS

Emulsions are generally made up of two or more immiscible liquids (typically water and oil, though not always), one of which is spread in the other as minute spherical droplets. The liquid that makes up the droplet compounds is the dispersed or discontinuous phase, while the surrounding liquid is a continuous phase (McClements, 2016). Inclusion of anthocyanins in the dispersed phase of an emulsion helps protect it from external environmental degradation such as mechanical, chemical, and enzymatic, and to manage its controlled release as well as enhance the bioavailability of anthocyanin in the human gut (Mao et al., 2017). Depending on the approach utilized, emulsiontechnology can be employed to develop systems to deliver bioactive components in the form of liquid or powder particles (Okuro et al., 2015). Emulsions can be prepared in two ways: low-energy and high-energy. Most of the traditional emulsification techniques utilized in the commercial food industry comprises high-energy-based system such as microfluidizers, high-pressure, or ultrasonic homogenizers which applies strong shear force in the macroscopic phase, resulting in their breakdown into micro or nano-sized emulsions (Santana et al., 2013). Depending on the phase composition and method utilized, high-energy emulsification can generate emulsions of varying particle sizes and degrees of viscosity and dispersity. Nonetheless, controlling particle size distribution is difficult, resulting in more polydisperse emulsions than emulsions created using a low-energy technique (Ushikubo et al., 2014).

Additional procedures such as coating, drying, and gelation may be used to enhance the durability of the emulsions once they have been formed (Okuro et al., 2015). To improve the stability and control the degradation of bioactive compounds in the gastrointestinal tract, several emulsion techniques have been used as delivery systems, such as conventional, double, multilayer, nano emulsions, and hydrogel (McClements & Li, 2010). Paula et al. (2018) developed a double emulsion comprising commercial anthocyanin extract with added guar gum (1.25%). They reported an encapsulation efficiency of 90.6% with enhanced thermal stability and decreased degradation. Guar gum was used to improve color stability. Although emulsion delivery techniques do a decent job of safeguarding anthocyanins, they are thermodynamically unstable and tend to break down with time. As a result, the main difficulty of emulsion technology is to give emulsions adequate kinetic stability so that they are not easily disrupted (McClements & Li, 2010).

5.4.5 COACERVATION

Recently, there has been a lot of interest in the coacervation technique among all the microencapsulation technologies because of its high loading capacity, low temperature operating requirements, reduced evaporation losses of volatile compounds and thermal degradation, and greater competence in regulating the release rate of nutraceuticals (Mohammadalinejhad & Kurek, 2021). Coacervates are a spherical collection of organic macromolecules kept together by hydrophobic forces in the shape of an inclusion. Simple organic molecules can be selectively adsorbed from the surrounding medium because of the boundaries of the coacervates. Frequently, the drop that forms the coacervate particle is a liquid oil. The drops will be encased in a shell of the hydrocolloid mixture; under the correct conditions, this will set to form a solid or semipermeable shell. Because both entropy and enthalpy control the formation of coacervates, a minimum temperature must be maintained (Johnson & Wang, 2014). In a stirred-liquid phase, coacervation entails generating minute droplets of the coacervate phase. This usually results in a colloidal dispersion that is unstable and prone to coalescence. The coacervate technique retains and sustains droplets by crosslinking or mechanical gelation of the polymer. Simple and complex coacervation are the two types of coacervation mechanisms (Sarkar et al., 2021). In medical technology, simple coacervation is often employed to entrap medicines in microcapsules. The use of coacervates for the microencapsulation of anthocyanins has been studied in a number of ways (Mohammadalinejhad & Kurek, 2021). Raspberry water extracts were microencapsulated utilizing a double-emulsion technology prior to complex coacervation with gelatin and gum Arabic in one study to lessen anthocyanin instability as a water-soluble compound, especially under harsh processing and storage conditions. The studies revealed the thermostability of microcapsules (Shaddel et al., 2018). The microcapsules in the best condition maintained a deep red color over time, indicating that the strategy for preserving anthocyanins was successful. Controlled protein–polysaccharide interactions via complex coacervation can point to a path to improve their functional position as ingredients not requiring chemical or enzymatic alterations. Guo et al. (2018) employed blueberry- and purple corn-derived anthocyanins to encapsulate particles based on hydrogels with alginate and pectin in an investigation utilizing coacervation. Microencapsulation efficiency was shown to be quite high. At the same time, this approach allows anthocyanins to be preserved for longer periods of time. Coacervation, like any other approach, has inherent drawbacks, mostly in terms of particle agglomeration and particle size management. Further, the developed particles are sensitive to pH and ionic strength, thus restricting their application in a variety of matrices (Joye & McClements, 2014).

5.5 TECHNOLOGICAL CHALLENGES IN COMMERCIAL PRODUCTION

Currently, a large range of strategies for encapsulating bioactive substances has been documented, despite the fact that just a few approaches, such as freeze drying and spray drying, are widely used in the food as well as pharma industries (Đorđević et al., 2016). Recently reported novel, emergent, or improved conventional procedures for industrial application should be investigated further in terms of their high throughput, good product quality, low production costs, fewer handling hassles, and safety concerns. Despite the fact that each technique has its own set of constraints that make it more difficult, they should be evaluated for future research in order to overcome those limitations and raise their level from lab to pilot to industrial scale. As a result, advice addressing the difficulties in micro- and nano-encapsulation can be summarized as follows (Shishir et al., 2018):

- Determine which generally recognized as safe (GRAS) polymers are best for each encapsulation process and compare their efficacy.
- Investigate the best processing parameters, such as technical conditions and formulations, to improve product stability, encapsulation efficiency, and target release profile.
- Several encapsulation techniques can be compared to determine which technique is best for a certain bioactive ingredient.
- Find alternatives to certain organic solvents (such as those used for anti-solvent precipitation) that are not suitable for human consumption.
- Improve the mechanical design of the process to boost, minimize energy consumption, and improve process efficiency.
- Develop new technologies and emerging nanoparticles having specific functionality within food and biological systems.
- Multi-compartment systems can be used to deliver numerous bioactive substances at once.
- In food and biological systems, investigate the toxicity and synergistic effects of co-delivery or various delivery mechanisms.

5.6 CONCLUSION

Anthocyanins play a significant role in human nutrition and in the food industry as colorants. Anthocyanins are extremely sensitive compounds that may degrade during commercial processing. So, there is a need to protect anthocyanins from external factors. For this reason, researchers are paying more attention to their microencapsulation. Generally, encapsulation techniques are categorized into five groups: spray drying, freeze drying, ion gelation method, emulsion method, and coacervation. Spray drying was found to be the most investigated method for encapsulating anthocyanins. Maltodextrin, an economic and efficient drying agent, was the most commonly reported biopolymer in this method. Freeze-dried microcapsules, for example, have been demonstrated to have greater anthocyanin stability and antioxidant activity than spray-dried microcapsules. Other techniques, including emulsion and ionic gelation, can also be utilized to encapsulate anthocyanins. Berries were the most researched and investigated sources of anthocyanins, regardless of the technique used. Nonetheless, further research is needed to attain the finest protection method, anthocyanin conservation, environmental care, high yield, and low prices.

REFERENCES

Adams, J. B. (1973). Thermal degradation of anthocyanins with particular reference to the 3-glycosides of cyanidin. I. In acidified aqueous solution at 100°C. *Journal of the Science of Food and Agriculture, 24*(7), 747–762. https://doi.org/10.1002/jsfa.2740240702

Aiyer, H. S., Li, Y., Losso, J. N., Gao, C. F., Schiffman, S. C., Slone, S. P., & Martin, R. C. G. (2011). Effect of freeze-dried berries on the development of reflux-induced esophageal adenocarcinoma. *Nutrition and Cancer, 63*, 1256–1262. https://doi.org/10.1080/01635581.2011.609307

Akhavan Mahdavi, S., Jafari, S. M., Assadpour, E., & Ghorbani, M. (2016). Storage stability of encapsulated barberry's anthocyanin and its application in jelly formulation. *Journal of Food Engineering, 181*, 59–66. https://doi.org/10.1016/j.jfoodeng.2016.03.003

Alvarez-Suarez, J. M., Dekanski, D., Ristić, S., Radonjić, N. V., Petronijević, N. D., Giampieri, F., Astolfi, P., Gonzalez-Paramás, A. M., Santos-Buelga, C., Tulipani, S., Quiles, J. L., Mezzetti, B., & Battino, M. (2011). Strawberry polyphenols attenuate ethanol-induced gastric lesions in rats by activation of antioxidant enzymes and attenuation of MDA increase. *PloS ONE, 6*, 25878. https://doi.org/10.1371/journal.pone.0025878

Arpagaus, C., Collenberg, A., Rütti, D., Assadpour, E., & Jafari, S. M. (2018). Nano spray drying for encapsulation of pharmaceuticals. *International Journal of Pharmaceutics, 546*(1–2), 194–214. https://doi.org/10.1016/j.ijpharm.2018.05.037

Arriola, N. D. A., Chater, P. I., Wilcox, M., Lucini, L., Rocchetti, G., Dalmina, M., . . . & Amboni, R. D. d. M. C. (2019). Encapsulation of *Stevia rebaudiana* Bertoni aqueous crude extracts by ionic gelation – Effects of alginate blends and gelling solutions on the polyphenolic profile. *Food Chemistry*, *275*, 123–134. https://doi.org/10.1016/j.foodchem.2018.09.086

Arroyo-Maya, I. J., & McClements, D. J. (2015). Biopolymer nanoparticles as potential delivery systems for anthocyanins: Fabrication and properties. *Food Research International*, *69*, 1–8. https://doi.org/10.1016/j.foodres.2014.12.005

Assadpour, E., & Jafari, S. M. (2019). A systematic review on nanoencapsulation of food bioactive ingredients and nutraceuticals by various nanocarriers. *Critical Reviews in Food Science and Nutrition*, *59*(19), 3129–3151. https://doi.org/10.1080/10408398.2018.1484687

Barea-Álvarez, M., Delgado-Andrade, C., Haro, A., Olalla, M., Seiquer, I., & Rufián-Henares, J. Á. (2016). Subtropical fruits grown in Spain and elsewhere: A comparison of mineral profiles. *Journal of Food Composition and Analysis*, *48*, 34–40. https://doi.org/10.1016/j.jfca.2016.02.001

Bermúdez-Soto, M.-J., Tomás-Barberán, F.-A., & García-Conesa, M.-T. (2007). Stability of polyphenols in chokeberry (*Aronia melanocarpa*) subjected to *in vitro* gastric and pancreatic digestion. *Food Chemistry*, *102*, 865–874. https://doi.org/10.1016/j.foodchem.2006.06.025

Betz, M., & Kulozik, U. (2011). Whey protein gels for the entrapment of bioactive anthocyanins from bilberry extract. *International Dairy Journal*, *21*, 703–710. https://doi.org/10.1016/j.idairyj.2011.04.003

Betz, M., Steiner, B., Schantz, M., Oidtmann, J., Mäder, K., Richling, E., & Kulozik, U. (2012). Antioxidant capacity of bilberry extract microencapsulated in whey protein hydrogels. *Food Research International*, *47*, 51–57. https://doi.org/10.1016/j.foodres.2012.01.010

Bornsek, S. M., Ziberna, L., Polak, T., Vanzo, A., Ulrih, N. P., Abram, V., Tramer, F., & Passamonti, S. (2012). Bilberry and blueberry anthocyanins act as powerful intracellular antioxidants in mammalian cells. *Food Chemistry*, *134*, 1878–1884. https://doi.org/10.1016/j.foodchem.2012.03.092

Bouayed, J., Hoffmann, L., & Bohn, T. (2011). Total phenolics, flavonoids, anthocyanins and antioxidant activity following simulated gastro-intestinal digestion and dialysis of apple varieties: Bioaccessibility and potential uptake. *Food Chemistry*, *128*, 14–21. https://doi.org/10.1016/j.foodchem.2011.02.052

Cai, X., Du, X., Cui, D., Wang, X., Yang, Z., & Zhu, G. (2019). Improvement of stability of blueberry anthocyanins by carboxymethyl starch/xanthan gum combinations microencapsulation. *Food Hydrocolloids*, *91*, 238–245. https://doi.org/10.1016/j.foodhyd.2019.01.034

Carvalho, A. G. da S., Machado, M. T. da C., da Silva, V. M., Sartoratto, A., Rodrigues, R. A. F., & Hubinger, M. D. (2016). Physical properties and morphology of spray dried microparticles containing anthocyanins of jussara (*Euterpe edulis* Martius) extract. *Powder Technology*, *294*, 421–428. https://doi.org/10.1016/j.powtec.2016.03.007.

Chiang, C. J., Kadouh, H., & Zhou, K. Q. (2013). Phenolic compounds and antioxidant properties of gooseberry as affected by in vitro digestion. *LWT-Food Science & Technology*, *51*, 417–422. https://doi.org/10.1016/j.lwt.2012.11.014

Chi, J., Ge, J., Yue, X., Liang, J., Sun, Y., Gao, X., & Yue, P. (2019). Preparation of nanoliposomal carriers to improve the stability of anthocyanins. *LWT – Food Science & Technology*, *109*, 101–107. https://doi.org/10.1016/j.lwt.2019.03.070

da Fonseca Machado, A. P., Rezende, C. A., Rodrigues, R. A., Barbero, G. F., e Rosa, P. d. T. V., & Martínez, J. (2018). Encapsulation of anthocyanin-rich extract from blackberry residues by spray-drying, freeze-drying and supercritical antisolvent. *Powder Technology*, *340*, 553–562. https://doi.org/10.1016/j.powtec.2018.09.063

da Rosa, J. R., Nunes, G. L., Motta, M. H., Fortes, J. P., Weis, G. C. C., Hecktheuer, L. H. R., ..., & da Rosa, C. S. (2019). Microencapsulation of anthocyanin compounds extracted from blueberry (*Vaccinium* spp.) by spray drying: Characterization, stability and simulated gastrointestinal conditions. *Food Hydrocolloids*, *89*, 742–748. https://doi.org/10.1016/j.foodhyd.2018.11.042

da Silva, F. L., Escribano-Bailón, M. T., Alonso, J. J. P., Rivas-Gonzalo, J. C., & Santos-Buelga, C. (2007). Anthocyanin pigments in strawberry. *LWT-Food Science & Technology*, *40*(2), 374–382. https://doi.org/10.1016/j.lwt.2005.09.018

da Silva Carvalho, A. G., da Costa Machado, M. T., Barros, H. D. d. F. Q., Cazarin, C. B. B., Junior, M. R. M., & Hubinger, M. D. (2019). Anthocyanins from jussara (*Euterpe edulis* Martius) extract carried by calcium alginate beads pre-prepared using ionic gelation. *Powder Technology*, *345*, 283–291. https://doi.org/10.1016/j.powtec.2019.01.016

de Moura, S. C. S. R., Berling, C. L., Germer, S. P. M., Alvim, I. D., & Hubinger, M. D. (2018). Encapsulating anthocyanins from *Hibiscus sabdariffa* L. calyces by ionic gelation: Pigment stability during storage of microparticles. *Food Chemistry*, *241(August 2017)*, 317–327. https://doi.org/10.1016/j.foodc hem.2017.08.095.

De Moura, S. C., Berling, C. L., Garcia, A. O., Queiroz, M. B., Alvim, I. D., & Hubinger, M. D. (2019). Release of anthocyanins from the hibiscus extract encapsulated by ionic gelation and application of microparticles in jelly candy. *Food Research International, 121*, 542–552. https://doi.org/10.1016/j.food res.2018.12.010

Del Bo, C., Kristo, A. S., Kalea, A. Z., Ciappellano, S., Riso, P., Porrini, M., & Klimis-Zacas, D. (2012). The temporal effect of a wild blueberry (*Vaccinium angustifolium*)-enriched diet on vasomotor tone in the Sprague-Dawley rat. *Nutrition, Metabolism & Cardiovascular Diseases*, *22*, 127–132. https://doi.org/ 10.1016/j.numecd.2010.05.004

Đorđević, V., Paraskevopoulou, A., Mantzouridou, F., Lalou, S., Panti, C. M., & Bugarski, B. (2016). Encapsulation technologies for food industry. In Nedovic, V., Raspor, P., Levic, J., Saponjac, V. T., & Barbosa-Canovas, G. V. (Eds.), *Emerging and Traditional Technologies for Safe, Healthy and Quality Food* (pp. 329–380). Switzerland: Springer International. https://doi.org/10.1007/978-3-319-24040-4_18

Esposito, D., Chen, A., Grace, M. H., Komarnytsky, S., & Lila, M. A. (2014). Inhibitory effects of wild blueberry anthocyanins and other flavonoids on biomarkers of acute and chronic inflammation *in vitro*. *Journal of Agricultural and Food Chemistry*, *62*, 7022–7028. https://doi.org/10.1021/jf4051599

Ezhilarasi, P. N., Karthik, P., Chhanwal, N., & Anandharamakrishnan, C. (2013). Nanoencapsulation techniques for food bioactive components: A review. *Food and Bioprocess Technology, 6*(3), 628–647. https://doi.org/10.1007/s11947-012-0944-0

Fan-Chiang, H. J., & Wrolstad, R. E. (2005). Anthocyanin pigment composition of blackberries. *Journal of Food Science*, *70*(3), C198–C202. https://doi.org/10.1111/j.1365-2621.2005.tb07125.x

Fischer, U. A., Carle, R., & Kammerer, D. R. (2013). Thermal stability of anthocyanins and colourless phenolics in pomegranate (*Punica granatum* L.) juices and model solutions. *Food Chemistry*, *138*, 1800–1809. https://doi.org/10.1016/j.foodchem.2012.10.072

Flores, F. P., Singh, R. K., Kerr, W. L., Pegg, R. B., & Kong, F. (2014). Total phenolics content and antioxidant capacities of microencapsulated blueberry anthocyanins during *in vitro* digestion. *Food Chemistry*, *153*, 272–278. https://doi.org/10.1016/j.foodchem.2013.12.063

Fredes, C., Osorio, M. J., Parada, J., & Robert, P. (2018). Stability and bioaccessibility of anthocyanins from maqui (*Aristotelia chilensis* [Mol.] *Stuntz*) juice microparticles. *LWT – Food Science & Technology, 91*, 549–556. https://doi.org/10.1016/j.lwt.2018.01.090

Garavand, F., Rahaee, S., Vahedikia, N., & Jafari, S. M. (2019). Different techniques for extraction and micro/nanoencapsulation of saffron bioactive ingredients. *Trends in Food Science and Technology, 89*, 26–44. https://doi.org/10.1016/j.tifs.2019.05.005

George, M., & Abraham, T. E. (2006). Polyionic hydrocolloids for the intestinal delivery of protein drugs: Alginate and chitosan – a review. *Journal of Controlled Release, 114*(1), 1–14. https://doi.org/ 10.1016/j.jconrel.2006.04.017

Ghasemi, S., Jafari, S. M., Assadpour, E., & Khomeiri, M. (2018). Nanoencapsulation of D-limonene within nanocarriers produced by pectin–whey protein complexes. *Food Hydrocolloids, 77*, 152–162. https://doi.org/10.1016/j.foodhyd.2017.09.030

Gibbs, B. F., Kermasha, S., Alli, I., & Mulligan, C. N. (1999). Encapsulation in the food industry: A review. *International Journal of Food Sciences and Nutrition, 50*(3), 213–224. https://doi.org/10.1080/0963 74899101256

Giusti, M. M., & Wrolstad, R. E. (1996). Characterization of red radish anthocyanins. *Journal of Food Science*, *61*(2), 322–326. https://doi.org/10.1111/j.1365-2621.1996.tb14186.x

Guo, J., Giusti, M. M., & Kaletunç, G. (2018). Encapsulation of purple corn and blueberry extracts in alginate-pectin hydrogel particles: Impact of processing and storage parameters on encapsulation efficiency. *Food Research International*, *107*, 414–422. https://doi.org/10.1016/j.foodres.2018.02.035

Hager, A., Howard, L. R., Prior, R. L., & Brownmiller, C. (2008). Processing and storage effects on monomeric anthocyanins, percent polymeric color, and antioxidant capacity of processed black raspberry products. *Journal of Food Science*, *73*, H134–H140. https://doi.org/10.1111/j.1750-3841.2008.00855.x

He, J., & Giusti, M. M. (2010). Anthocyanins: Natural colorants with health-promoting properties. *Annual Review of Food Science and Technology, 1*(1), 163–187. https://doi. org/10.1146/annurev.food.080708.100754

Hébrard, G., Hoffart, V., Cardot, J.-M., Subirade, M., & Beyssac, E. (2013). Development and characterization of coated-microparticles based on whey protein/alginate using the Encapsulator device. *Drug Development and Industrial Pharmacy, 39*(1), 128–137. https://doi.org/10.3109/03639045.2012.660950

Hogan, S., Zhang, L., Li, J. R., Sun, S., Canning, C., & Zhou, K. Q. (2010a). Antioxidant rich grape pomace extract suppresses postprandial hyperglycemia in diabetic mice by specifically inhibiting alpha glucosidase. *Nutrition & Metabolism., 7*, 71–79. https://doi.org/10.1186/1743-7075-7-71

Hogan, S., Canning, C., Sun, S., Sun, X. X., & Zhou, K. Q. (2010b). Effects of grape pomace antioxidant extract on oxidative stress and inflammation in diet induced obese mice. *Journal of Agricultural and Food Chemistry, 58*, 11250–11256. https://doi.org/10.1021/jf102759e

Huang, Z., Wang, B., Williams, P., & Pace, R. D. (2009). Identification of anthocyanins in muscadine grapes with HPLC-ESI-MS. *LWT-Food Science & Technology, 42*(4), 819–824. https://doi.org/10.1016/j.lwt.2008.11.005

Idham, Z., Muhamad, I. I., & Sarmidi, M. R. (2012). Degradation kinetics and color stability of spray-dried encapsulated anthocyanins from *Hibiscus sabdariffa* L. *Journal of Food Process Engineering, 35*(4), 522–542. https://doi.org/10.1111/j.1745-4530. 2010.00605.x

Jackman, R. L., Yada, R. Y., & Tung, M. A. (1987). A review: Separation and chemical properties of anthocyanins used for their qualitative and quantitative analysis. *Journal of Food Biochemistry, 11*(4), 279–308. https://doi.org/10.1111/j.1745-4514.1987.tb00128.x

Jafari, S. M., Assadpoor, E., He, Y., & Bhandari, B. (2008). Encapsulation efficiency of food flavours and oils during spray drying. *Drying Technology, 26*(7), 816–835. https://doi.org/10.1080/07373930802135972

Jafari, S. M., Mahdavi-Khazaei, K., & Hemmati-Kakhki, A. (2016). Microencapsulation of saffron petal anthocyanins with cress seed gum compared with Arabic gum through freeze drying. *Carbohydrate Polymers, 140*, 20–25. https://doi.org/10.1016/j.carbpol.2015.11.079

Jiménez-Aguilar, D. M., Ortega-Regules, A. E., Lozada-Ramírez, J. D., Pérez-Pérez, M. C. I., Vernon-Carter, E. J., & Welti-Chanes, J. (2011). Color and chemical stability of spray-dried blueberry extract using mesquite gum as wall material. *Journal of Food Composition and Analysis, 24*, 889–894. https://doi.org/10.1016/j.jfca.2011.04.012

Jing, P., & Giusti, M. M. (2005). Characterization of anthocyanin-rich waste from purple corncobs (*Zea mays* L.) and its application to color milk. *Journal of Agricultural and Food Chemistry, 53*(22), 8775–8781. https://doi.org/10.1021/jf051247o

Jing, P., Zhao, S. J., Ruan, S. Y., Xie, Z. H., Dong, Y., & Yu, L. L. (2012). Anthocyanin and glucosinolate occurrences in the roots of Chinese red radish (*Raphanus sativus* L.), and their stability to heat and pH. *Food Chemistry, 133*(4), 1569–1576. https://doi.org/10.1016/j.foodchem.2012.02.051

Johnson, N. R., & Wang, Y. (2014). Coacervate delivery systems for proteins and small molecule drugs. *Expert Opinion on Drug Delivery, 11*, 1829–1832. https://doi.org/10.1517/17425247.2014.941355

Joye, I. J., & McClements, D. J. (2014). Biopolymer-based nanoparticles and microparticles: Fabrication, characterization, and application. *Current Opinion in Colloid & Interface Science, 19*, 417–427. https://doi.org/10.1016/j.cocis.2014.07.002

Kader, F., Irmouli, M., Zitouni, N., Nicolas, J. P., & Metche, M. (1999). Degradation of cyanidin 3-glucoside by caffeic acid o-quinone. Determination of the stoichiometry and characterization of the degradation products. *Journal of Agricultural and Food Chemistry, 47*(11), 4625–4630. https://doi.org/10.1021/jf981400x

Kader, F., Irmouli, M., Nicolas, J. P., & Metche, M. (2001). Proposed mechanism for the degradation of pelargonidin 3-glucoside by caffeic acid o-quinone. *Food Chemistry, 75*(2), 139–144. https://doi.org/10.1016/S0308-8146(00)00301-0

Li, Y., Han, L., Ma, R., Xu, X., Zhao, C., Wang, Z., ... & Hu, X. (2012). Effect of energy density and citric acid concentration on anthocyanins yield and solution temperature of grape peel in microwave-assisted extraction process. *Journal of Food Engineering, 109*(2), 274–280. https://doi.org/10.1016/j.jfoodeng.2011.09.021

Liang, L. H., Wu, X. Y., Zhao, T., Zhao, J. L., Li, F., Zou, Y., Mao, G. H., & Yang, L. Q. (2012). *In vitro* bioaccessibility and antioxidant activity of anthocyanins from mulberry (*Morus atropurpurea Roxb.*) following simulated gastro-intestinal digestion. *Food Research International, 46*, 76–82. https://doi.org/10.1016/j.foodres.2011.11.024

Liang, Y., Chen, J., Zuo, Y., Ma, K. Y., Jiang, Y., Huang, Y., & Chen, Z.-Y. (2013). Blueberry anthocyanins at doses of 0.5 and 1% lowered plasma cholesterol by increasing fecal excretion of acidic and neutral sterols in hamsters fed a cholesterol-enriched diet. *European Journal of Nutrition*, *52*, 869–875. https://doi.org/10.1007/s00394-012-0393-6

Longo, L., & Vasapollo, G. (2006). Extraction and identification of anthocyanins from *Smilax aspera* L. berries. *Food Chemistry*, *94*(2), 226–231. https://doi.org/10.1016/j.foodchem.2004.11.008

Mahdavi, S. A., Jafari, S. M., Ghorbani, M., & Assadpoor, E. (2014). Spray-drying microencapsulation of anthocyanins by natural biopolymers: A review. *Drying Technology*, *32*(5), 509–518. https://doi.org/10.1080/07373937.2013.839562

Mao, L., Roos, Y. H., Biliaderis, C. G., & Miao, S. (2017). Food emulsions as delivery systems for flavor compounds: A review. *Critical Reviews in Food Science and Nutrition*, *57*(15), 3173–3187. https://doi.org/10.1080/10408398.2015.1098586

Markaris, P., Livingston, G. E., & Fellers, C. R. (1957). Quantitative aspects of strawberry pigment degradation. *Journal of Food Science*, *22*(2), 117–130. https://doi.org/10.1111/j.1365-2621.1957.tb16991.x

Maza, G., & Miniatti, E. (1993). *Anthocyanins in Fruits, Vegetable and Grains*. New York, NY: CRC.

McClements, D. J. (2016). *Food Emulsions: Principles, Practice, and Techniques* (3rd ed.). Boca Raton, FL: CRC Press.

McClements, D. J., & Li, Y. (2010). Structured emulsion-based delivery systems: Controlling the digestion and release of lipophilic food components. *Advances in Colloid and Interface Science*, *159*(2), 213–228. https://doi.org/10.1016/j.cis.2010.06.010

McDougall, G. J., Dobson, P., Smith, P., Blake, A., & Stewart, D. (2005). Assessing potential bioavailability of raspberry anthocyanins using an *in vitro* digestion system. *Journal of Agricultural and Food Chemistry*, *53*, 5896–5904.

McGhie, T. K., & Walton, M. C. (2007). The bioavailability and absorption of anthocyanins: Towards a better understanding. *Molecular Nutrition & Food Research*, *51*(6), 702–713. https://doi.org/10.1002/mnfr.200700092

Mohammadalinejhad, S., & Kurek, M. A. (2021). Microencapsulation of anthocyanins – Critical review of techniques and wall materials. *Applied Sciences*, *11*(9), 3936. https://doi.org/10.3390/app11093936

Morais, C. A., de Rosso, V. V., Estadella, D., & Pisani, L. P. (2016). Anthocyanins as inflammatory modulators and the role of the gut microbiota. *Journal of Nutritional Biochemistry*, *33*, 1–7. https://doi.org/10.1016/j.jnutbio.2015.11.008

Nayak, C. A., & Rastogi, N. K. (2010). Effect of selected additives on microencapsulation of anthocyanin by spray drying. *Drying Technology*, *28*(12), 1396–1404. https://doi.org/10.1080/07373937.2010.482705

Nayak, C. A., Srinivas, P., & Rastogi, N. K. (2010). Characterisation of anthocyanins from *Garcinia indica* Choisy. *Food Chemistry*, *118*(3), 719–724.

Ngo, T., Wrolstad, R. E., & Zhao, Y. (2007). Color quality of Oregon strawberries – impact of genotype, composition, and processing. *Journal of Food Science*, *72*(1), C025–C032. https://doi.org/10.1111/j.1750-3841.2006.00200.x

Norkaew, O., Thitisut, P., Mahatheeranont, S., Pawin, B., Sookwong, P., Yodpitak, S., & Lungkaphin, A. (2019). Effect of wall materials on some physicochemical properties and release characteristics of encapsulated black rice anthocyanin microcapsules. *Food Chemistry*, *294*, 493–502. https://doi.org/10.1016/j.foodchem.2019.05.086

Oidtmann, J., Schantz, M., Mäder, K., Baum, M., Berg, S., Betz, M., Kulozik, U., Leick, S., Rehage, H., Schwarz, K., & Richling, E. (2012). Preparation and comparative release characteristics of three anthocyanin encapsulation systems. *Journal of Agricultural and Food Chemistry*, *60*, 844–851. https://doi.org/10.1021/jf2047515

Okuro, P. K., Furtado, G. F., Sato, A. C. K., & Cunha, R. L. (2015). Structures design for protection and vehiculation of bioactives. *Current Opinion in Food Science*, *5*, 67–75. https://doi.org/10.1016/j.cofs.2015.09.003

Ordaz-Galindo, A., Wesche-Ebeling, P., Wrolstad, R. E., Rodriguez-Saona, L., & Argaiz-Jamet, A. (1999). Purification and identification of Capulin (*Prunus serotina Ehrh*) anthocyanins. *Food Chemistry*, *65*(2), 201–206. https://doi.org/10.1016/S0308-8146(98)00196-4

Ozkan, G., Franco, P., De Marco, I., Xiao, J., & Capanoglu, E. (2019). A review of microencapsulation methods for food antioxidants: Principles, advantages, drawbacks and applications. *Food Chemistry*, *272*, 494–506. https://doi.org/10.1016/j.foodchem.2018.07.205

Palamidis, N., & Markakis, P. (1975). Stability of grape anthocyanin in a carbonated beverage. *Journal of Food Science, 40*(5), 1047–1049. https://doi.org/10.1111/j.1365-2621.1975.tb02264.x

Paula, D. de A., Ramos, A. M., de Oliveira, E. B., Martins, E. M. F., de Barros, F. A. R., Vidigal, M. C. T. R., ... & da Rocha, C. T. (2018). Increased thermal stability of anthocyanins at pH 4.0 by guar gum in aqueous dispersions and in double emulsions W/O/W. *International Journal of Biological Macromolecules, 117*, 665–672. https://doi.org/10.1016/j.ijbiomac.2018.05.219

Pazmiño-Durán, E. A., Giusti, M. M., Wrolstad, R. E., & Glória, M. B. A. (2001). Anthocyanins from banana bracts (*Musa X paradisiaca*) as potential food colorants. *Food Chemistry, 73*(3), 327–332. https://doi.org/10.1016/S0308-8146(00)00305-8

Pieczykolan, E., & Kurek, M. A. (2019). Use of guar gum, gum arabic, pectin, beta-glucan and inulin for microencapsulation of anthocyanins from chokeberry. *International Journal of Biological Macromolecules, 129*, 665–671. https://doi.org/10.1016/j.ijbiomac.2019.02.073

Prietto, L., Pinto, V. Z., El Halal, S. L. M., de Morais, M. G., Costa, J. A. V., Lim, L. T., ... & Zavareze, d. R. (2018). Ultrafine fibers of zein and anthocyanins as natural pH indicator. *Journal of the Science of Food and Agriculture, 98*(7), 2735–2741. https://doi.org/10.1002/jsfa.8769

Prior, R. L., Wu, X. L., Gu, L. W., Hager, T. J., Hager, A., & Howard, L. R. (2008). Whole berries versus berry anthocyanins: Interactions with dietary fat levels in the C57BL/6J mouse model of obesity. *Journal of Agricultural and Food Chemistry*, 56, 647–653. https://doi.org/10.1021/jf071993o

Rajabi, H., Jafari, S. M., Rajabzadeh, G., Sarfarazi, M., & Sedaghati, S. (2019). Chitosan-gum Arabic complex nanocarriers for encapsulation of saffron bioactive components. *Colloids and Surfaces A: Physicochemical and Engineering Aspects, 578*, 123644. https://doi.org/10.1016/j.colsurfa.2019.123644

Riaz, M., Zia-Ul-Haq, M., & Saad, B. (2016). Springer briefs in food, health, and nutrition anthocyanins and human health: Biomolecular and therapeutic aspects. In Hartel, R. W., Finley, J. W., Rodriguez-Lazaro, D., Roos, Y. H., & Topping, D. (Eds.), Basel: Springer Nature. https://doi.org/10.1007/s10803-015-2603-6.

Rocha, J. d. C. G., de Barros, F. A. R., Perrone, Í. T., Viana, K. W. C., Tavares, G. M., Stephani, R., & Stringheta, P. C. (2019). Microencapsulation by atomization of the mixture of phenolic extracts. *Powder Technology, 343*, 317–325. https://doi.org/10.1016/j.powtec.2018.11.040

Rosenberg, M., Kopelman, I., & Talmon, Y. (1990). Factors affecting retention in spray drying microencapsulation of volatile materials. *Journal of Agricultural and Food Chemistry, 38*(5), 1288–1294. https://doi.org/10.1021/jf00095a030

Rosso, V. V. D., & Mercadante, A. Z. (2007). The high ascorbic acid content is the main cause of the low stability of anthocyanin extracts from acerola. *Food Chemistry, 103*, 935–943. https://doi.org/10.1016/j.foodchem.2006.09.047

Sadilova, E., Carle, R., & Stintzing, F. C. (2007). Thermal degradation of anthocyanins and its impact on color and *in vitro* antioxidant capacity. *Molecular Nutrition & Food Research, 51*(12), 1461–1471. https://doi.org/10.1002/mnfr.200700179

Santana, R. C., Perrechil, F. A., & Cunha, R. L. (2013). High- and low-energy emulsifications for food applications: A focus on process parameters. *Food Engineering Reviews, 5*(2), 107–122. https://doi.org/10.1007/s12393-013-9065-4

Santos-Buelga, C., & González-Paramás, A. M. (2018). Anthocyanins. *Reference Module in Food Science*, 1–12. https://doi.org/10.1016/B978-0-08-100596-5.21609-0

Šaponjac, V. T., Ćetković, G., Čanadanović-Brunet, J., Djilas, S., Pajin, B., Petrović, J., ... & Vulić, J. (2017). Encapsulation of sour cherry pomace extract by freeze drying: Characterization and storage stability. *Acta Chimica Slovenica, 64*(2), 283–289. http://dx.doi.org/10.17344/acsi.2016.2789

Sarkar, R., Dutta, A., Patra, A., & Saha, S. (2021). Bio-inspired biopolymeric coacervation for entrapment and targeted release of anthocyanin. *Céllulose, 28*, 377–388. https://doi.org/10.1007/s10570-020-03523-w

Seeram, N. P., Bourquin, L. D., & Nair, M. G. (2001). Degradation products of cyanidin glycosides from tart cherries and their bioactivities. *Journal of Agricultural and Food Chemistry, 49*(10), 4924–4929. https://doi.org/10.1021/jf0107508

Seymour, E. M., Singer, A. A., Kirakosyan, A., Urcuyo-Llanes, D. E., Kaufman, P. B., & Bolling, S. F. (2008). Altered hyperlipidemia, hepatic steatosis, and hepatic peroxisome proliferator-activated receptors in rats with intake of tart cherry. *Journal of Medicinal Food, 11*, 252 https://doi.org/10.1089/jmf.2007.658

Shaddel, R., Hesari, J., Azadmard-Damirchi, S., Hamishehkar, H., Fathi-Achachlouei, B., & Huang, Q. (2018). Use of gelatin and gum Arabic for encapsulation of black raspberry anthocyanins by complex coacervation. *International Journal of Biological Macromolecules, 107*, 1800–1810. https://doi.org/10.1016/j.ijbiomac.2017.10.044

Sharif, N., Khoshnoudi-Nia, S., & Jafari, S. M. (2020). Nano/microencapsulation of anthocyanins: A systematic review and meta-analysis. *Food Research International, 132*, 109077. https://doi.org/10.1016/j.foodres.2020.109077

Shiga, H., Yoshii, H., Ohe, H., Yasuda, M., Furuta, T., Kuwahara, H., ... & Linko, P. (2004). Encapsulation of shiitake (*Lenthinus edodes*) flavors by spray drying. *Bioscience, Biotechnology, and Biochemistry, 68*(1), 66–71. https://doi.org/10.1271/bbb.68.66

Shishir, M. R. I., Xie, L., Sun, C., Zheng, X., & Chen, W. (2018). Advances in micro and nano-encapsulation of bioactive compounds using biopolymer and lipid-based transporters. *Trends in Food Science & Technology, 78*, 34–60. https://doi.org/10.1016/j.tifs.2018.05.018

Simões, L. de S., Madalena, D. A., Pinheiro, A. C., Teixeira, J. A., Vicente, A. A., & Ramos, Ó. L. (2017). Micro- and nano bio-based delivery systems for food applications: *In vitro* behavior. *Advances in Colloid and Interface Science, 243*, 23–45. https://doi.org/10.1016/j.cis.2017.02.010

Souza, A. C. P., Gurak, P. D., & Marczak, L. D. F. (2017). Maltodextrin, pectin and soy protein isolate as carrier agents in the encapsulation of anthocyanins-rich extract from jaboticaba pomace. *Food and Bioproducts Processing, 102*, 186–194. https://doi.org/10.1016/j.fbp.2016.12.012

Srivastava, A., Greenspan, P., Hartle, D. K., Hargrove, J. L., Amarowicz, R., & Pegg, R. B. (2010). Antioxidant and anti-inflammatory activities of polyphenolics from southeastern U.S. range blackberry cultivars. *Journal of Agricultural and Food Chemistry, 58*, 6102–6109. https://doi.org/10.1021/jf1004836

Stoll, L., Costa, T. M. H., Jablonski, A., Flôres, S. H., & de Oliveira Rios, A. (2016). Microencapsulation of anthocyanins with different wall materials and its application in active biodegradable films. *Food and Bioprocess Technology, 9*(1), 172–181. https://doi.org/10.1007/s11947-015-1610-0

Suravanichnirachorn, W., Haruthaithanasan, V., Suwonsichon, S., Sukatta, U., Maneeboon, T., & Chantrapornchai, W. (2018). Effect of carrier type and concentration on the properties, anthocyanins and antioxidant activity of freeze-dried mao [*Antidesma bunius* (*L.*) *Spreng*] powders. *Agriculture and Natural Resources, 52*(4), 354–360. https://doi.org/10.1016/j.anres.2018.09.011

Tagliazucchi, D., Verzelloni, E., Bertolini, D., & Conte, A. (2010). *In vitro* bio-accessibility and antioxidant activity of grape polyphenols. *Food Chemistry, 120*, 599–606. https://doi.org/10.1016/j.foodchem.2009.10.030

Tarone, A. G., Cazarin, C. B. B., & Junior, M. R. M. (2020). Anthocyanins: New techniques and challenges in microencapsulation. *Food Research International, 133*, 109092. https://doi.org/10.1016/j.foodres.2020.109092

Tian, Q., Giusti, M. M., Stoner, G. D., & Schwartz, S. J. (2006). Characterization of a new anthocyanin in black raspberries (*Rubus occidentalis*) by liquid chromatography electrospray ionization tandem mass spectrometry. *Food Chemistry, 94*(3), 465–468. https://doi.org/10.1016/j.foodchem.2005.01.020

Tsai, P. J., McIntosh, J., Pearce, P., Camden, B., & Jordan, B. R. (2002). Anthocyanin and antioxidant capacity in Roselle (*Hibiscus sabdariffa* L.) extract. *Food Research International, 35*(4), 351–356. https://doi.org/10.1016/S0963-9969(01)00129-6

Tsuda, T. (2012). Dietary anthocyanin-rich plants: Biochemical basis and recent progress in health benefits studies. *Molecular Nutrition and Food Research, 56*(1), 159–170. https://doi.org/10.1002/mnfr.201100526.

Ushikubo, F. Y., Oliveira, D. R. B., Michelon, M., & Cunha, R. L. (2014). Designing food structure using microfluidics. *Food Engineering Reviews, 7*(4), 393–416. https://doi.org/10.1007/s12393-014-9100-0

Von Elbe, J. H., & Schwartz, S. J. (1996). Colorants. In Fennema, O. R. (Ed.), *Food Chemistry* (3rd ed.) (pp. 651–722). New York, NY: Marcel Dekker.

Wang, Y., Li, J., & Li, B. (2017). Chitin microspheres: A fascinating material with high loading capacity of anthocyanins for colon specific delivery. *Food Hydrocolloids*, 63, 293–300. https://doi.org/10.1016/j.foodhyd.2016.09.003

Xiong, S. Y., Melton, L. D., Easteal, A. J., & Siew, D. (2006). Stability and antioxidant activity of black currant anthocyanins in solution and encapsulated in glucan gel. *Journal of Agricultural and Food Chemistry*, 54, 6201–6208. https://doi.org/10.1021/jf060889o

Yi, W. G., Fischer, J., Krewer, G., & Akoh, C. C. (2005). Phenolic compounds from blueberries can inhibit colon cancer cell proliferation and induce apoptosis. *Journal of Agricultural and Food Chemistry, 53*, 7320–7329. https://doi.org/10.1021/jf0513330

Yousuf, B., Gul, K., Wani, A. A., & Singh, P. (2016). Health benefits of anthocyanins and their encapsulation for potential use in food systems: A review. *Critical Reviews in Food Science and Nutrition, 56*(13), 2223–2230. https://doi.org/10.1080/10408398.2013.805316

Zhao, L., Temelli, F., & Chen, L. (2017). Encapsulation of anthocyanin in liposomes using supercritical carbon dioxide: Effects of anthocyanin and sterol concentrations. *Journal of Functional Foods, 34*, 159–167. https://doi.org/10.1016/j.jff.2017.04.021

6 A Short Glimpse of Synthetic and Natural Anthocyanins Obtained from Various Subtropical Fruits

M. Selvamuthukumaran

CONTENTS

6.1   INTRODUCTION

Anthocyanin is a major water-soluble pigment that has a wide variety of colors, like purple, blue, and brilliant red. It is present in various subtropical fruits, like blood orange, berries, especially those grown in subtropical climateric conditions, like blueberries, craneberries, as well as in fruits like figs, dates, and litchis. Anthocyanin compounds are widely used as food additives in functional food-processing industries on account of their colorant properties. In addition, these colorants have therapeutic properties (Krishnan et al., 2017).

6.2   NATURAL ANTHOCYANINS AND SYNTHETIC ANTHOCYANINS

Various subtropical fruits, like blood orange, fig, dates, berries such as blueberries, raspberries, cranberries, litchi, and passion fruits are an ample source of natural anthocyanin. Advanced processing techniques can be successfully deployed to extract colorants from such fruits. No synthetic anthocyanin is yet available, so there is great scope for the production of natural colorants from dark-colored subtropical fruits. The main advantages of producing color from subtropical fruit are

Highly Toxic

Not environmentally friendly

Has adverse health effects

Banned in most of the country for its regular use

Can cause serious health hazards for consumers

FIGURE 6.1   Limitations of using synthetic food dyes.

TABLE 6.1
Anthocyanin Content of Various Subtropical Fruits

Name of Fruit	Type of Anthocyanin
Blood ranges	Cyanidin-3-glucosides, cyanidin-3-(6"-melonyl) glucosides
Plums	Cyanidin-3-glucosides, cyanidin-3-rutinoside
Grapes	Malvidin-3-glucosides, peonidin-3-glucosides, petunidin-3-glucosides, malvidin-3-coumaryl-glucoside
Jamun	Delphinidin-3,5-diglucoside, malvidin-3,5-diglucoside
Blueberries	Malvidin-3-glucosides, malvidin-3-galactosides, malvidin-3-arabinosides, petunidin-3-arabinoside
Acai berries	Cyanidin-3-rutinosides, cyanidin-3-glucosides
Sour cherries	Cyanidin-3-rutinosides, cyanidin-3-glucosides
Strawberries	Cyanidin-3-glucosides, pelargonidin-3-glucosides, pelargonidin-3-rutinosides

that natural colorants are non-toxic, environmentally friendly and have potential health benefits. Therefore, natural colorants obtained from various subtropical fruit sources can be used as a substitute for synthetic food dyes. The use of synthetic dye in food has several disadvantages (Figure 6.1). Dye obtained from natural sources has higher water solubility and can be easily incorporated into aqueous food systems (Melo et al., 2009; Mercadante and Bobbio, 2008). In addition to colorant properties, anthocyanins possess additional health benefits, i.e. preventing several degenerative diseases (Fraga, 2010; Ghosh and Konishi, 2007; Jaganath and Crozier, 2010; Pascual-Teresa and Sanchez-Ballesta, 2008). They work synergistically when used with other phytochemicals present in berries (Jing and Giusti, 2011; Lila, 2009; Lila et al., 2005; Pascual-Teresa et al., 2010; Schwinn and Davies, 2004).

6.3   ANTHOCYANIN PROFILE OF VARIOUS SUBTROPICAL FRUITS

A wide variety of anthocyanin compounds can be extracted and purified from subtropical fruits (Table 6.1).

6.4   MEMBRANE TECHNIQUE FOR PURIFICATION OF ANTHOCYANINS

During extraction of anthocyanins by non-thermal processing methods or traditional methods of extraction, other compounds like proteins, sugars and acids are extracted together with the anthocyanin compound. Therefore, it is essential to separate the anthocyanin compound from other compound mixtures. This can be achieved by a membrane separation technique.

This membrane separation technique can use both natural as well as synthetic membranes to separate and purify the anthocyanin molecule. A target compound such as anthocyanin can be effectively separated from other molecules based on molecular weight (Martin et al., 2017). Membranes used to separate the anthocyanin molecule include ultra-filtration membrane and micro-filtration membrane. There are several advantages associated with using membrane technology, including lower energy consumption, non-involvement of chemical reaction, non-phase change during the separation process, resistance to alkali and acid. In cranberry pomace, anthocyanin compounds were effectively recovered using a membrane-based separation technique (Woo et al., 1980). Concentrated powder was found to contain 0.11% anthocyanins using a membrane-processing technique like infiltration, ultrafiltration and subsequent concentration (Woo et al., 1980).

6.5 ANTHOCYANIN AS A FOOD ADDITIVE

Food-processing industries have used numerous synthetic colorants, which has harmful effects for consumers. A wide variety of food colorants obtained from synthetic sources, like amaranth, orange II, rhodamine, and fast red, have been banned due to their harmful effect. Therefore, there is a great demand for using anthocyanin, especially that obtained from subtropical fruits. The high-intensity anthocyanin color was ascribed to the higher resonance of the completely conjugated 10-electron A–C ring system, with a lesser contribution of the B ring (Krishnan et al., 2017). The anthocyanin color will be destroyed if the resonance structure is disrupted. Anthocyanin compounds will condense with polyphenolic compounds to give colored polymeric pigments. Anthocyanin tannin complex molecules are highly resistant to bisulfite bleaching; this needs to be considered when using anthocyanin as a food additive. Anthocyanin in monomeric form also undergoes a bleaching reaction.

Anthocyanin extracted from subtropical fruits possesses less stability, which is a major challenge when it is used as a coloring agent during food processing. Factors affecting anthocyanin stability include light, temperature, presence of polyphenolic compounds, ascorbic acid, and metal ions and as well as their chemical structure.

6.6 FACTORS AFFECTING ANTHOCYANIN COLOR STABILITY

6.6.1 Effect of Processing and Storage

Anthocyanin is highly influenced by processing and storage temperatures. Stability can be greatly enhanced at low temperature, i.e. at 4°C.

6.6.2 Effect of pH

Anthocyanin can undergo structural transformation when subjected to differential or change in pH, which can create a significant change in color. Red coloration for foods was achieved when anthocyanin-rich foods have a pH value of 3; when the pH is increased to 4–5 the red intensity is reduced and further enhancement of pH may lead to anthocyanin coloring food purple and blue (Krishnan et al., 2017). Based on these color changes it is desirable to add anthocyanin colorant to acidic foods, contributing an intense red color, which can add aesthetic appeal to food products.

6.6.3 Effect of Oxygen

Anthocyanin molecule had a detrimental effect on oxygen either through direct or indirect oxidative mechanism. With oxidation either brownish colored pigments or colorless products may resulr.

6.6.4   EFFECT OF ENZYMES

The presence of polyphenol oxidase enzymes in subtropical fruits degrades anthocyanin. Consumer acceptability is affected by the decompartmentalization of phenolic compounds and the occurrence of polyphenol oxidase; contact between these polyphenol oxidase and phenolic compounds can produce brownish pigments which affect nutritional value and marketability.

6.6.5   EFFECT OF PRESERVATIVES

The use of preservatives, especially sulfur dioxide, can stabilize the anthocyanin compound even at lower concentrations as they lead to the formation of anthocyanin–bisulfite complexes.

6.6.6   EFFECT OF FERMENTATION

Anthocyanin degradation was also noticed in the fermentation process, especially during lactic fermentation. This process can result in phenols of unstable nature with glycosides of undetermined structure and more reactive aglycones (Wang et al., 1996).

6.7   STRATEGIES TO PREVENT THE COLOR LOSS OF ANTHOCYANIN

The strategies below can be adopted to retain the colorant properties of anthocyanin (Krishnan et al., 2017).

- The use of preservatives, particularly metabisulfite, leads the anthocyanin moiety to develop intermediate adducts with HSO_3 which can leach out the color of the anthocyanin. This color degradation can be prevented by inactivating the enzyme, i.e. polyphenol oxidase, also prohibiting higher-temperature treatment, and also using co-pigments, which can further help the anthocyanin molecule remain stable in food complexes during processing.
- Adding these anthocyanin colorants along with co-pigments incorporation during fermentation can stabilize the anthocyanin, reducing the anthocyanin amenability to hydration, and leading to a greater hydration constant (pK_H value) (Stintzing et al., 2002).
- Chung et al. (2016) noted that the use of either stabilizers or emulsifiers, i.e. gum Arabic at 1.5%, in commercial beverage-based products can retain anthocyanin color stability as well as retain it up to 50% (Benech, 2008). Even in the presence of vitamin C, anthocyanin hydrogen bonding with gum Arabic glycoprotein fractions can make anthocyanin color more stable.

6.8   MARKET POTENTIAL OF USING ANTHOCYANIN AS A COLORANT

The production of anthocyanin dates back 40 years. It is among the preferred natural colorants in the food and beverage processing industries at a global level. In addition to its use as a colorant it has therapeutic properties that can be applied in pharmaceutical product manufacturing, ascribed to its antioxidant, anticancer and neuroprotective roles. Anthocyanin can be used in cosmetic industries, especially as an additive in personal care products because it improves the appearance of the skin.

The market value for anthocyanin at global level in the year 2015 was US$300 mn and it was forecast to rise to US$387.4 mn by 2021 according to a Transparency Market Research (2021) report, demonstrating that there will be a huge demand for its use as a colorant in food, pharma, cosmetic and other industries.

6.9   CONCLUSION

The natural colorant obtained from the subtropical fruits can be effectively used in food and beverage processing. In addition to its use as a colorant it has health benefits like the prevention and cure of degenerative diseases. It has good market value when compared to synthetic colorants, which are commonly used in the food and beverage-processing industries. Therefore, there is a great demand and huge scope for producing natural anthocyanin from various subtropical fruits.

REFERENCES

Benech, A. (2008). Gum Arabic – A functional hydrocolloid for beverages. *Agro Food Industry Hi-Tech.* 19(3): 58–59.

Chung, C., Rojanasasithara, T., Mutilangi, W., & Julian, M. D. (2016). Enhancement of colour stability of anthocyanins in model beverages by gum arabic addition. *Food Chem.* 201: 14–22.

Fraga, C. G., ed. (2010). *Plant Phenolics and Human Health: Biochemistry, Nutrition and Pharmacology.* New York: Wiley.

Ghosh, D., & Konishi, T. (2007). Anthocyanin and anthocyanin-rich extracts: Role in diabetes and eye function. *Asia Pac. J. Clin. Nutr.* 16(2): 200–208.

Jaganath, I. B., & Crozier, A. (2010). Dietary flavonoids and phenolic compounds. In Fraga, G. G. (Ed.), *Plant Phenolics and Human Health: Biochemistry, Nutrition and Pharmacology.* New York: Wiley; pp. 1–50.

Jing, P., & Giusti, M. M. (2011). Contribution of berry anthocyanins to their chemopreventive properties. In Stoner, G. D., & Seeram, N. P. (Eds.), *Berries and Cancer Protection.* New York: Springer; pp. 1–38.

Krishnan, V., Singh, S., Kaur, C., Dahuja, A., & Praveen, S. (2017). *Anthocyanin – A Premium Functional Superfood Supplement.* New Delhi: Division of Biochemistry, ICAR-Indian Agricultural Research Institute.

Lila, M. A. (2009). Interactions between flavonoids that benefit human health. In Gould, K., Davies, K., & Winifield, C. (Eds.), *Anthocyanins: Biosynthesis, Functions and Applications.* New York: Springer Science.

Lila, M. A., Yousef, G. G., Jiang, Y., & Weaver, C. M. (2005). Sorting out bioactivity in flavonoid mixtures. *J. Nutr.,* 135 (5): 1231–1235.

Martin, J., Enrique, J., Diaz, M., & Asuero, A. G. (2017). Recovery of anthocyanins using membrane technologies: A review. *Crit. Rev. Anal. Chem.* 48: 143–175.

Melo, M. J., Pina, F., & Andary, C. (2009). Anthocyanins: Nature's glamorous palete. In Bechtold, T., & Mussak, R. (Eds.), *Handbook of Natural Colorants.* Chichester, UK: Wiley; pp. 134–150.

Mercadante, A. Z., & Bobbio, F. O. (2008). Anthocyanins in foods: Occurrence and physicochemical properties. In Sosaciu, C. (Ed.), *Food Colorants: Chemical and Functional Properties.* Boca Raton, FL: CRC Press.

Pascual-Teresa, S. de, & Sanchez-Ballesta, M. T. (2008). Anthocyanins: From plant to health. *Phytochem. Rev.* 7: 281–299.

Pascual-Teresa, S. de, Moreno, D. A., & García-Viguera, C. (2010). Flavanols and anthocyanins in cardiovascular health: A review of current evidence. *Int. J. Mol. Sci.* 11(4): 1679–1703.

Schwinn, K. E., & Davies, K. M. (2004). Flavonoids. In *Plant Pigments and Their Manipulation; Annual Plant Reviews 14.* Oxford: Blackwell; p. 92149.

Stintzing, F. C., Stintzing, A. S., Carle, R., Frei, B., & Wrolstad, R. E. (2002). Color and antioxidant properties of cyanidin-based anthocyanin pigments. *J. Agric. Food Chem.* 50(21): 6172–6181.

Transparency market report: *Anthocyanins Market (Product Type – Cyanidin, Malvidin, Delphinidin, and Peonidin; Application – Food Beverage, Pharmaceuticals Products, and Personal Care) – Global Industry Analysis, Size, Share, Growth, Trends, and Forecast 2015 – 2021.* www.transparencymarketr esearch.com/anthocyaninmarket.html

Wang, H., Cao, G., & Prior, R. L. (1996). Total antioxidant capacity of fruits. *J. Agric. Food Chem.* 44(3): 701–705.

Woo, A. H., Elbe, J. H. V., & Amundson, C. H. (1980). Anthocyanin recovery from cranberry pulp wastes by membrane technology. *J. Food Sci.* 45: 875–879.

7 Natural Anthocyanins from Subtropical Fruits for Cancer Prevention

M. Selvamuthukumaran

CONTENTS

7.1 INTRODUCTION

Anthocyanins play a major role in preventing cancer, as they can interact with cancerous cells, modifying cellular response (Table 7.1). Anthocyanins were regarded as potent antioxidants because of their phenolic structure and their hydroxyl groups. They prohibit cell mutagenesis in the cells by quenching the reactive oxygen species (ROS), and are toxic for various cancerous cells (Wang and Stoner, 2008; Cvorovic et al., 2010).

Several research studies have suggested that dietary habits were solely responsible for the occurrence of cancer. This can be prevented by incorporating more fruit and vegetables into the diet, as they are the natural source of several polyphenol complex molecules, and they have a potent chemopreventive effect on the biological system. Subtropical fruits like grapes, plums, and a wide array of berries grown in subtropical climacteric conditions contain high concentrations of anthocyanin components. Therefore the consumption of such anthocyanin-rich subtropical fruits helps to prevent cancer (Figure 7.1). In addition anthocyanin's cancer-preventing effects, it can reduce diabetes, heart disease, and arthritis, thanks to its antioxidative and anti-inflammatory activities (Prior and Wu, 2006).

DOI: 10.1201/9781003242598-7

TABLE 7.1
Chemopreventive Activity of Anthocyanins Obtained from Subtropical Fruits

Name of Subtropical Fruit	Type of Cancer Studied	Effect Achieved	Reference
Tart cherry	Colon	Reduced cecal tumor was shown in an animal model fed with tart cherry extract containing anthocyanin at 375–3000 mg/kg diet. The tumor effect was reduced to 74% for the animals that received this extract	Kang et al. (2003)
Bilberry	Colon	The adenoma number was reduced to 30–45% for animals fed with bilberry extract containing anthocyanin, cyanidin-3-glucoside and its mixture at 0.3%	Cooke et al. (2006)
Pomegranate	Skin	The skin application of pomegranate extract containing anthocyanin and tannin in CD-1 mice model highly prevented hyperplasia, ornithine decarboxylase (ODC) activity, tissue-type plasminogen activator-mediated increases in skin edema and protein expression of both ODC and cyclooxygenase--2	Afaq et al. (2005)
Black raspberry	Esophageal cancer	Esophageal cancer was prevented: esophageal tumors were reduced by 42–47% as a result of feeding black raspberries; anthocyanins had a chemopreventive effect	Stoner et al. (2007)

7.2 ANTICANCER ACTIVITY OF VARIOUS SUBTROPICAL FRUITS

7.2.1 PLUMS

Symonds et al. (2013) reported that the polyphenol extract obtained from Illawarra plum had successfully suppressed the growth of cells, enhanced apoptosis, and diminished the activity of telomerase as well as leading to arrest of the S-cell phase. Similar findings were also confirmed by Charepalli et al. (2016) in Java plum. The anthocyanin extract obtained from Java plum enhanced colon cancer stem cell proportions, and in HCT-116 cells, apoptosis was promoted with suppressed proliferation of cells.

7.2.2 GRAPES

Grapes contain abundant anthocyanins. Several research studies have proved that these anthocyanin extracts obtained from grapes promoted apoptosis and suppressed cell growth in CaCo-2 cells and HT-29 cells (Zhao et al., 2004; Yi et al., 2005a; Mazewski et al., 2018). Esselen et al. (2011) reported that extract obtained from grapes minimized the side effects of chemotherapy drugs, reducing the activity of topoisomerase I and also DNA strand breaks. Grape contains phenolic acids, flavonol compounds, tannins and anthocyanins. Among all these components, anthocyanins exhibited significant antiproliferation activity. Lala et al. (2006) reported that grape extract successfully diminished the burden of cyclooxygenase-2 (COX-2) mRNA expression, which was proved in the AOM rat model. Cai et al. (2010) observed that grape pomace reduced the burden of adenoma, adenoma number, Ki-67 and AKT expression in APCm in mice.

7.2.3 BERRIES

Some of the berries that can be successfully harvested under subtropical climate conditions possess potent anticarcinogenic properties. They are briefly described below.

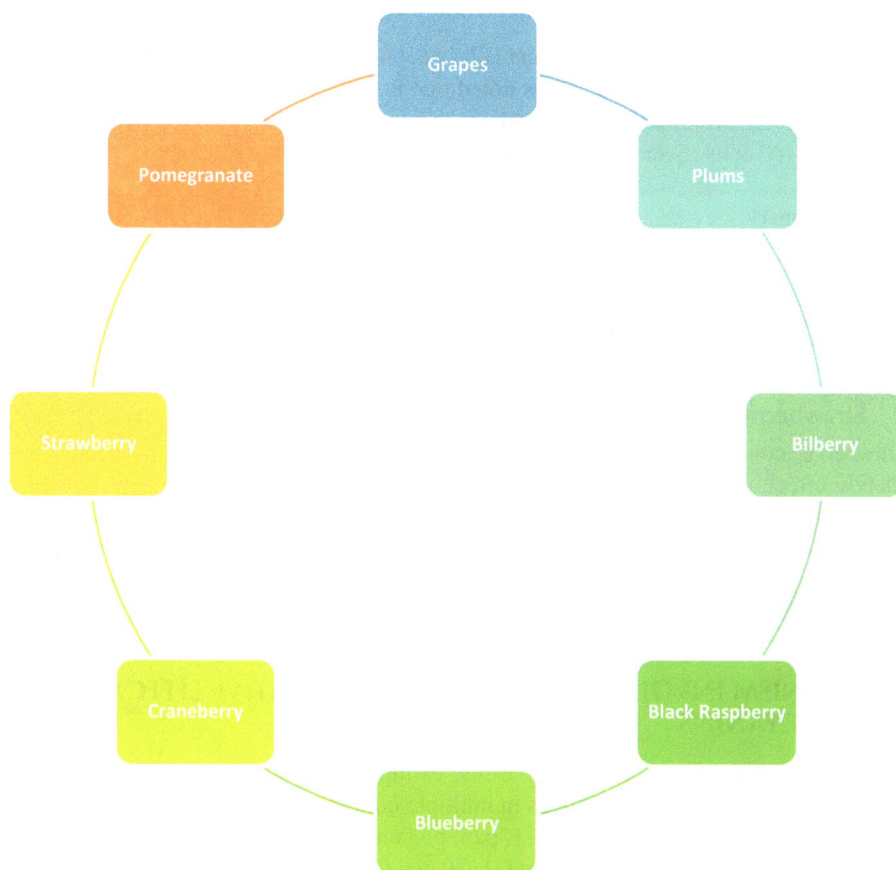

FIGURE 7.1 Various subtropical fruits that exhibit antiproliferative activity.

7.2.3.1 Blueberry

It was claimed that the extract obtained from blueberry fruit suppressed Caco-2 cell proliferation with reported half maximal inhibitory concentration (IC_{50}) value of 0.78 (Yi et al., 2005b; Bornsek et al., 2012). Seeram et al. (2006) and Yi et al. (2005b) explained that extract obtained from blueberry had significantly inhibited the proliferation of cells and promoted apoptosis in HT-29 and HCT116 cells. The higher antiproliferation effect was pronounced in anthocyanin extract obtained from blueberry when compared to flavonol components, tannins and phenolic acids. Zu et al. (2010) also reported that the anthocyanin obtained from a Chinese blueberry variety had suppressed the proliferation of COLO205 and DLD-1 cells.

7.2.3.2 Black Raspberry

Wang et al. (2013) and Seeram et al. (2006) found that the anthocyanin obtained from black raspberry had suppressed the proliferation of cells, thereby inducing apoptosis. It also diminished the activity of SFRP2, SFRP5, demethylate CDKN2A, DNMT3B, DNMT1 and Wnt inhibitory factor 1 (WIF1) in the Wnt pathway of CRC cell lines. In AOM/DSS colon cancer-induced mouse model, the incorporation of black raspberry anthocyanin extracts reduced the multiplicity level of tumor, thereby modulating the gut commensal microbiota composition. It also changes the methylation status of the secreted frizzled-related protein 2 (*SFRP2*) gene and inflammation index (Chen et al., 2018).

7.2.3.3 Cranberry

The cell proliferation of SW620, SW480 and HT-29 cells was suppressed by cranberry anthocyanin extracts (Seeram et al., 2004, 2006). It was noted that the anthocyanin extract obtained from cranberry has successfully exhibited inhibitory effects in metastatic and progressive SW620 cells, when compared to SW480 cells (Seeram et al., 2004).

7.2.3.4 Blackberry

The anthocyanin extract obtained from blackberry had exhibited potent antiproliferative, anti-inflammatory and antioxidant activity in Caco-2 cells and HT-29 cells, thereby suppressing cell growth as well as interleukin (IL)-12 release from dendritic cells derived from mouse bone marrow (Seeram et al., 2006; Dai et al., 2007; Elisia and Kitts, 2008).

7.2.3.5 Strawberries

The strawberry contains anthocyanin, i.e. cyanidin-3-glucoside, which possesses a potent antioxidative effect, enabling it to inhibit HT-116 and HT-29 cell growth. It was justified that the cyanidin-3-glucoside exhibited the best antioxidative effect compared to other anthocyanins extracted from strawberries (Zhang et al., 2008). Anthocyanin components like pelargonidin-3-glucoside, pelargonidin, and pelargonidin-3-rutinoside obtained from strawberries suppressed the cell growth of HCT-116 and HT-29 cells (Zhang et al., 2008).

7.3 MECHANISM INVOLVED IN THE CHEMOPREVENTIVE EFFECT OF ANTHOCYANINS

Anthocyanin obtained from subtropical fruits exhibited potent antiproliferative activity, as demonstrated in several animals as well as in multiple colon cancer cell types. The cell cycle gene (cyclin A, cyclin B, cyclin E, cyclin D1, p21, p25, p53 and cyclin-dependent kinase inhibitors) was modulated and the cell cycle was arrested at G1/G0 and G2/M by anthocyanins (Kamei et al., 1998; Malik et al., 2003; Lazze et al., 2004; Yun et al., 2009; Symonds et al., 2013). During carcinogenesis the MAPK pathway (ERK, c-Jun-NH2 kinase and p38), represents proteins regulating cell proliferation, differentiation and survival during carcinogenesis (Shi et al., 2015). In colon cancer, MAPK pathway-blocking activation has been reported as one of the mechanisms involved in exhibiting anticarcinogenic activities of anthocyanidins in colon cancer (Shi et al., 2011, 2015; Sido et al., 2017). Teller et al. (2009 a, b) reported that another molecular mechanism involved in exhibiting anthocyanin antiproliferative activity may be ascribed to suppression of insulin-like growth factor 1 receptor (IGF1R), epidermal growth factor receptor (EGFR), ErbB2, ErbB3, vascular endothelial growth factor receptor-2 (VEGFR-2) and VEGFR-3. Mazewski et al. (2018) reported that delphinidin-3-O-glucoside and cyanidin-3-O-glucoside significantly prohibited EGFR in HT-29 and HCT-116 cells. Chen et al. (2019) reported that cell proliferation and survival were greatly influenced by AKT/ERK/NF-κB signaling cascades.

7.3.1 APOPTOSIS INDUCTION

Apoptosis is one of the targets of antitumor activity. Several research studies have shown that anthocyanin components can exhibit apoptosis-induced activities in animal models and colon cancer cell types. The anthocyanin pro-apoptotic effect can be ascribed to an ROS-mediated mitochondrial caspase-independent pathway (Wang and Stoner, 2008; Charepalli et al., 2016; Zhang et al., 2021). Extracts obtained from anthocyanin successfully promoted apoptosis by reducing the expression of anti-apoptotic proteins (survivin, cIAP-2 and XIAP) (Mazewski et al., 2018). This further shows that anthocyanins can successfully promote apoptosis in a p53-independent manner in colon cancer stem cells by elevating protein-mediating mitochondrial apoptosis, Bax and cytochrome c (Charepalli et al., 2015; Zhang et al., 2021).

7.3.2 GENE DEMETHYLATION

Chen et al. (2019) reported that in the AOM-DSS mouse model, black raspberry anthocyanins successfully diminished the DNA methyltransferase expression and demethylated the hypermethylated promoters of the *SFRP2* gene in order to induce the colon type of cancer. The methylation of tumor suppressor genes, *SFRP2*, paired box 6a (PAX6a) and WIF1 were reduced due to incorporation of black raspberry anthocyanins in a human test model (Wang et al., 2011). The anthocyanin extract obtained from black raspberry reduced the activity and protein expression of DNA methyltransferase enzymes, DNMT1 and DNMT3B, and demethylated WNT upstream regulators, CDKN2A, SFRP2, SFRP5 and WIF1 (Wang et al., 2013).

7.4 CONCLUSION

Incorporation of subtropical fruits in one's diet can help augment the bioavailability of anthocyanin, which can exhibit an anticancer preventive effects in addition to scavenging the formation of free radicals in the biological system.

REFERENCES

Afaq, F., Saleem, M., Kueger, C. G., Reed, J. D., & Mukhtar, H. Anthocyanin- and hydrolyzable tannin-rich pomegranate fruit extract modulates MAPK and NF-kappaB pathways and inhibits skin tumorigenesis in CD-1 mice. Int. J. Cancer 2005;113: 423–433.

Bornsek, S. M., Ziberna, L., Polak, T., Vanzo, A., Ulrih, N. P., Abram, V.,Tramer, F., & Passamonti, S. Bilberry and blueberry anthocyanins act as powerful intracellular antioxidants in mammalian cells. Food Chem. 2012;134: 1878–1884.

Cai, H., Marczylo, T. H., Teller, N., Brown, K., Steward, W. P., Marko, D., & Gescher, A. J. Anthocyanin-rich red grape extract impedes adenoma development in the Apc(Min) mouse: Pharmacodynamic changes and anthocyanin levels in the murine biophase. Eur. J. Cancer 2010;46: 811–817.

Charepalli, V., Reddivari, L., Radhakrishnan, S., Vadde, R., Agarwal, R., & Vanamala, J. K. Anthocyanin-containing purple-fleshed potatoes suppress colon tumorigenesis via elimination of colon cancer stem cells. J. Nutr. Biochem. 2015;26: 1641–1649.

Charepalli, V., Reddivari, L., Vadde, R., Walia, S., Radhakrishnan, S., Vanamala, J. K. *Eugenia jambolana* (java plum) fruit extract exhibits anti-cancer activity against early stage human HCT-116 colon cancer cells and colon cancer stem cells. Cancers 2016;8: 29.

Chen, L., Jiang, B., Zhong, C., Guo, J., Zhang, L., Mu, T., Zhang, Q., & Bi, X. Chemoprevention of colorectal cancer by black raspberry anthocyanins involved the modulation of gut microbiota and SFRP2 demethylation. Carcinogenesis 2018;39: 471–481.

Chen, T., Shi, N., & Afzali, A. Chemopreventive effects of strawberry and black raspberry on colorectal cancer in inflammatory bowel disease. Nutrients 2019;11: 1261–1281.

Cooke, D., Schwarz, M., Boocock, D., Winterhalter, P., Steward, W. P., Gescher, A. J., & Marczylo, T. H. Effect of cyanidin-3-glucoside and an anthocyanin mixture from bilberry on adenoma development in the ApcMin mouse model of intestinal carcinogenesis – relationship with tissue anthocyanin levels. Int. J. Cancer 2006;119:2213–2220.

Cvorovic, J., Tramer, F., Granzotto, M., Candussio, L., Decorti, G., & Passamonti, S. Oxidative stress-based cytotoxicity of delphinidin and cyanidin in colon cancer cells. Arch. Biochem. Biophys. 2010;501: 151–157. doi: 10.1016/j. abb.2010.05.019

Dai, J., Patel, J. D., & Mumper, R. J. Characterization of blackberry extract and its antiproliferative and anti-inflammatory properties. J. Med. Food 2007;10: 258–265.

Elisia, I., & Kitts, D. D. Anthocyanins inhibit peroxyl radical-induced apoptosis in Caco-2 cells. Mol. Cell Biochem. 2008;312: 139–145.

Esselen, M., Fritz, J., Hutter, M., Teller, N., Baechler, S., Boettler, U., Marczylo, T. H., Gescher, A. J., & Marko, D. Anthocyanin-rich extracts suppress the DNA-damaging effects of topoisomerase poisons in human colon cancer cells. Mol. Nutr. Food Res. 2011;55(Suppl. 1): S143–S153.

Kamei, H., Hashimoto, Y., Koide, T., Kojima, T., & Hasegawa, M. Anti-tumor effect of methanol extracts from red and white wines. Cancer Biother. Radiopharm. 1998;13: 447–452.

Kang, S. Y., Seeram, N. P., Nair, M. G., & Bourquin, L. D. Tart cherry anthocyanins inhibit tumor development in Apc(Min) mice and reduce proliferation of human colon cancer cells. Cancer Lett. 2003;194:13–19.

Lala, G., Malik, M., Zhao, C., He, J., Kwon, Y., Giusti, M. M., & Magnuson, B. A. Anthocyanin-rich extracts inhibit multiple biomarkers of colon cancer in rats. Nutr. Cancer 2006;54: 84–93.

Lazze, M. C., Savio, M., Pizzala, R., Cazzalini, O., Perucca, P., Scovassi, A. I., Stivala, L. A., & Bianchi, L. Anthocyanins induce cell cycle perturbations and apoptosis in different human cell lines. Carcinogenesis 2004;25: 1427–1433.

Malik, M., Zhao, C., Schoene, N., Guisti, M. M., Moyer, M. P., & Magnuson, B. A. Anthocyanin-rich extract from *Aronia meloncarpa* E induces a cell cycle block in colon cancer but not normal colonic cells. Nutr. Cancer 2003;46: 186–196.

Mazewski, C., Liang, K., & Gonzalez de Mejia, E. Comparison of the effect of chemical composition of anthocyanin-rich plant extracts on colon cancer cell proliferation and their potential mechanism of action using in vitro, in silico, and biochemical assays. Food Chem. 2018;242: 378–388.

Prior, R. L., & Wu, X. Anthocyanins: Structural characteristics that result in unique metabolic patterns and biological activities. Free Radic. Res. 2006;40:1014–1028.

Seeram, N. P., Adams, L. S., Hardy, M. L., & Heber, D. Total cranberry extract versus its phytochemical constituents: Antiproliferative and synergistic effects against human tumor cell lines. J. Agric. Food Chem. 2004;52: 2512–2517.

Seeram, N. P., Adams, L. S., Zhang, Y., Lee, R., Sand, D., Scheuller, H. S., & Heber, D. Blackberry, black raspberry, blueberry, cranberry, red raspberry, and strawberry extracts inhibit growth and stimulate apoptosis of human cancer cells in vitro. J. Agric. Food Chem. 2006;54: 9329–9339.

Shi, N., Clinton, S. K., Liu, Z., Wang, Y., Riedl, K. M., Schwartz, S. J., Zhang, X., Pan, Z., & Chen, T. Strawberry phytochemicals inhibit azoxymethane/dextran sodium sulfate-induced colorectal carcinogenesis in Crj: CD-1 mice. Nutrients 2015;7: 1696–1715.

Shin, D. Y., Lu, J. N., Kim, G. Y., Jung, J. M., Kang, H. S., Lee, W. S., Choi, Y. H. Anti-invasive activities of anthocyanins through modulation of tight junctions and suppression of matrix metalloproteinase activities in HCT-116 human colon carcinoma cells. Oncol. Rep. 2011;25: 567–572.

Sido, A., Radhakrishnan, S., Kim, S. W., Eriksson, E., Shen, F., Li, Q., Bhat, V., Reddivari, L., & Vanamala, J. K. P. A food-based approach that targets interleukin-6, a key regulator of chronic intestinal inflammation and colon carcinogenesis. J. Nutr. Biochem. 2017;43: 11–17.

Stoner, G. D., Wang, L. S., & Chen, T. Chemoprevention of esophageal squamous cell carcinoma. Toxicol Appl Pharmacol 2007;224:337–349.

Symonds, E. L., Konczak, I., & Fenech, M. The Australian fruit Illawarra plum (*Podocarpus elatus* Endl., Podocarpaceae) inhibits telomerase, increases histone deacetylase activity and decreases proliferation of colon cancer cells. Br. J. Nutr. 2013;109: 2117–2125.

Teller, N., Thiele, W., Boettler, U., Sleeman, J., & Marko, D. Delphinidin inhibits a broad spectrum of receptor tyrosine kinases of the ErbB and VEGFR family. Mol. Nutr. Food Res. 2009a;53: 1075–1083.

Teller, N., Thiele, W., Marczylo, T. H., Gescher, A. J., Boettler, U., Sleeman, J., & Marko, D. Suppression of the kinase activity of receptor tyrosine kinases by anthocyanin-rich mixtures extracted from bilberries and grapes. J. Agric. Food Chem. 2009b;57: 3094–3101.

Wang, L. S., & Stoner, G. D. Anthocyanins and their role in cancer prevention. Cancer Lett. 2008;269: 281–290.

Wang, L. S., Arnold, M., Huang, Y. W., Sardo, C., Seguin, C., Martin, E., Huang, T. H., Riedl, K., Schwartz, S., Frankel, W., et al. Modulation of genetic and epigenetic biomarkers of colorectal cancer in humans by black raspberries: A phase I pilot study. Clin. Cancer Res. 2011;17: 598–610.

Wang, L. S., Kuo, C. T., Cho, S. J., Seguin, C., Siddiqui, J., Stoner, K., Weng, Y. I., Huang, T. H., Tichelaar, J., Yearsley, M., et al. Black raspberry-derived anthocyanins demethylate tumor suppressor genes through the inhibition of DNMT1 and DNMT3B in colon cancer cells. Nutr. Cancer 2013;65: 118–125.

Yi, W., Fischer, J., & Akoh, C. C. Study of anticancer activities of muscadine grape phenolics in vitro. J. Agric. Food Chem. 2005a;53: 8804–8812.

Yi, W., Fischer, J., Krewer, G., & Akoh, C. C. Phenolic compounds from blueberries can inhibit colon cancer cell proliferation and induce apoptosis. J. Agric. Food Chem. 2005b;53: 7320–7329.

Yun, J. M., Afaq, F., Khan, N., & Mukhtar, H. Delphinidin, an anthocyanidin in pigmented fruits and vegetables, induces apoptosis and cell cycle arrest in human colon cancer HCT116 cells. Mol. Carcinog. 2009;48: 260–270.

Zhang, Y., Seeram, N. P., Lee, R., Feng, L., & Heber, D. Isolation and identification of strawberry phenolics with antioxidant and human cancer cell antiproliferative properties. J. Agric. Food Chem. 2008;56: 670–675.

Zhang, Z., Pan, Y., Zhao, Y., Ren, M., Li, Y., Lu, G., Wu, K., & He, S. Delphinidin modulates JAK/STAT3 and MAPKinase signaling to induce apoptosis in HCT116 cells. Environ. Toxicol. 2021;36: 1557–1566.

Zhao, C., Giusti, M. M., Malik, M., Moyer, M. P., & Magnuson, B. A. Effects of commercial anthocyanin-rich extracts on colonic cancer and nontumorigenic colonic cell growth. J. Agric. Food Chem. 2004;52: 6122–6128.

Zu, X. Y., Zhang, Z. Y., Zhang, X. W., Yoshioka, M., Yang, Y. N., & Li, J. Anthocyanin extracted from Chinese blueberry (Vaccinium uliginosum L.) and its anticancer effects on DLD-1 and COLO205 cells. Chin. Med. J. 2010;123: 2714–2719.

8 Natural Anthocyanins from Subtropical Fruits for Cardiac Disease Prevention

Sidra Kazmi, Mehvish Habib, Khalid Bashir, Shumaila Jan, Sachin Kumar, and Kulsum Jan

CONTENTS

8.1   INTRODUCTION

Dietary phytochemicals such as phenols, flavonoids and carotenoids are in high demand because of their health benefits. It has been found that these phytochemicals reduce the risk of chronic diseases such as cardiovascular disease (CVD) (Boyer & Liu, 2004). Polyphenols are the most numerous of bioactive molecules (Abbas et al., 2017), among which "phenolics" represents the largest group of phytochemicals (Kennedy, 2014). Polyphenols are natural compounds found in fruit and vegetables. Anthocyanins belong to the polyphenolic group and are colored water-soluble pigments. The pigments are present in glycosylated form (Khoo et al., 2017). Anthocyanins impart the colors red, purple, and blue. Fruits and vegetables like blueberries, pomegranate, beetroot, grapes and some tropical fruits have high anthocyanin content. Anthocyanins contain two aromatic benzene rings. In the 20 types of anthocyanin pigments, only a few are found in plants where cyanidin-3-glucoside is the major anthocyanin. They range in color from orange to violet. Anthocyanins have antioxidant properties, i.e., they are free radical scavengers, which in turn benefits human health. Other factors also help to decrease health-related risks (Tsuda et al., 1996, 2002, 2003; Wang & Jiao, 2000). Isolates of anthocyanins protect against unwanted DNA cleavage and altered development of hormone-dependent disease symptoms. Factors like chemical structure also influence the bioactive properties of anthocyanins (Lila, 2004).

Anthocyanins are also used as a natural food colorant. Their color and color stability can be influenced by pH, light (as they are light-sensitive), and temperature. The color of anthocyanin remains red at lower pH but changes as the pH increases (Khoo et al., 2017). Liquid chromatography is commonly used for the extraction, separation, and quantification of anthocyanin components (Singh et al., 2020).

Approximately 70% of daily dietary intake of anthocyanins can be provided by fruits such as apples, berries, grapes, and pears (Zamora-Ros et al., 2011). The main dietary source of anthocyanins in Northern Europe and the United States is berries (Cassidy, 2018; Heinonen, 2007). Although a recommended daily intake has not been established for anthocyanins as they are not considered essential nutrients, China has suggested 50 mg of daily anthocyanin intake (Wallace & Giusti, 2015). Due to incomplete and insufficient data, it is difficult to evaluate daily intake of anthocyanins. However, according to estimates from various countries, the daily intake of anthocyanins ranges from 19 to 65 mg/day for men and from 18 to 44 mg/day for women in Europe (Zamora-Ros et al., 2011), 12.5 mg/day in the United States (Wu et al., 2006), and about 24 mg/day in Australia (Igwe et al., 2019). Moreover, in Finland, berries are consumed as a primary source; thus daily intake is up to 150 mg/day (Heinonen, 2007).

Encouraging the consumption of a daily anthocyanin-rich diet guards against various chronic and degenerative diseases. Moreover, it assures an adequate amount of antioxidant and protecting substances at plasma level. Adhering to a Mediterranean diet which is rich in anthocyanin content is therefore associated with reduced inflammation markers and decreased risk of diseases such as cancer, diabetes, CVD and obesity (Tosti et al., 2018). In contrast, consuming fruit and vegetables in lesser quantities results in millions of deaths worldwide: major causes include gastrointestinal cancer (14%), ischemic heart disease (11%), stroke (9%), and many others.

In order to improve the stability of anthocyanins, techniques and methods such as chemical derivatization, co-pigment addition with colorless molecules in solution, oxygen exclusion, and microencapsulation have been incorporated. Efficient protection is provided by microencapsulation with biopolymer, specifically by the application of beta-glucans and maltodextrins (Lourith & Kanlayavattanakul, 2020; Mahdavi et al., 2016; Pieczykolan & Kurek, 2019; Yousuf et al., 2016). Freeze drying or spray drying can be produced by microcarriers. Nanoformulation, including nanoemulsions or nanoliposomes, is considered to be an alternative method for microencapsulation (Chen & Inbaraj, 2019).

8.2   ANTHOCYANINS: CHEMICAL STRUCTURE AND TYPES

Anthocyanin structure contains flavylium cations, i.e., 2-phenyl-1-benzopyrilium. They lack ketone oxygen at the fourth position. Anthocyanins are present in the form of anthocyanidins (aglycone) glycosides and acylated anthocyanins (Khoo et al., 2017). They contain conjugated bonds which result in different plant colors, such as red, orange, blue.

Pelargonidin, cyanidin, delphinidin, petunidin, peonidin, and malvidin are the most common types of anthocyanidins (Reyes et al., 2018). These pigments contain carbohydrates where glucose is the most common type. Other than glucose, fructose, xylose, and galactose are also present in small quantities. These sugars are mainly linked to the C3 and C5 position as 3-monoglycosides and diglycosides respectively. Glycosylations have been also found at C7, C3′, and C5′ positions (Khoo et al., 2017).

The potential role of anthocyanin is determined by its chemical structure. Since they have a specific chemical structure, anthocyanins are known to be electron-deficient; furthermore, they are highly reactive towards reactive oxygen species (ROS), also known as free radicals. Therefore, anthocyanins are called natural antioxidants. Various plant species have different types of structure and anthocyanin profile (Pervaiz et al., 2017).

On the basis of 31 anthocyanidins, more than 500 different anthocyanins have been discovered depending on type of methoxylation, hydroxylation and glycosylation, and various substituents linked to sugar units (McCallum et al., 2007; Tsao & McCallum, 2009).

From these 31 anthocyanidin monomers, 18% are from pelargonidin, 22% from delphinidin, and 30% from cyanidin. Furthermore, their methylated derivatives in total represent 20% of anthocyanins, including malvidin, petunidin, and peonidin. Thus, it can be concluded that the most frequently found anthocyanins (up to 90%) are associated with cyanidin, delphinidin, and pelargonidin, as well as their methylated derivatives (Goulas et al., 2012).

The majorly distributed pigment in plants is cyanidin. Cyanidin consists of two hydroxyl groups on the B-ring. Cyanidin-3-O-β-glucoside is mostly found in anthocyanin in edible plants, followed by delphinidin, peonidin glucosides, and pelargonidin (Khoo et al., 2017). Generally, methylation induces red color and enhances stability, whereas blue color and decrease in stability are caused by hydroxylation. Acylation or glycosylation may be helpful in modifying hydroxyl groups. Chemical and physical properties of anthocyanins are affected by these modifications. Moreover, molecule polarity and chemical reactivity are also modified (He & Giusti, 2010).

8.3 EXTRACTION AND IDENTIFICATION OF ANTHOCYANINS

The stability of anthocyanins is usually influenced by various factors such as temperature, pH, light, oxygen, metal ions, sugars, ascorbic acid, and complex formation (Cavalcanti et al., 2011). Thus, analytical chemistry has a significant role in advances in separation techniques.

Identification of anthocyanins plays a significant role in adulteration (Gardana et al., 2014) and taxonomic (Picariello et al., 2014) studies. Extensively used separation techniques for anthocyanins include high-performance liquid chromatography (HPLC), specifically in reversed phase. Different molecular forms of anthocyanins have pH-dependent interconversions, and as a result it is necessary to have a highly acidic mobile phase of pH < 2. This helps to ensure that anthocyanins are predominantly maintained in flavylium cationic form for maximum efficiency of chromatography. But in some cases, interconversion between carbinol pseudobasic forms and anthocyanin flavylium cationic takes place at low pH as well (Alberts et al., 2012; Borrás-Linares et al., 2015; Pina, 2014; Pina et al., 2012).

For the detection of anthocyanins, frequently used chromatographic detection techniques include mass spectrometry (MS) or tandem mass spectrometry (MS/MS) and diode array detection (DAD). However, advancements in nuclear magnetic resonance (NMR) also hold promise in the analysis of anthocyanin and complex mixtures of phenolic compounds (Corradini et al., 2011; Rupasinghe, 2015). NMR is primarily based on analysis of ^{1}H NMR spectra, although ^{13}C NMR also provides necessary structural information. ^{13}C NMR especially provides information about compounds having many quaternary carbons, combining hetero- and homonuclear two-dimensional (2D) and 3D techniques (Kirby et al., 2013). But the barrier to their routine use in laboratories is its high capital cost (Scotter, 2011).

Depending on the pH of the medium, anthocyanins can either form part of the complexes or can be present in matrices of a complex nature or in distinct equilibrium form (Pina et al., 2012, 2015; Silva et al., 2016). It is of the utmost important to have acid dissociation, tautomerization constants, and a rate constant for the purpose of bioactive compound analysis.

Prior to separation and quantification of compounds, extraction and semipreparative isolation methods are generally applied (Figure 8.1). For the isolation, purification, and determination of anthocyanin structure, time-consuming processes are involved, and these should be properly completed (Jordheim, 2007).

Extraction of anthocyanin pigments using organic solvents such as ethanol and methanol leads to toxicity issues. Extraction on the basis of water is considered to be a greener method. Furthermore, a method of subcritical water-based extraction is considered to be a highly effective method to extract anthocyanins from fruits such as berries. In this method, acidified water of pH ~2.3 and 0.01% hydrochloric acid is subjected to temperatures ranging between 110 and 160°C under a constant

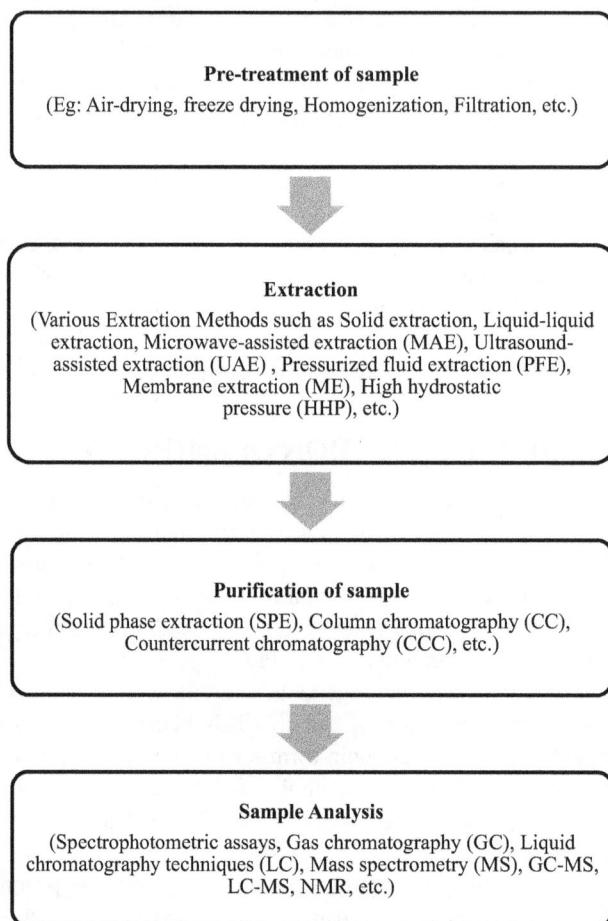

```
┌─────────────────────────────────────────────┐
│         Pre-treatment of sample               │
│  (Eg: Air-drying, freeze drying,              │
│   Homogenization, Filtration, etc.)           │
└─────────────────────────────────────────────┘
                      ↓
┌─────────────────────────────────────────────┐
│                Extraction                     │
│ (Various Extraction Methods such as Solid     │
│  extraction, Liquid-liquid extraction,        │
│  Microwave-assisted extraction (MAE),         │
│  Ultrasound-assisted extraction (UAE),        │
│  Pressurized fluid extraction (PFE),          │
│  Membrane extraction (ME), High hydrostatic   │
│  pressure (HHP), etc.)                         │
└─────────────────────────────────────────────┘
                      ↓
┌─────────────────────────────────────────────┐
│           Purification of sample              │
│ (Solid phase extraction (SPE), Column         │
│  chromatography (CC), Countercurrent          │
│  chromatography (CCC), etc.)                   │
└─────────────────────────────────────────────┘
                      ↓
┌─────────────────────────────────────────────┐
│              Sample Analysis                  │
│ (Spectrophotometric assays, Gas               │
│  chromatography (GC), Liquid chromatography   │
│  techniques (LC), Mass spectrometry (MS),     │
│  GC-MS, LC-MS, NMR, etc.)                      │
└─────────────────────────────────────────────┘
```

FIGURE 8.1   General flow chart for the extraction, preparation, and characterization of anthocyanin samples from fruit.

pressure of 40 bars (King et al., 2003). Moreover, sulfur dioxide can added to water to extract anthocyanin pigments. This helps to stabilize the structure of anthocyanins with increased diffusion coefficient anthocyanin molecules through solid (Ju & Howard, 2005), resulting in increased solubility of anthocyanins from plant during extraction with water.

Chromatographic methods such as cellulose column chromatography, chromatography based on high-speed countercurrent forces, HPLC, gas chromatography (GC), thin-layer chromatography (TLC), and reversed-phase ion-pair chromatography are used for the separation and identification of anthocyanins. Isolation of specific types of anthocyanin is needed for a specific purpose since these pigments are extracted from plants in the form of a crude mixture (Schwarz et al., 2003). Furthermore, anthocyanin compounds are also quantified by HPLC and GC (Petersson, 2009). Specially invented for the purpose of purification of flavonoid, AB-8 resin is a type of macroporous resin used in a study for the purification of phenolic pigments (Liu et al., 2010).

8.4   SUBTROPICAL FRUITS

Subtropical fruits are the crops grown in climatic conditions between tropical and temperate. They can be evergreen or deciduous. They are capable of withstanding low temperature, although not frost.

Moreover, during the day and night they adopt fluctuations of dark and light. Common examples of subtropical fruit include berries, pomegranates, dates, varieties of avocado, figs, litchi, citrus fruit such as oranges, lemons, grapefruits, etc. They can be subdivided into three groups: (1) less perishable; for example, dried figs, pomegranate, date, citrus fruit, kiwi fruit; (2) moderately perishable; for example, olive, persimmon, cherimoya; and (3) highly perishable; for example, litchi, fresh figs, loquat (Yahia et al., 2011).

Anthocyanin-rich fruits that are most commonly consumed belong to the Rosaceace family (for example, raspberries, cherries, strawberries, plums, apples), Ericaceae family (for example, blueberries, bilberries, cranberries, blackcurrants), and Vitaceae family, including the common grape (*Vitis vinifera*) (Konczak & Zhang, 2004). Providing a considerable anthocyanin dose in one serving, the concentration of anthocyanin in these fruit sources ranges from several hundreds of milligrams in 100 g of fresh weight (FW). Elderberries have the highest anthocyanin concentration, ranging between 664 and 1816 mg/100 g FW, followed by chokeberries, ranging between 410 and 1480 mg/100 g FW, bilberries, ranging between 300 and 698 mg/100 g FW, raspberries, ranging between 20 and 687 mg/100 g FW, blackcurrants, ranging between 130 and 476 mg/100 g FW, blackberries, ranging between 82.5 and 325.9 mg/100 g FW, and blueberries, ranging between 61.8 and 299.6 mg/100 g FW (Konczak & Zhang, 2004). Moreover, anthocyanin concentration is influenced by food-processing and storage conditions, environmental and genetic factors, humidity, temperature, light, and fertilization. A comparative study of various types of processed blueberries shows preservation of anthocyanins is better in canned fruits (~70%) and purées (~57%) compared to clarified juice (~31%) (Konczak & Zhang, 2004).

8.5 PRESENCE OF ANTHOCYANINS IN SUBTROPICAL FRUITS

Table 8.1 lists types of fruit and anthocyanins, as reported in the literature.

8.6 ANTHOCYANINS FOR HYPERTENSION

Studies have showed results of protection against hypertension and improvement of cognition by the consumption of anthocyanins from fruit sources. Plum juice with high anthocyanin content significantly decreased blood pressure, although dose timing did not appear to be a significant factor in the potential acute blood pressure-lowering effect of Queen Garnet plum juice (Igwe et al., 2017).

According to many researchers, anthocyanins present mainly in fruit and vegetables such as strawberries, blackcurrants, blueberries, cranberries, raspberries, blood oranges, and eggplant tend to provide protection against high blood pressure (Igwe et al., 2019).

A study has shown the vasodilator activities of polyphenols extracts from grape and wine because of increased expression and phosphorylation of endothelial nitric oxide synthase (eNOS), and as a result vasodilator nitric oxide (NO) is produced. There is a large amount of anthocyanin flavonoids in berries which helps to induce NO production and decrease inflammation and oxidative stress. As a result, vascular function was improved (Festa et al., 2021).

A study of animal models investigated the vasodilatory properties of preparations containing only anthocyanins. It was discovered that coronary vessel relaxation results from the presence of anthocyanins (Bell & Gochenaur, 2006).

Hypertension and atherosclerosis are two major conditions associated with reduced blood flow and that have the potential to cause adverse cardiac-related diseases. The effect of anthocyanins in the prevention of these cardiac diseases has been studied. One study shows an anti-atherosclerosis effect with increase in high-density lipoprotein cholesterol (HDL-C) (Sanossian et al., 2007). Another study showed increased HDL-C levels in pre-hypertensive men who had been taking 4-week anthocyanin intervention capsules (Hassellund et al., 2013). The intervention capsule

TABLE 8.1

Types of Fruit and the Anthocyanins Present in Them

Subtropical Fruit	Major Types of Anthocyanins Present	Reference
Blueberry	Monoarabinosides, monogalactosides, and monoglucosides of delphinidins, cyanidins, petunidins, peonidins, and malvidins	Spinardi et al. (2019)
Litchi	Cyanidin-3-rutinoside, cyanidin-3-glucoside, cyanidin-3-galactoside, malvidin-3-acetyl-glucoside, pelargonidin-3-glycoside, and quercetin 3-rutinoside	Li and Jiang (2007)
Blood orange	Cyanidin 3-glucoside (C3G) and cyanidin 3-(6″-malonyl)-β-glucoside (C3-6MG)	Scordino et al. (2015)
Bilberry	Delphinidins, petunidins, cyanidins, peonidins, malvidins	Upton (2001)
Cranberry	Glucosides of cyanidin and peonidin (small European cranberries) 3-Galactosides and 3-arabinosides (American cranberries)	Mazza and Miniati (2018)
Blackberry	Cyanidin-3-glucoside, cya-3-rutinoside delphinidin, malvidin, pelargonidin, pelargonidin-3-glucoside, peonidin	Ivanovic et al. (2014); Lee et al. (2016)
Strawberry	Cyanidin-3-glucoside, pelargonidin-3-glucoside, pelargonidin-3-rutinoside, and pelargonidin-3-acetylglucoside	Karaaslan and Yaman (2017)
Red grape	Malvidin 3-O-glucoside, paeonidin 3-O-glucoside, petunidin 3-O-glucoside, cyanidin 3-O-glucoside, delphinidin 3-O-glucoside	Kallithraka et al. (2009)
Grape	Mono- and di-glucosylated cyanidin, delphinidin, petunidin, peonidin, and malvidin	Mazza and Miniati (2018)
Pomegranate	Cyanidin-3,5-diglucoside, cyanidin-3-glucoside, cyanidin-pentoside, delphinidin-3,5-diglucoside, delphinidin-3-glucoside, pelargonidin-3,5-diglucoside, pelargonidin-3-glucoside	Cano-Lamadrid et al. (2017)
Apple	Cyanidin-3-galactoside	Bars-Cortina et al. (2017)
Cherry	Cyanidin-3-glucoside and cyanidin-3-rutinoside	Karaaslan and Yaman (2016)
Haskap berry	Cyanidin-3-glucoside	Caprioli et al. (2016); Chaovanalikit et al. (2004); Zhao et al. (2015)
Mulberry	Cyanidin-3-glucoside, cyanidin-3-galactoside, cyanidin-3-rutinoside, cyanidin-3-(6″-rhaminosyl) glucoside, and delphinidin-3-rutinoside	Chen et al. (2017); Khalifa et al. (2018)
Plum	Ccyanidin-3-rutinoside, cyanidin-3-glucoside	Sahamishirazi et al. (2017); Venter et al. (2014)
Blackcurrant	3-Rutinoside and 3-glucoside of cyanidin and delphinidin	Chen et al. (2014)
Chokeberry	Cyanidin 3-galactoside, cyanidin 3-glucoside, cyanidin 3-arabinoside, and cyanidin 3-xyloside	Slimestad et al. (2005)

consisted of 17 types of anthocyanins from blackcurrants and blueberries; the maximum amount of capsule containing delphinidin 3-O-β-glucosides and cyanidin 3-O-β-glucosides and total weight of the capsule was 80 mg (Hassellund et al., 2013).

Evidence shows supplementation of anthocyanins may have a beneficial effect on reversing hypertension and atherosclerosis in an aging population and subsequently can help in the treatment and prevention of cardiac-related diseases. Anthocyanin-rich extracts of bilberry and chokeberry have vasorelaxation properties (Bell & Gochenaur, 2006). A clinical trial in healthy volunteers suggested that platelet function and lipid profile were improved if anthocyanin-rich strawberries had been consumed for 1 month (Alvarez-Suarez et al., 2014). However, non-anthocyanin compounds such as phenolic compounds and vitamin C in strawberries may also play a part in these effects.

In one study, healthy humans were treated with anthocyanin-rich fruits. Participants were advised to avoid consuming strawberry and other polyphenols for 10 days. After 10 days, these participants were instructed to consume strawberries for 30 days, i.e., 500 g/day. After this, samples of participants' blood and urine were collected and analyzed. Furthermore, after 30 days of strawberry consumption, participants were again recommended to avoid strawberries for 15 days. After all these periods, samples of participants' blood and urine were collected again for analysis. The results after supplementing strawberries to participants were observed to decrease levels of low-density lipoproteins (LDL), triglycerides, and cholesterol. After 15 days of supplementing strawberries, these parameters returned to baseline levels. Spontaneous and oxidative hemolysis was also observed in the study. This study provides evidence that consumption of a strawberry-rich diet could partially help in protecting against and preventing CVD (Alvarez-Suarez et al., 2014).

In a study on blueberry, 66 obese participating patients with metabolic syndrome were divided into two groups. Participants included both men and women. Freeze-dried blueberry (50 g), reconstituted and dissolved in 480 mL of water and vanilla, was given to one group. In another group (control group) patients were recommended to drink water as control. These patients were treated for 8 weeks and evaluated after each 4-week period. The results showed significant reduction in blood pressure, oxidized LDL (ox-LDL), serum malondialdehyde, and hydroxynonenal. No significant changes were observed in lipid profile and serum glucose concentration. These results show that blueberries have a correlation with CVD markers and improve aspects of metabolic syndrome (Basu et al., 2010).

Several clinical studies have investigated the effects of different varieties of berries such as raspberries, strawberries, cranberries, acai berries, blueberries, and bilberries on patients with CVD risk factors as well as on healthy humans. The results of berry interventions showed an increased antioxidant capacity of plasma and increased HDL cholesterol. Furthermore, there was a decrease in lipid peroxidation, LDL oxidation, and total cholesterol. Coronary artery disease (CAD) is associated with lipids, lipid oxidation, and elevated plasma glucose; therefore, the addition of berries to the diet is beneficial in counteracting CAD-associated risk factors such as postprandial metabolic and oxidative stress. These data suggest the importance of berries as a potential dietary source for hypertension management (Basu et al., 2009, 2010).

Higher systolic blood pressure in individuals is reduced by daily consumption of boysenberry juice; moreover, it was found to be beneficial in reducing the risk of CVD (Matsusima et al., 2013). The antihypertensive effect of this fruit is due to the presence of anthocyanins in roselle. It was observed in patients with metabolic syndrome (125 mg/kg/day for 4 weeks) and also in a double-blind, lisinopril-controlled clinical trial of 171 hypertensive patients for 4 weeks (Herrera-Arellano et al., 2007). To summarize: studies on anthocyanins present in fruits found them to be beneficial in reducing oxidative stress and ameliorating inflammation, and suitable candidates for the treatment and prevention of cardiac diseases (Joven et al., 2014).

8.7 MECHANISM OF ACTION

In order to understand the underlying mechanism of the cardioprotective effects of anthocyanins, many in vivo and in vitro studies have been done. These studies have suggested that the action of anthocyanin is mostly restricted at the endothelial level, which contributes to vascular homeostasis. Atherosclerosis develops due to impaired endothelial function and is linked with cardiovascular risk factors such as hypertension. These changes are characterized by various properties of the endothelium, including decreased endothelial vasodilatation, proinflammatory and prothrombic properties (Hadi et al., 2005).

For the purpose of the treatment, prevention, or reduction of various diseases, anthocyanins are considered to be an excellent source of antioxidants and therefore they are capable of effectively scavenging free radicals. Moreover, the mechanism of action for disease prevention by anthocyanins involves two pathways: direct and indirect. Decreasing the risk of various chronic diseases by scavenging free radicals and thereby further reducing oxidative stress is known as the direct pathway. In contrast, in the indirect pathway, through reduction of oxidative stress and lipid peroxidation, there is a reduction or downregulation of cell proliferation and apoptosis (Khoo et al., 2017). Therefore, by improving blood lipid profile and biomarkers, anthocyanins decrease the risk of cardiovascular diseases.

It was suggested that anthocyanins help to prevent LDL oxidation and associated inflammatory responses by exerting a direct antioxidant effect against ROS, thereby reducing the development of atherosclerosis. Moreover, recent studies have also proposed more complex mechanisms of action involved in disease prevention, such as miRNA expression, gene expression, and cell signaling (Krga & Milenkovic, 2019).

Mechanistic analysis provides backing to the constructive effects of flavonoids, along with anthocyanins, on the rooted biomarkers of CVD risk as in the case of NO, inflammation, and endothelial dysfunction (Loke et al., 2008; Pergola et al., 2006; Steffen et al., 2008). Inflammation, described by *calor* (heat), *rubor* (redness), and *tumor* (swelling), plays a dominant role in the expansion of CVD. The role of anthocyanins in CVD prohibition is closely linked to safeguard against oxidative stress. In vivo anti-inflammatory responses were intended to clarify through diverse mechanisms/ structures. The combination of anthocyanin isolates and anthocyanin-rich flavonoids may provide security from DNA cleavage, estrogenic activity (altering the development of hormone-dependent disease symptoms), enzyme restriction, expanded cytokine production (thus monitoring immune responses), anti-inflammatory activity, lipid peroxidation, reduced capillary penetration and brittleness, and membrane strengthening (Acquaviva et al., 2003; Lazze et al., 2003; Lefevre et al., 2004; Ramirez-Tortosa et al., 2001; Rossi et al., 2003).

The chemical arrangement (types of replacement or substitution, position, and number) of an anthocyanin performs a vital role in the biological action exercised. Dietary anthocyanins have accumulated in the tissues of pigs on long-term feeding, remaining in tissues rather than in the blood stream (Kalt et al., 2008). Although it is not yet known whether anthocyanins assemble in cardiac or vascular tissues during long-term feeding, evidence from animal investigations has shown that anthocyanins affect vascular reactivity (Kalea et al., 2009). Clinically diagnosed patients with vascular disease were administered with relatively low-dose anthocyanin. As a result, there was a significant reduction in lipid level, blood pressure, inflammatory status, and ischemia.

Inhibited platelet activity and experimental coronary thrombosis in vivo occurred as a result of significant impact of commercial grape juice (10 mL/kg) (Demrow et al., 1995).

In healthy human participants, there is clinical evidence of less effect of proinflammatory markers; however, in research by Karlsen et al. (2007) supplementation with anthocyanins resulted in a significant improvement in plasma risk biomarkers.

However, to understand better the role of anthocyanin in cardiac disease prevention and its underlying molecular mechanism, more in vitro studies on the effects of anthocyanins and their metabolites on miRNA, gene and protein expressions need to be done using a proper holistic approach and relevant design. Furthermore, to provide solid evidence and understand the exact role of anthocyanin-rich foods in their cardioprotective effect, it is of the utmost importance to perform well-developed clinical trials for future studies (Krga & Milenkovic, 2019).

8.8   CONCLUSION AND FUTURE PROSPECTS

Anthocyanins are water-soluble colored pigments present in various plants that appear blue in alkaline solution and red in acidic solution. They are principally responsible for the purple, red, and blue colors found in various tropical and subtropical fruit, vegetables, and flowers as well in cereal grains. Their structure in different pH conditions is maintained by highly heat-stable acylated and co-pigmented anthocyanidins. Research has shown that anthocyanins have health benefits for humans. They are value-added colorants that can be used to treat and prevent diseases such as CVD, diabetes, microbial infections, some metabolic diseases, cancers, and obesity. Anthocyanidins and anthocyanins have various mechanisms of action in preventing these diseases, such as free-radical scavenging, cyclooxygenase, and MAPK pathways, and changes in blood biomarkers. Moreover, they have a neuroprotective effect and are capable of improving visual ability. Therefore, research studies should focus on the development of dietary supplements and nutraceuticals based on anthocyanin-rich extracts that will confer health benefits to humans. Further studies should also emphasize better understanding of the mechanisms involved by which food components achieve their effects and their pharmacokinetic characteristics.

REFERENCES

Abbas, M., Saeed, F., Anjum, F. M., Afzaal, M., Tufail, T., Bashir, M. S., ... & Suleria, H. A. R. (2017). Natural polyphenols: An overview. International Journal of Food Properties, 20(8), 1689–1699. https://doi.org/10.1080/10942912.2016.1220393

Acquaviva, R., Russo, A., Galvano, F., Galvano, G., Barcellona, M. L., Volti, G. L., & Vanella, A. (2003). Cyanidin and cyanidin 3-O-β-D-glucoside as DNA cleavage protectors and antioxidants. Cell Biology and Toxicology, 19(4), 243–252. https://doi.org/10.1023/B:CBTO.0000003974.27349.4e

Alberts, P., Stander, M. A., & de Villiers, A. (2012). Advanced ultra high pressure liquid chromatography–tandem mass spectrometric methods for the screening of red wine anthocyanins and derived pigments. Journal of Chromatography A, 1235, 92–102.

Alvarez-Suarez, J. M., Giampieri, F., Tulipani, S., Casoli, T., Di Stefano, G., González-Paramás, A. M., ... & Battino, M. (2014). One-month strawberry-rich anthocyanin supplementation ameliorates cardiovascular risk, oxidative stress markers and platelet activation in humans. The Journal of Nutritional Biochemistry, 25(3), 289–294.

Bars-Cortina, D., Macia, A., Iglesias, I., Romero, M. P., & Motilva, M. J. (2017). Phytochemical profiles of new red-fleshed apple varieties compared with traditional and new white-fleshed varieties. Journal of Agricultural and Food Chemistry, 65(8), 1684–1696. https://doi.org/10.1021/acs.jafc.6b02931

Basu, A., Wilkinson, M., Penugonda, K., Simmons, B., Betts, N. M., & Lyons, T. J. (2009). Freeze-dried strawberry powder improves lipid profile and lipid peroxidation in women with metabolic syndrome: baseline and post intervention effects. Nutrition Journal, 8(1), 1–7.

Basu, A., Du, M., Leyva, M. J., Sanchez, K., Betts, N. M., Wu, M., ... & Lyons, T. J. (2010). Blueberries decrease cardiovascular risk factors in obese men and women with metabolic syndrome. Journal of Nutrition, 140(9), 1582–1587.

Bell, D. R., & Gochenaur, K. (2006). Direct vasoactive and vasoprotective properties of anthocyanin-rich extracts. Journal of Applied Physiology, 100(4), 1164–1170. https://doi.org/10.1152/japplphysiol.00626.2005

Borrás-Linares, I., Fernández-Arroyo, S., Arráez-Roman, D., Palmeros-Suárez, P. A., Del Val-Díaz, R., Andrade-Gonzáles, I., ... & Segura-Carretero, A. (2015). Characterization of phenolic compounds, anthocyanidin, antioxidant and antimicrobial activity of 25 varieties of Mexican Roselle (Hibiscus sabdariffa). Industrial Crops and Products, 69, 385–394.

Boyer, J., & Liu, R.H. (2004). Apple phytochemicals and their health benefits. Nutrition Journal, 3, 5. https://doi.org/10.1186/1475-2891-3-5

Cano-Lamadrid, M., Trigueros, L., Wojdyło, A., Carbonell-Barrachina, Á. A., & Sendra, E. (2017). Anthocyanins decay in pomegranate enriched fermented milks as a function of bacterial strain and processing conditions. LWT, 80, 193–199. DOI: 10.1016/j.lwt.2017.02.023

Caprioli, G., Iannarelli, R., Innocenti, M., Bellumori, M., Fiorini, D., Sagratini, G., ... & Maggi, F. (2016). Blue honeysuckle fruit (Lonicera caerulea L.) from eastern Russia: Phenolic composition, nutritional value and biological activities of its polar extracts. Food & Function, 7(4), 1892–1903.

Cassidy, A. (2018). Berry anthocyanin intake and cardiovascular health. Molecular Aspects of Medicine, 61, 76–82. https://doi.org/10.1016/j.mam.2017.05.002

Cavalcanti, R. N., Santos, D. T., & Meireles, M. A. A. (2011). Non-thermal stabilization mechanisms of anthocyanins in model and food systems – An overview. Food Research International, 44(2), 499–509.

Chaovanalikit, A., Thompson, M. M., & Wrolstad, R. E. (2004). Characterization and quantification of anthocyanins and polyphenolics in blue honeysuckle (Lonicera caerulea L.). Journal of Agricultural and Food Chemistry, 52(4), 848–852. https://doi.org/10.1021/jf030509o

Chen, B. H., & Inbaraj, S. B. (2019). Nanoemulsion and nanoliposome based strategies for improving anthocyanin stability and bioavailability. Nutrients, 11(5), 1052.

Chen, X., Parker, J., Krueger, C. G., Shanmuganayagam, D., & Reed, J. D. (2014). Validation of HPLC assay for the identification and quantification of anthocyanins in black currants. Analytical Methods, 6(20), 8141–8147.

Chen, Y., Du, F., Wang, W., Li, Q., Zheng, D., Zhang, W., ... & Yang, L. (2017). Large-scale isolation of high-purity anthocyanin monomers from mulberry fruits by combined chromatographic techniques. Journal of Separation Science, 40(17), 3506–3512. https://doi.org/10.1002/jssc.201700471

Corradini, E., Foglia, P., Giansanti, P., Gubbiotti, R., Samperi, R., & Lagana, A. (2011). Flavonoids: Chemical properties and analytical methodologies of identification and quantitation in foods and plants. Natural Product Research, 25(5), 469–495.

Demrow, H. S., Slane, P. R., & Folts, J. D. (1995). Administration of wine and grape juice inhibits in vivo platelet activity and thrombosis in stenosed canine coronary arteries. Circulation, 91(4), 1182–1188. https://doi.org/10.1161/01.CIR.91.4.1182

Festa, J., Da Boit, M., Hussain, A., & Singh, H. (2021). Potential benefits of berry anthocyanins on vascular function. Molecular Nutrition & Food Research, 2100170. https://doi.org/10.1002/mnfr.202100170

Gardana, C., Ciappellano, S., Marinoni, L., Fachechi, C., & Simonetti, P. (2014). Bilberry adulteration: Identification and chemical profiling of anthocyanins by different analytical methods. Journal of Agricultural and Food Chemistry, 62(45), 10998–11004.

Goulas, V., Vicente, A. R., & Manganaris, G. A. (2012). Structural Diversity of Anthocyanins in Fruits. Nova Science https://ktisis.cut.ac.cy/handle/10488/3619

Hadi, H. A., Carr, C. S., & Al Suwaidi, J. (2005). Endothelial dysfunction: Cardiovascular risk factors, therapy, and outcome. Vascular Health and Risk Management, 1(3), 183.

Hassellund, S. S., Flaa, A., Kjeldsen, S. E., Seljeflot, I., Karlsen, A., Erlund, I., & Rostrup, M. (2013). Effects of anthocyanins on cardiovascular risk factors and inflammation in pre-hypertensive men: A double-blind randomized placebo-controlled crossover study. Journal of Human Hypertension, 27(2), 100–106. https://doi.org/10.1038/jhh.2012.4

He, J., & Giusti, M. M. (2010). Anthocyanins: Natural colorants with health-promoting properties. Annual Review of Food Science and Technology, 1, 163–187.

Heinonen, M. (2007). Antioxidant activity and antimicrobial effect of berry phenolics – A Finnish perspective. Molecular Nutrition & Food Research, 51(6), 684–691. https://doi.org/10.1002/mnfr.200700006

Herrera-Arellano, A., Miranda-Sánchez, J., Ávila-Castro, P., Herrera-Álvarez, S., Jiménez-Ferrer, J. E., Zamilpa, A., ... & Tortoriello, J. (2007). Clinical effects produced by a standardized herbal medicinal product of Hibiscus sabdariffa on patients with hypertension. A randomized, double-blind, lisinopril-controlled clinical trial. Planta Medica, 73(01), 6–12.

Igwe, E. O., Charlton, K. E., Roodenrys, S., Kent, K., Fanning, K., & Netzel, M. E. (2017). Anthocyanin-rich plum juice reduces ambulatory blood pressure but not acute cognitive function in younger and older adults: A pilot crossover dose-timing study. Nutrition Research, 47, 28–43. DOI: 10.1016/j.nutres.2017.08.006

Igwe, E. O., Charlton, K. E., & Probst, Y. C. (2019). Usual dietary anthocyanin intake, sources and their association with blood pressure in a representative sample of Australian adults. Journal of Human Nutrition and Dietetics, 32(5), 578–590. https://doi.org/10.1111/jhn.12647

Ivanovic, J., Tadic, V., Dimitrijevic, S., Stamenic, M., Petrovic, S., & Zizovic, I. (2014). Antioxidant properties of the anthocyanin-containing ultrasonic extract from blackberry cultivar "Čačanska Bestrna". Industrial Crops and Products, 53, 274–281. http://dx.doi.org/10.1016/j.indcrop.2013.12.048

Jordheim, M. (2007). Isolation, Identification and Properties of Pyranoanthocyanins and Anthocyanin Forms.

Joven, J., March, I., Espinel, E., Fernández-Arroyo, S., Rodríguez-Gallego, E., Aragonès, G., ... & Camps, J. (2014). Hibiscus sabdariffa extract lowers blood pressure and improves endothelial function. Molecular Nutrition & Food Research, 58(6), 1374–1378.

Ju, Z., & Howard, L. R. (2005). Subcritical water and sulfured water extraction of anthocyanins and other phenolics from dried red grape skin. Journal of Food Science, 70(4), S270–S276. https://doi.org/10.1111/j.1365-2621.2005.tb07202.x

Kalea, A. Z., Clark, K., Schuschke, D. A., & Klimis-Zacas, D. J. (2009). Vascular reactivity is affected by dietary consumption of wild blueberries in the Sprague-Dawley rat. Journal of Medicinal Food, 12(1), 21–28. https://doi.org/10.1089/jmf.2008.0078

Kallithraka, S., Aliaj, L., Makris, D. P., & Kefalas, P. (2009). Anthocyanin profiles of major red grape (*Vitis vinifera* L.) varieties cultivated in Greece and their relationship with in vitro antioxidant characteristics. International Journal of Food Science & Technology, 44(12), 2385–2393. 10.1111/j.1365-2621.2008.01869.x

Kalt, W., Blumberg, J. B., McDonald, J. E., Vinqvist-Tymchuk, M. R., Fillmore, S. A., Graf, B. A., ... & Milbury, P. E. (2008). Identification of anthocyanins in the liver, eye, and brain of blueberry-fed pigs. Journal of Agricultural and Food Chemistry, 56(3), 705–712. https://doi.org/10.1021/jf071998l

Karaaslan, N. M., & Yaman, M. (2016). Determination of anthocyanins in cherry and cranberry by high-performance liquid chromatography–electrospray ionization–mass spectrometry. European Food Research and Technology, 242(1), 127–135.

Karaaslan, N. M., & Yaman, M. (2017). Anthocyanin profile of strawberry fruit as affected by extraction conditions. International Journal of Food Properties, 20(suppl. 3), S2313–S2322. https://doi.org/10.1080/10942912.2017.1368548

Kennedy, D.O. (2014). Polyphenols and the human brain: plant "secondary metabolite" ecologic roles and endogenous signaling functions drive benefits. Advances in Nutrition, 5(5), 515–533. https://doi.org/10.3945/an.114.006320

Karlsen, A., Retterstøl, L., Laake, P., Paur, I., Kjølsrud-Bøhn, S., Sandvik, L., & Blomhoff, R. (2007). Anthocyanins inhibit nuclear factor-κ B activation in monocytes and reduce plasma concentrations of pro-inflammatory mediators in healthy adults. Journal of Nutrition, 137(8), 1951–1954. https://doi.org/10.1093/jn/137.8.1951

Khalifa, I., Zhu, W., Li, K. K., & Li, C. M. (2018). Polyphenols of mulberry fruits as multifaceted compounds: Compositions, metabolism, health benefits, and stability – A structural review. Journal of Functional Foods, 40, 28–43. https://doi.org/10.1016/j.jff.2017.10.041

Khoo, H. E., Azlan, A., Tang, S. T., & Lim, S. M. (2017). Anthocyanidins and anthocyanins: Colored pigments as food, pharmaceutical ingredients, and the potential health benefits. Food & Nutrition Research, 61(1), 1361779. https://doi.org/10.1080/16546628.2017.1361779

King, J. W., Grabiel, R. D., & Wightman, J. D. (2003). Subcritical water extraction of anthocyanins from fruit berry substrates. In Proceedings of the 6th International Symposium on Supercritical Fluids; Apr 28–30; Lorraine, France: National Polytechnic Institute of Lorraine; pp. 28–30.

Kirby, C. W., Wu, T., Tsao, R., & McCallum, J. L. (2013). Isolation and structural characterization of unusual pyranoanthocyanins and related anthocyanins from Staghorn sumac (*Rhus typhina* L.) via UPLC–ESI-MS, 1H, 13C, and 2D NMR spectroscopy. Phytochemistry, 94, 284–293.

Konczak, I., & Zhang, W. (2004). Anthocyanins – More than nature's colours. Journal of Biomedicine and Biotechnology, 2004(5), 239.

Krga, I., & Milenkovic, D. (2019). Anthocyanins: From sources and bioavailability to cardiovascular-health benefits and molecular mechanisms of action. Journal of Agricultural and Food Chemistry, 67(7), 1771–1783.

Lazze, M. C., Pizzala, R., Savio, M., Stivala, L. A., Prosperi, E., & Bianchi, L. (2003). Anthocyanins protect against DNA damage induced by tert-butyl-hydroperoxide in rat smooth muscle and hepatoma cells. Mutation Research/Genetic Toxicology and Environmental Mutagenesis, 535(1), 103–115. https://doi.org/10.1016/S1383-5718(02)00285-1

Lee, S. G., Vance, T. M., Nam, T. G., Kim, D. O., Koo, S. I., & Chun, O. K. (2016). Evaluation of pH differential and HPLC methods expressed as cyanidin-3-glucoside equivalent for measuring the total anthocyanin contents of berries. Journal of Food Measurement and Characterization, 10(3), 562–568. https://doi.org/10.1007/s11694-016-9337-9

Lefevre, M., Howard, L., Most, M. M., Ju, Z. Y., & Delany, J. (2004, March). Microarray analysis of the effects of grape anthocyanins on hepatic gene expression in mice. FASEB Journal, 18(5), A851–A851.

Li, J., & Jiang, Y. (2007). Litchi flavonoids: Isolation, identification and biological activity. Molecules, 12(4), 745–758. https://doi.org/10.3390/12040745

Lila, M. A. (2004). Anthocyanins and human health: an in vitro investigative approach. Journal of Biomedicine and Biotechnology, 2004(5), 306. https://doi.org/10.1155/S111072430440401X

Liu, Y., Liu, J., Chen, X., Liu, Y., & Di, D. (2010). Preparative separation and purification of lycopene from tomato skins extracts by macroporous adsorption resins. Food Chemistry, 123(4), 1027–1034. https://doi.org/10.1016/j.foodchem.2010.05.055

Loke, W. M., Hodgson, J. M., Proudfoot, J. M., McKinley, A. J., Puddey, I. B., & Croft, K. D. (2008). Pure dietary flavonoids quercetin and (−)-epicatechin augment nitric oxide products and reduce endothelin-1 acutely in healthy men. The American Journal of Clinical Nutrition, 88(4), 1018–1025. https://doi.org/10.1093/ajcn/88.4.1018

Lourith, N., & Kanlayavattanakul, M. (2020). Improved stability of butterfly pea anthocyanins with biopolymeric walls. Journal of Cosmetic Science, 71(1), 1–10.

Mahdavi, S. A., Jafari, S. M., Assadpoor, E., & Dehnad, D. (2016). Microencapsulation optimization of natural anthocyanins with maltodextrin, gum Arabic and gelatin. International Journal of Biological Macromolecules, 85, 379–385.

Matsusima, A., Furuuchi, R., Sakaguchi, Y., Goto, H., Yokoyama, T., Nishida, H., & Hirayama, M. (2013). Acute and chronic flow-mediated dilation and blood pressure responses to daily intake of boysenberry juice: A preliminary study. International Journal of Food Sciences and Nutrition, 64(8), 988–992.

Mazza, G., & Miniati, E. (2018). Anthocyanins in Fruits, Vegetables, and Grains. CRC Press.

McCallum, J. L., Yang, R., Young, J. C., Strommer, J. N., & Tsao, R. (2007). Improved high performance liquid chromatographic separation of anthocyanin compounds from grapes using a novel mixed-mode ion-exchange reversed-phase column. Journal of Chromatography A, 1148(1), 38–45. https://doi.org/10.1016/j.chroma.2007.02.088

Pergola, C., Rossi, A., Dugo, P., Cuzzocrea, S., & Sautebin, L. (2006). Inhibition of nitric oxide biosynthesis by anthocyanin fraction of blackberry extract. Nitric Oxide, 15(1), 30–39. https://doi.org/10.1016/j.niox.2005.10.003

Pervaiz, T., Songtao, J., Faghihi, F., Haider, M. S., & Fang, J. (2017). Naturally occurring anthocyanin, structure, functions and biosynthetic pathway in fruit plants. Journal of Plant Biochemistry and Physiology, 5(2), 1–9.

Petersson, E. V. (2009). Analysis of Acrylamide and Anthocyanins in Foods: Extraction Optimization for Challenging Analytes. Doctoral dissertation, Acta Universitatis Upsaliensis.

Picariello, G., Ferranti, P., Garro, G., Manganiello, G., Chianese, L., Coppola, R., & Addeo, F. (2014). Profiling of anthocyanins for the taxonomic assessment of ancient purebred V. vinifera red grape varieties. Food Chemistry, 146, 15–22.

Pieczykolan, E., & Kurek, M. A. (2019). Use of guar gum, gum arabic, pectin, beta-glucan and inulin for microencapsulation of anthocyanins from chokeberry. International Journal of Biological Macromolecules, 129, 665–671.

Pina, F. (2014). Chemical applications of anthocyanins and related compounds. A source of bioinspiration. Journal of Agricultural and Food Chemistry, 62(29), 6885–6897.

Pina, F., Melo, M. J., Laia, C. A., Parola, A. J., & Lima, J. C. (2012). Chemistry and applications of flavylium compounds: A handful of colours. Chemical Society Reviews, 41(2), 869–908.

Pina, F., Oliveira, J., & de Freitas, V. (2015). Anthocyanins and derivatives are more than flavylium cations. Tetrahedron, 71(20), 3107–3114.

Ramirez-Tortosa, C., Andersen, Ø. M., Gardner, P. T., Morrice, P. C., Wood, S. G., Duthie, S. J., ... & Duthie, G. G. (2001). Anthocyanin-rich extract decreases indices of lipid peroxidation and DNA damage in vitamin E-depleted rats. Free Radical Biology and Medicine, 31(9), 1033–1037. https://doi.org/10.1016/S0891-5849(01)00618-9

Reyes, B. A., Dufourt, E. C., Ross, J., Warner, M. J., Tanquilut, N. C., & Leung, A. B. (2018). Selected phyto and marine bioactive compounds: Alternatives for the treatment of type 2 diabetes. Studies in Natural Products Chemistry, 55, 111–143. https://doi.org/10.1016/B978-0-444-64068-0.00004-8

Rossi, A., Serraino, I., Dugo, P., Di Paola, R., Mondello, L., Genovese, T., ... & Cuzzocrea, S. (2003). Protective effects of anthocyanins from blackberry in a rat model of acute lung inflammation. Free Radical Research, 37(8), 891–900. https://doi.org/10.1080/1071576031000112690

Rupasinghe, H. V. (2015). Application of NMR spectroscopy in plant polyphenols associated with human health. In Applications of NMR Spectroscopy (pp. 3–92). Bentham Science.

Sahamishirazi, S., Moehring, J., Claupein, W., & Graeff-Hoenninger, S. (2017). Quality assessment of 178 cultivars of plum regarding phenolic, anthocyanin and sugar content. Food Chemistry, 214, 694–701. https://doi.org/10.1016/j.foodchem.2016.07.070

Sanossian, N., Saver, J. L., Navab, M., & Ovbiagele, B. (2007). High-density lipoprotein cholesterol: An emerging target for stroke treatment. *Stroke*, *38*(3), 1104–1109.

Schwarz, M., Hillebrand, S., Habben, S., Degenhardt, A., & Winterhalter, P. (2003). Application of high-speed countercurrent chromatography to the large-scale isolation of anthocyanins. Biochemical Engineering Journal, 14(3), 179–189. https://doi.org/10.1016/S1369-703X(02)00219-X

Scordino, M., Sabatino, L., Lazzaro, F., Borzì, M. A., Gargano, M., Traulo, P., & Gagliano, G. (2015). Blood orange anthocyanins in fruit beverages: How the commercial shelf life reflects the quality parameter. Beverages, 1(2), 82–94. https://doi.org/10.3390/beverages1020082

Scotter, M. J. (2011). Emerging and persistent issues with artificial food colours: Natural colour additives as alternatives to synthetic colours in food and drink. Quality Assurance and Safety of Crops & Foods, 3(1), 28–39.

Silva, V. O., Freitas, A. A., Maçanita, A. L., & Quina, F. H. (2016). Chemistry and photochemistry of natural plant pigments: The anthocyanins. Journal of Physical Organic Chemistry, 29(11), 594–599.

Singh, M. C., Kelso, C., Price, W. E., & Probst, Y. (2020). Validated liquid chromatography separation methods for identification and quantification of anthocyanins in fruit and vegetables: A systematic review. Food Research International, 138, 109754. https://doi.org/10.1016/j.foodres.2020.109754

Slimestad, R., Torskangerpoll, K., Nateland, H. S., Johannessen, T., & Giske, N. H. (2005). Flavonoids from black chokeberries, *Aronia melanocarpa*. Journal of Food Composition and Analysis, 18(1), 61–68. https://doi.org/10.1016/j.jfca.2003.12.003

Spinardi, A., Cola, G., Gardana, C. S., & Mignani, I. (2019). Variation of anthocyanin content and profile throughout fruit development and ripening of highbush blueberry cultivars grown at two different altitudes. Frontiers in Plant Science, 10, 1045.

Steffen, Y., Gruber, C., Schewe, T., & Sies, H. (2008). Mono-O-methylated flavanols and other flavonoids as inhibitors of endothelial NADPH oxidase. Archives of Biochemistry and Biophysics, 469(2), 209–219. https://doi.org/10.1016/j.abb.2007.10.012

Tosti, V., Bertozzi, B., & Fontana, L. (2018). Health benefits of the Mediterranean diet: Metabolic and molecular mechanisms. The Journals of Gerontology: Series A, 73(3), 318–326. https://doi.org/10.1093/gerona/glx227

Tsao, R., & McCallum, J. (2009). Chemistry of Flavonoids. In Fruit and Vegetable Phytochemicals: Chemistry, Nutritional Value and Stability. John Wiley & Sons.

Tsuda, T., Shiga, K., Ohshima, K., Kawakishi, S., & Osawa, T. (1996). Inhibition of lipid peroxidation and the active oxygen radical scavenging effect of anthocyanin pigments isolated from *Phaseolus vulgaris* L. Biochemical Pharmacology, 52(7), 1033–1039.https://doi.org/10.1016/0006-2952(96)00421-2

Tsuda, T., Horio, F., & Osawa, T. (2002). Cyanidin 3-O-β-D-glucoside suppresses nitric oxide production during a zymosan treatment in rats. Journal of Nutritional Science and Vitaminology, 48(4), 305–310. https://doi.org/10.3177/jnsv.48.305

Tsuda, T., Horio, F., Uchida, K., Aoki, H., & Osawa, T. (2003). Dietary cyanidin 3-O-β-d-glucoside-rich purple corn color prevents obesity and ameliorates hyperglycemia in mice. The Journal of Nutrition, 133(7), 2125–2130. https://doi.org/10.1093/jn/133.7.2125

Upton, R. (2001). Bilberry Fruit *Vaccinium myrtillus* L. Standards of Analysis, Quality Control, and Therapeutics. Santa Cruz, CA: American Herbal Pharmacopoeia and Therapeutic Compendium.

Venter, A., Joubert, E., & De Beer, D. (2014). Nutraceutical value of yellow-and red-fleshed South African plums (*Prunus salicina* Lindl.): Evaluation of total antioxidant capacity and phenolic composition. Molecules, 19(3), 3084–3109. https://doi.org/10.3390/molecules19033084

Wallace, T. C., & Giusti, M. M. (2015). Anthocyanins. Advances in Nutrition, 6(5), 620–622. https://doi.org/10.3945/an.115.009233

Wang, S. Y., & Jiao, H. (2000). Scavenging capacity of berry crops on superoxide radicals, hydrogen peroxide, hydroxyl radicals, and singlet oxygen. Journal of Agricultural and Food Chemistry, 48(11), 5677–5684.

Wu, X., Beecher, G. R., Holden, J. M., Haytowitz, D. B., Gebhardt, S. E., & Prior, R. L. (2006). Concentrations of anthocyanins in common foods in the United States and estimation of normal consumption. Journal of Agricultural and Food Chemistry, 54(11), 4069–4075. https://doi.org/10.1021/jf060300l

Yahia, E. M., Ornelas-Paz, J. D. J., & Gonzalez-Aguilar, G. A. (2011). Nutritional and health-promoting properties of tropical and subtropical fruits. In Postharvest Biology and Technology of Tropical and Subtropical Fruits (pp. 21–78). Woodhead. https://doi.org/10.1533/9780857093622.21

Yousuf, B., Gul, K., Wani, A. A., & Singh, P. (2016). Health benefits of anthocyanins and their encapsulation for potential use in food systems: A review. Critical Reviews in Food Science and Nutrition, 56(13), 2223–2230.

Zamora-Ros, R., Knaze, V., Lujan-Barroso, L., Slimani, N., Romieu, I., Touillaud, M., ... & Gonzalez, C. A. (2011). Estimation of the intake of anthocyanidins and their food sources in the European Prospective Investigation into Cancer and Nutrition (EPIC) study. British Journal of Nutrition, 106(7), 1090–1099. DOI: https://doi.org/10.1017/S0007114511001437

Zhao, L., Li, S., Zhao, L., Zhu, Y., & Hao, T. (2015). Antioxidant activities and major bioactive components of consecutive extracts from blue honeysuckle (Lonicera caerulea L.) cultivated in China. Journal of Food Biochemistry, 39(6), 653–662. https://doi.org/10.1111/jfbc.12173

9 Natural Anthocyanins from Subtropical Fruits for Hypertension

Mehvish Habib, Kulsum Jan, Khalid Bashir, Savita Rani, Abishek Chandra, and Sachin Kumar

CONTENTS

9.1   INTRODUCTION

Anthocyanin originates from the Greek *anthos*, signifying flower, and *kianos*, signifying blue. These are purple, blue, or red pigments that are present in plants, particularly in fruits, flowers, and shrubs. Anthocyanin is a pigment found in various plants (Table 9.1). The flavonoid family comprises anthocyanins, which are naturally occurring chemical compounds with two benzene rings attached to three equal chains of C6–C3–C6 (Wilska, 2007). The fundamental flavonoid structure's oxygen C-ring has a favorable atomic charge, and another name is the flavylium ion (2-phenylchromenylium). The molecular geometry of anthocyanin is demonstrated in Figure 9.1. Anthocyanin stability can be regulated by its design, pH, temperature, and exposure to light (Laleh, Frydoonfar H. et al., 2006). So far, over 600 anthocyanins together with their alternatives have been identified (Grayer and Veitch, 2008). Primarily blue, purple, and red flowers possess anthocyanins. Pink bloom, red clover, red pineapple sage, red rose, and red hibiscus are types of red flowers. The delectable crimson blooms are available in a range of colors. Blue and purple flowers (blue rosemary, blue chicory, and cornflower) are the most typical edible flowers (lavender, common violet, purple sage, purple passionflower, and purple mint). Most of the flowers mentioned above have been used as medicine, food coloring, and food in the past. Along with their traditional uses, blue, purple, and red-colored

TABLE 9.1
Anthocyanins Present in Selected Plant Sources

Name of Anthocyanin	Plant Source
Cyanidin	Apple, elderberry, blackberry
Cyanidin and delphinidin	Blackcurrant, pomegranate
Pelargonidin	Strawberry, red radish, banana
Cyanidin and peonidin	Sweet cherry, plum
Delphinidin	Pomegranate, passion fruit, eggplant
Petunidin and malvidin	Red grape, bilberry
Peonidin	Mango

Source: Bueno et al. (2012).

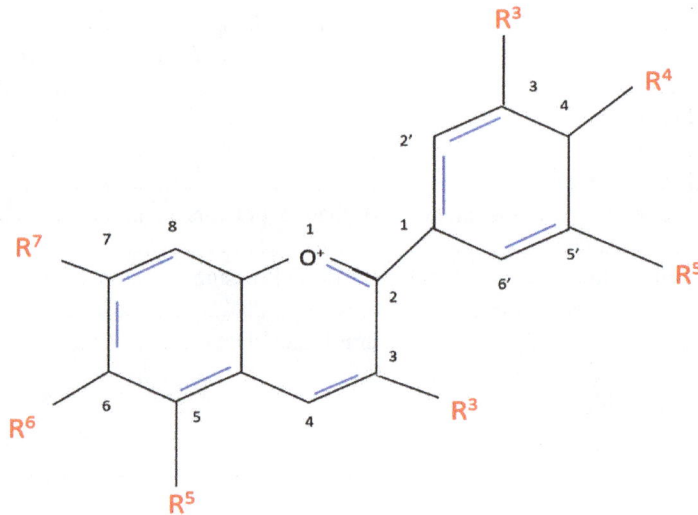

FIGURE 9.1   Molecular structure of anthocyanin (Yi et al., 2009).

fruits can also be used for numerous health advantages. Anthocyanins are potent antioxidants that can be discovered in blackcurrants and berries as well as other blue-color food items. Purple potatoes, red cabbage, and black carrots, that have high quantities of anthocyanins, can also be used to cure specific diseases and as a nutritional supplement.

Anthocyanidins are modified with aromatic, glycosyl, or aliphatic acyl moieties to produce anthocyanins (Castaneda et al., 2009; Oren, 2009) and are responsible for the color of processed and fresh vegetables, fruits, as well as other natural food items. Often anthocyanins are dissolved vacuolar solutions of epidermal cells (Rogez et al., 2011). Extracts that are rich in anthocyanin are frequently used as natural alternatives for Food, Drug, and Cosmetic synthetic lakes and dyes due to their coloring characteristics (Bueno et al., 2012). Because such compounds are water-soluble and non-toxic it is simple to mix them with food systems (Pazmino et al., 2001). As opposed to this, anthocyanins from plants have also been extensively studied for their pharmacological uses, having anti-cancer, anti-bacterial, anti-diabetic, anti-obesity, and anti-inflammatory characteristics and protecting against heart problems (He et al., 2011).

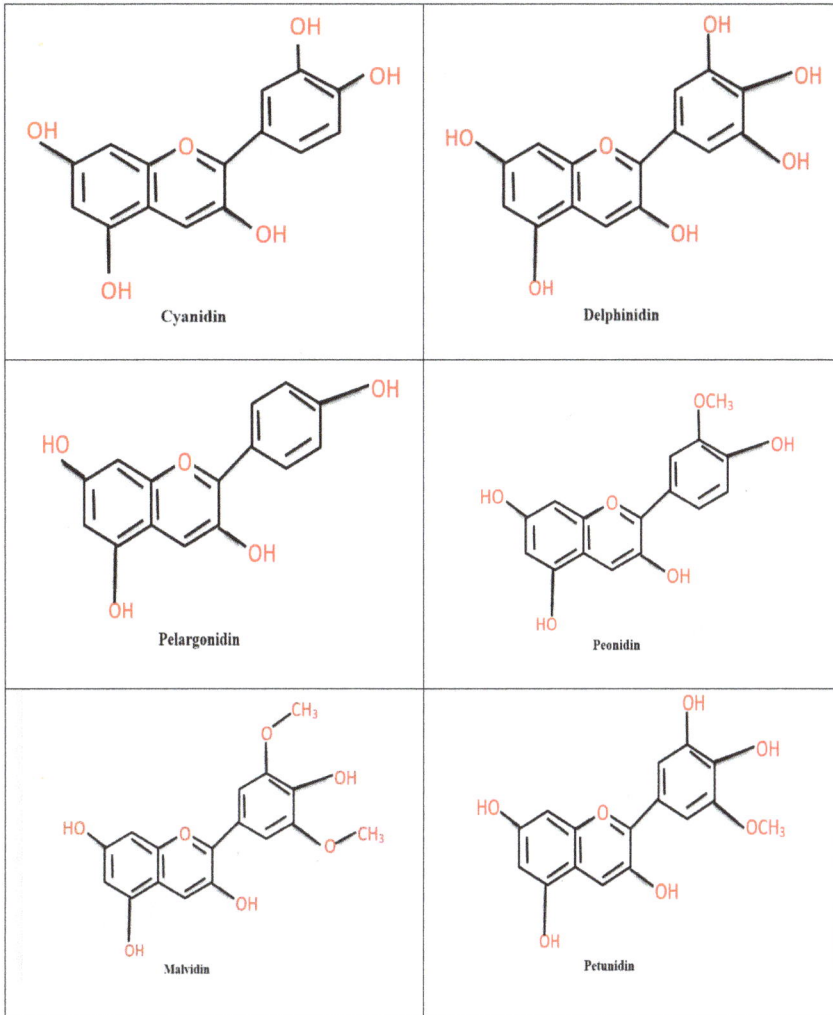

FIGURE 9.2 Major anthocyanidins in a plant (Khoo et al., 2012).

9.2 CHEMISTRY

Anthocyanins are flavonoid phenolic compounds that give many plants, flowers, and fruits their color and the color of commodities made from colored vegetable matrices like wine (Strack and Wray, 1989; Mazza, 1995). The structure of anthocyanin is shown in Figure 9.1; it consists of two benzene rings coupled to a series of three carbon chains, implying the fundamental C6–C3–C6 bone structure (Wilska, 2007). Flavylium or 2-phenylbenzopyrilium cations are used to make anthocyanins, which are anthocyanidins possessing glycoside moieties. The six most abundant and ubiquitous anthocyanidins discovered in nature are malvidin, petunidin, peonidin, pelargonidin, delphinidin, and cyanidin. Their molecular structure is shown in Figure 9.2. As anthocyanins are polar, they are soluble in polar solvents like water, methanol, and ethanol; that is why such solvents are required in many solvent extraction procedures. These solvents are transformed into acids to increase the anthocyanins in the flavylium cation. The proportion of OH molecules, the pace at which OH moieties are methylated, the kind and number of sugar particles embedded in the aglycone molecule,

TABLE 9.2
Quantity of Anthocyanins Reported in Different Studies

Sources	Total Anthocyanin Content (mg/100g)	References
Apple	1.7–3.2	Wu et al., 2006
Strawberry	13–315	Silva et.al., 2007
Fresh blackberrys	75	Ju et al., 2005
Black raspberries	145–607	Ngo et al., 2007
Capulin	31.7	Ordaz-Galindo et al., 1999
Red rice	7.9–34.4	Chen et al., 2012
Red vine grapes	30–750	Mazza & Miniati, 1993

and the precise location of these particles all play a role in anthocyanin structural variations. Furthermore, the number of anthocyanins varies depending on their source (Table 9.2).

9.3 ANTHOCYANIN STABILITY

As a natural food coloring, anthocyanin has limitations which restrict its usage in the diet. Compared to a few manufacturing settings, formulations, and storage circumstances, they anthocyanins have limited stability that causes the finished product to have an undesirable odor or flavor. Extreme heat, pressure, and pH conditions, on the other hand, may result in the loss of anthocyanins occurring naturally, or impair their anti-oxidant activity. Anthocyanins can be unstable after release and can also be highly sensitive (Giusti and Wrolstad, 2003). Anthocyanin stability can change with different parameters, including storage temperature, pH, metallic ions, enzymes, flavonoids, proteins, solvents, oxygen, light, concentration, and chemical structure (Rein, 2005). The chemical composition of anthocyanin is a primary determinant of pigment stability.

The heat conductivity of extracts from radish anthocyanin from the Tou Hong area was also studied over 24 h at a temperature of 90 and 100°C. In an acidic environment, the presence of hydroxycinnamic acid and strong acylation contributed to the considerable stability of radish anthocyanins with regard to heat (Jing et al., 2012).

The presence of pelargonidin 3-glucoside, which contains a considerable amount of anthocyanin, is responsible for the vivid red, appealing strawberry hue. The stability of anthocyanins based on pelargonidin was investigated at various degrees of water activity (Garzona and Wrolstad, 2001). Following these findings, degradation of anthocyanin increases with rising water levels and follows the original kinetics. Anthocyanin cells have been found to have a half-life ranging from 56 to 934 days.

A study of the degradation kinetics and color density of anthocyanins incorporated into a spray from *Hibiscus sabdariffa* concluded that adding polysaccharide-containing anthocyanins after adequate processing could improve anthocyanin stability and efficiency in dietary systems (Idham et al. 2012). The authors also discovered that maltodextrin and gum Arabic have very high encapsulation effectiveness.

When ascorbic acid is present, the stability of anthocyanin reduces. Because of the high content of ascorbic acid, the anthocyanin extract of acerola has lower strength. Because acerola is the most natural and wealthiest source of ascorbic acid, its impact on synthesizing anthocyanins from acerola extract was discovered when compared with acai, that does not include ascorbic acid. It was also discovered that color fading worsened as ascorbic acid is mixed with the acai anthocyanin solution.

Anthocyanins are derived from black carrots and used in various drinks and fruits (Kirca et al., 2006). Anthocyanin degradation follows the first order of kinetics in all colored juices and nectars.

9.4 USE OF ANTHOCYANIN AS FOOD ADDITIVE AND COLORANT

Natural color and additives play an essential role in boosting the consumer acceptability of processed foods and beverages. Anthocyanins are brilliantly colored pigments produced from plants that have a pleasing appearance and low toxicity. Natural colors are slightly safer than synthetic colorants and can be consumed in high doses. Anthocyanins have value-added capabilities as natural colorants (Bridle and Timberlake, 1997). They have antioxidant characteristics, act as a nutraceutical, and provide many health benefits, including anti-bacterial activity and the prevention of chronic conditions.

9.5 ANTHOCYANIN BIOAVAILABILITY

The number of specific nutrients absorbed, digested, or metabolized in the usual way is called the bioavailability of a particular nutrient. Bioavailability is a significant problem in terms of its biological effects. The bioavailability of the anthocyanin substance found in different vegetables and fruits is essential to maximize the results, but it remains a lightly studied field.

Anthocyanins must be successfully absorbed and circulated throughout the body. However, anthocyanins, are usually incorporated in varying combinations of food sources, and the impact of the food matrix on its composition makes research on bioavailability extremely difficult.

Multiple experiments have been conducted to investigate the presence of anthocyanins. Anthocyanins and their bioavailability were found to be less than 1% (Bub et al., 2001; Manach et al., 2005). Anthocyanins contain a number of glycosides with variable bioavailability.

Generally, absorption of acylated anthocyanins is less than non-acylated anthocyanins. Bub et al. (2001) conducted an experiment on malvidin-3-glucoside (M3G) anthocyanin, which occurs in red grape juice and wine. It was found that M3G absorbs poorly and not even anthocyanins on their own or even the defined polyphenols or anthocyanin metabolites are responsible for the health effects, as observed antioxidants. Anthocyanins are efficiently absorbed and also rapidly eliminated (Manach et al., 2005). Therefore, a steady diet of anthocyanin may be required for systematic benefit.

9.6 EXTRACTION OF ANTHOCYANINS

Extracting anthocyanins from plant sources is done in various ways. Selecting an effective method is influenced by many factors, including economic feasibility and the suitability of the procedure in a given situation.

As previously stated, polar molecules such as anthocyanins are isolated from various vegetables and fruits as well as from other waste products (Clifford, 2000). Anthocyanins can be extracted using an aqueous mixture of compounds like ethanol, methanol, or acetone (Kahkonen et al., 2001). According to previous studies, solvent extraction is the most prevalent approach for anthocyanin extraction, with hydrochloric acid (HCl) and methanol being the most commonly employed solvents (Durana et al., 2001). Using agitation or stirring techniques can help improve extraction. After that, the extract is vacuum-concentrated and filtered with the help of a rotary evaporator. Thermal degradation of anthocyanins can be avoided using membrane methods like nanofiltration and ultrafiltration. These technologies can concentrate anthocyanins, resulting in a comparable concentrate to the original extract (Cisse et al., 2011). These membrane technologies could be helpful for preconcentrating extracts without causing thermal damage before the final concentration (spray drying, osmotic evaporation, as well as vacuum evaporation).

An acidified methanol solution (also known as concentrated HCl/methanol = 0.01:100 mL) was used to extract radish anthocyanins (Jing et al., 2012). Separation or extraction of natural compounds from betalain or other similar crude extracts is done using aqueous two-phase extraction (ATPE) method in the case of mulberry (Chethana et al., 2007). ATPE is an effective, adaptable, and powerful developing technology for biomolecule downstream processing. Anthocyanins from mulberry have recently been extracted using ATPE. Compared to standard extraction procedures, the ATPE extract had a comparatively significant antioxidant activity that also affected the configuration of anthocyanin compound (Wu et al., 2011).

Supercritical fluid extraction (SFE) is a valuable technique for extracting anthocyanins. SFE could be used as a substitute for organic solvent extraction. SFE is quite helpful because it is a quick and automated process. SFE procedures do not require the use of significant amounts of harmful solvents and are a selective process. Compared to traditional extraction techniques, SFE does not require air or light during extraction, which decreases the likelihood of the degradation process occurring during extraction. Due to the polar nature of anthocyanins, SFE utilizing CO_2 necessitates high pressure and a high concentration of organic solvent during extraction (Mantell et al., 2003).

Extraction of anthocyanins could also be done using microwave-assisted extraction (MAE). With a persistent dipole, MAE creates energy from microwaves and causes molecular motion and the rotation of liquids so that the material heats up quickly. MAE has several advantages, including reduced solvent usage, increased efficiency, and reduced extraction time. MAE was used to extract anthocyanin from purple corn (*Zea mays* L.) cobs, and it was shown that MAE extraction was effective and quick compared to solvent extraction (Yang and Zhai, 2010). The same outcomes were also obtained by Zou et al. (2012) and Liazid et al. (2011) for mulberry and grape skin using MAE.

Ultrasound-assisted extraction (UAE) is a secondary approach for extracting anthocyanins. This method could be a viable alternative to the time-consuming and inefficient traditional solvent extraction method. Acoustic cavitation is used to create molecular mobility in the solvent and sample. In comparison to conventional extraction procedures, UAE delivers higher automation level and higher outcomes for the target molecule. Anthocyanins from red raspberries were extracted using UAE, and the process parameters were adjusted using the response surface technique (Chen et al., 2007). UAE's efficacy and speed in extracting anthocyanins make it a viable alternative to the traditional solvent extraction approach. The significant disruption of the structure of fruit tissue caused by ultrasonic acoustic cavitation is the reason behind this (Chen et al., 2007).

9.7   ENCAPSULATION OF ANTHOCYANINS

Anthocyanins are promising natural food colorants with potential health benefits, but their unstable nature impedes their practical applications. During production and storage, anthocyanins have a poor level of environmental stability. Anthocyanin extracts are volatile and vulnerable to deterioration at the moment of isolation (Giusti and Wrolstad, 2003). As a result of their poor strength, using anthocyanin pigment in foods was limited, and its application in medical and food items gives the impression of a complex undertaking. Encapsulation is another excellent technique to get these components into these items. Encapsulating agents, like the protective coat, protect against environmental hazards like oxygen, humidity, and light. Bioactive chemicals that are encapsulated are easy to handle and have better stability. Interactions of food and pharmaceutical components with environmental elements like light, moisture, and oxygen have already been reduced using encapsulation techniques.

Microencapsulation may be a valuable strategy for preserving the sensitivity of components such as anthocyanins and preventing their degradation. Maltodextrin is frequently utilized as a microencapsulation wall material. Maltodextrin and gum Arabic mixture were used during the microencapsulation of anthocyanins, resulting in the highest encapsulation efficiencies (Idham et al., 2012). This maltodextrin and gum Arabic combination was also the best wall material for microencapsulation, with the most extended shelf-life and most minor color pigment change. Maltodextrin and dextrose equivalents are utilized as wall materials for encapsulating anthocyanins and betacyanins and employed in amounts ranging from 10 to 25% (Ersus and Yurdagel, 2007). Berg et al. (2012) also worked on the impact of different pectins on powder properties of microencapsulated anthocyanins in a study on the microencapsulation of anthocyanins.

Microencapsulation can be done in many ways. Pan coating, spray drying, coacervation phase separation, air suspension, solvent evaporation, interfacial polymerization, and multi-orifice centrifugal processes are the most often utilized.

The most popular method for microencapsulating anthocyanins is spray drying. Because microencapsulation necessitates a wall material, the most widely utilized wall materials in this approach are citrus fiber, tapioca starch, gum Arabic, inulin, maltodextrin, as well as other matrix compounds like soy protein isolate and glucose syrup. Because starch is abundant in nature, it can be utilized to encapsulate flavor components as well as other essences by spray drying them in a way that protects them from oxidation and allows for controlled release over a set period of time (Wani et al., 2012). Natural polymers also improve anthocyanin stabilities and aid in the controlled release of these beneficial components for more efficient nutraceutical use in the human body. Light and oxygen are present when anthocyanins are spray-dried and during dry storage, which improves the stability and inhibits the degradation of these derived phenolics. Previous research has shown that encapsulation circumstances, like the technique used and the gelling agent, can impact the degradation of anthocyanin directly.

Microencapsulation of roselle (*Hibiscus sabdariffa*) was done using freeze drying and various materials used for coating, like gum Arabic, trehalose, and maltodextrin (Gradinaru et al., 2003; Duangmal et al., 2008; Selim et al., 2008), although it was suggested that freeze-drying procedures are more expensive than spray drying.

Coacervation is a costly method developed for food-grade encapsulation (Diaz et al., 2006). This method was created in the 1950s to provide a carbonless copy paper using a two-ink system (Shahidi and Han, 1993). This technique consists of three steps that should be carried out during continuous agitatation. The first stage creates three phases: liquid production vehicle, core material, and coating substance. The second phase coats the core material with a liquid polymer coating, which is then rigidized using cross-linking or heating processes to generate self-sustaining microcapsules, which is the last coacervation step. These particles were tossed in the pan while the coating material was slowly applied in the pan-coating method. It is an early industrial method for creating tiny and coated objects, and it is frequently used in the pharmaceutical industry (Tiwari et al., 2010).

In the method of coating with air suspension, the core solid particulate materials are distributed to support air streams before spray coating. The air suspension approach is used to apply submicron or micron particle core materials successfully, but an aggregation of the particles to enormous sizes may also occur (Bansode et al., 2010).

The coating for the microcapsules is mixed in an incompatible volatile solvent with a liquid production vehicle phase during the solvent evaporation process. In the solution of the coating polymer, the core material which is to be microencapsulated is disseminated. Such core coating mixtures are distributed in the vehicle phase of liquid manufacturing (Dubey et al., 2009). In the polycondensation method two reactants meet an interface and react fast; this is interfacial polymerization. As a result, the approach relies on polymerizing reactive monomers to build a capsule shell on a particle or droplet surface. Multifunctional monomers are the chemicals employed (Agnihotri et al., 2012).

In addition to spray drying, other anthocyanin encapsulation techniques are unknown, making this an exciting subject for study.

9.8 USE OF ANTHOCYANINS IN NUTRACEUTICALS AND PHARMACEUTICALS

Anthocyanin, a bioactive component, is extensively employed in traditional treatments and nutraceuticals. It has been used for centuries as an appetite stimulant, phytopharmaceutical, treatment, and choleretic agent for various ailments. Anthocyanins are effective nutraceuticals or medicinal compounds in the form of colorful pigments. Anthocyanins are a type of nutraceutical that aids in disease prevention and maintenance of good health.

Anthocyanins with low bioavailability absorb chemicals slowly into the blood circulation system and are quickly eliminated in urine and feces, limiting their efficacy in scavenging free radicals. On the other hand, those with high bioavailability effectively reduce lipid peroxidation of cells, lowering the risk of various disorders.

The bioavailability of M3G and cyanidin-3-glucoside is the highest of all anthocyanins. Peonidin, glycosylated cyanidin, malvidin, delphinidin, and petunidin have the highest relative bioavailability (Frank et al. 2003). However, because most red-wine anthocyanins has lower bioavailability, their efficacy can be enhanced by increasing their consumption. An increase in anthocyanin consumption may have unidentified adverse effects.

M3G absorption following red grape juice and wine consumption was investigated in a study (Bub et al., 2001). The authors concluded that M3G was detected in plasma as well as in urine within 3 or 6 h of intake of these liquids. These results also showed that the compound's availability was twice as high after drinking red grape juice as it was after drinking red wine. Urine samples of 6 healthy volunteers who had ingested 300 mL red wine (218 mg) were studied by Lapidot et al. in 1998, and it was discovered that anthocyanin levels in urine reached a high level after 6 h of intake. On an empty stomach, another study reported on the ingestion of 11 g of elderberry concentration with 17.2%w/w anthocyanins (primarily cyanidin-3-sambubioside and cyanidin-3-glucoside) (Mülleder et al. 2002). The researchers concluded that anthocyanins have a low bioavailability based on their findings.

Anthocyanins are, in fact, more bioavailable than previously thought. Within 48 h of oral intake, cyanidin-3-glucoside had a relative bioavailability of 12.38 + 1.38%, with 6.91 + 1.59% excreted in breath and 5.37.67% in urine, according to the literature (Cassidy et al. 2011). Phenylpropenoic acids, phenylacetic, hippuric, phenolic conjugates, and phenolic acid are the metabolites formed after the human body digests cyanidin-3-glucoside. Anthocyanins from blackcurrants including delphinidin-3-glucoside, cyanidin-3-glucoside, delphinidin-3-rutinoside, and cyanidin-3-rutinoside, are reported to be absorbed straight into the human blood stream and eliminated in urine in their glycosylated forms (Matsumoto et al., 2001). It was reported that the absorbed anthocyanins are not converted into aglycones or other types of metabolite in the body; this was tested 30 min after eating an anthocyanin combination from red fruits (Miyazawa et al., 1999).

9.9 ANTHOCYANINS IN HUMAN HEALTH

Anthocyanins have high antioxidant qualities and aid in preventing many diseases, such as heart disease, neurological disease, cancer, diabetes, and inflammation. According to the literature, anthocyanins have been linked to human nutrition (Stintzing and Carle, 2004) and cancer treatment (Nichenametla et al., 2006). They are thought to work by stopping cell division between the S and G2 phases of a cell cycle, decreasing tumor growth (Koide et al., 1996). Anthocyanin's ability to combat oxidants means it to can fight atherosclerosis. Anthocyanins have been discovered for relaxing blood vessels and protecting from damage the endothelial cells that line the walls of blood vessels. Strawberry has been found in animal studies to have an inhibitory impact in esophageal cancer and to reverse neural and behavioral aging in the animals tested (Torronen and

Maatta, 2002). Therapeutic activities of strawberry are linked to the anthocyanin component of the fruit. A healthy heart, allergy relief (Basu et al., 2010; Wallace, 2011), sharp eyesight (Ghosh and Konishi, 2007), cognitive function and ulcer treatment are some of the applications of anthocyanins for health benefits (Moskovitz et al., 2002). Anthocyanins have been discovered to be beneficial in treating disorders caused by capillary fragility (Tamura and Yamagami, 1994). The two main examples are decreasing platelet aggregation and preventing atherosclerosis induced with high cholesterol. As an outcome of their unique color and health benefits, scientists are investigating the natural features of anthocyanins. Many studies have been conducted on separation and purification methods of anthocyanins, using anthocyanins in food (Giusti and Wrolstad, 2003), identification and distribution in plants (Matera et al., 2012), quantitative analysis using electrophoretic and chromatographic method (Reyes and Zevallos, 2007), and stability (Cavalcanti et al., 2011; Durge et al., 2013).

Understanding of the importance of diet has grown as the incidence of chronic conditions like cardiovascular disorders and cancer has increased. A number of epidemiological researchers have implied that quitting smoking, boosting fruit and vegetable consumption, and avoiding infections can all help in lowering the incidence of chronic disorders like cancer and cardiovascular disorders (Percival et al., 2006; Yahia et al., 2011).

Fruits are high in key dietary fiber and micronutrients, which recently also have been recognized as the vital source of phytochemicals which may enhance health separately or in combination (Table 9.3) (Rechkemmer, 2001; Yahia et al., 2011). Tropical and subtropical fruits include various nutritional components and phytochemicals, including ascorbic acid phenolic compounds and carotenoids (Percival et al., 2006; Yahia et al., 2011). They exhibit significant *in vitro* antioxidative action, having significant health ramifications. A fruit- and vegetable-rich diet raises antioxidant levels in the blood and body tissues, potentially protecting cells and tissues from oxidative damage (Yahia et al., 2011).

Increasing fruit and vegetable consumption to 400–800 g/day is proposed as a desirable public health strategy. An average of 400 g of fruits and vegetables every day should be consumed, as recommended by the World Health Organization, and health authorities worldwide should follow this advice.

In terms of plant physiology, numerous factors have an influence on the metabolism and synthesis of bioactive chemicals in subtropical and tropical fruits. These include before and after harvest conditions and climatic factors, particularly temperature and light, which have a significant impact. Irrigation, soil type, fertilizer, rootstocks, and mulching, along with other cultural activities,

TABLE 9.3
Phytochemicals That May Have a Beneficial Effect on Human Health

Phytochemical	Plant Source	Proposed or Established Effects
Carotenoids	Papaya, mango, pineapple, citrus	Night blindness prevention
Tocopherols	Olives, avocados, nuts	Cancer, diabetes, immune system, low-density lipoprotein oxidation, and heart disease
Ascorbic acid	Papaya, mango, pineapple, guava, citrus fruits, cantaloupe	Aids wound healing and prevents scurvy
Folic acid	Most nuts and fruits	Heart disease, diabetes, congenital disabilities, cancer, heart disease
Phenolic compounds		
Proanthocyanin	Pomegranates	Cancer
Anthocyanidins	Pomegranate, grape	Allergies, blood pressure, cataracts, diabetes, heart disease

are equally critical (Goldmann et al., 1999). Many other aspects have a significant impact on health and nutrition, including harvest maturity, harvesting methods, precooling, the number of days between harvest and processing or consumption, and other post-harvesting methods (like cooking, processing, controlled or modified atmosphere, heat treatment, and refrigeration) (Kader, 2001; Yahia et al., 2011).

9.9.1 ROLE OF ANTHOCYANINS IN REGULATING BLOOD PRESSURE

Cardiovascular disease (CVD) is one of the most significant causes of mortality worldwide, and hypertension is a known risk factor (Clark et al., 2015). The importance of lifestyle changes, particularly nutrition, in reducing CVD risk is becoming better recognized, but it requires a deliberate, evidence-based approach (Cicero et al., 2017). It holds true for controlling or curing hypertension, affecting most Western and European nations (Cicero et al., 2019).

Anthocyanins belong to a broad family of plant metabolites soluble in water responsible for red, blue, and purple coloration of a number of vegetables, fruits, and flowers. Anthocyanins are glycosylated versions of anthocyanidins, which are extremely reactive and somewhat unstable compounds with a flavylium cation structure (De Ferrars et al., 2014). Even though there are over 600 naturally occurring anthocyanins, the vast majority are made up of different glycosylated components of six anthocyanidins: average malvidin, pelargonidin, petunidin, peonidin, cyanidin, and delphinidin (Stoner et al., 2005). The daily typical intake of anthocyanins is 200 mg, accounting for around a fifth of total phenolic intake (Wu et al., 2006). While plants manufacture anthocyanins to protect themselves from uncontrollable environmental factors like ultraviolet light, temperature, drought, and dietary anthocyanins intake which has been intensively researched for its health-enhancing potential, particularly in preventing CVD (Wallace, 2011).

Anthocyanin consumption lowers the risk of coronary heart disease and CVD according to solid epidemiological evidence (Mink et al., 2007). While the consumption of anthocyanin is linked to its anti-inflammatory, antioxidant, and lipid-lowering characteristics, its impact on blood pressure management has been less reliable, with inconsistent results (Vendrame, Klimis et al., 2015; Vendrame et al., 2016). A review of medical trials tested anthocyanin-rich or purified anthocyanin extracts, whole phenolic extracts, purées, powders, berry juices, or whole berries, rich in anthocyanins with anthocyanins as the predominant bioactive to see if any patterns could explain such variability in blood pressure results. Over 400 mg/100 g *Aronia melanocarpa* (chokeberries) (Jurikova et al., 2017), over 100 mg/100 g raspberries, blueberries, blackcurrants, and blackberries, and over 50 mg/100 g cherries are the berries with the highest anthocyanin content (Garcia-Conesa, M. T et al., 2018, Kelley et al., 2018). On the other hand, grapes, cranberries, and strawberries have less than 50 mg/100 g of cyanidin and other bioactive phytochemicals (Wu et al., 2006; Corona et al., 2011).

9.9.1.1 Anthocyanin and Blood Pressure

High blood pressure could be generated by a larger cardiac output or high systemic vascular resistance caused by a narrower diameter or reduced flexibility of blood vessels. Hypertension can be caused by lack of vasodilator stimuli or excess of vasoconstricting stimuli (Clark et al., 2015). Three critical pathways have been linked to anthocyanins and regulation of blood pressure, either indirectly or directly.

1. Endothelial-derived nitric oxide (NO) has been demonstrated to be increased by anthocyanins via modification of endothelial NO synthase (eNOS) production and activity. One of the essential factors in endothelium-dependent vasorelaxation is NO. Activation of soluble guanylate cyclase, that raises cyclic guanosine monophosphate (cGMP), produces vascular smooth-muscle relaxation and prevents intracellular calcium from making vascular smooth-muscle contractions by blocking its release (Bell and Gochenaur, 2006).

2. Reactive species of oxygen hence promote hypertension and vasoconstriction. Because of intense antioxidant activities, anthocyanins prevent NO oxidative damage and NO conversion induced with radicals, like the reactions caused due to NADPH oxidase (Kim et al., 2017).

3. Anthocyanins are also demonstrated to diminish vasoconstricting molecule synthesis, like endothelin-1, thromboxanes and angiotensin II, by inhibiting the cyclooxygenase pathway and inhibiting angiotensin-converting enzyme activity (Parichatikanond et al., 2012).

9.9.1.2 Meta-Analysis and Epidemiological Data

Epidemiological research suggests establishing a correlation between regulation of blood pressure and anthocyanins. A 12-year epidemiological prospective research conducted on 34,489 post-menopausal women from Iowa Women's Health Study resulted in a significantly inversed relation between coronary heart disease and anthocyanins; and CVD and total mortality, but so far, no significant inverse relationship has been found between other individual subclasses and total flavonoids (Beecher G.R., Wu, X; et al., 2006). In a group of 156,957 women and men tracked for 14 years from the Nurses' Health Study, and Follow-Up Study of Health Professionals, Cassidy and colleagues revealed an inverse interaction between anthocyanin intake and hypertension (Cassidy et al., 2011). This connection is not true for the total consumption of flavonoid or other flavonoid subclasses (flavanones, flavan-3-ols, flavanols, and flavones) except for two specific molecules (apigenin and catechin) (Cassidy et al., 2011). In a cross-sectional evaluation of adult women of 1989 from the registry of TwinsUK, increased consumption of anthocyanin was also linked to considerably low central systolic blood pressure (SBP) and mean arterial pressure (MAP). There was no unfavorable correlation between total intake of flavanones, flavonoid, flavonols, flavones, and flavan-3-ols (Jennings et al., 2012).

Several meta-analyses using data from clinical studies involving anthocyanin have yielded varied outcomes. Red grapes, red wine, and berries are the primary anthocyanin sources considered for this research. Anthocyanin was seen to lower both diastolic blood pressure (DBP) and SBP) in a meta-analysis of 128 clinical studies conducted in different ellagitannins and anthocyanin sources, with 5538 participants (Garcia-Consea et al., 2018). When research on berries only and experiments on red grapes/wine exclusively are evaluated, the effect is more substantial (Garcia-Consea et al., 2018). Meta-analysis conducted on 22 medical trials (with 1251 respondents) indicated a considerable reduction in SBP but not in DPB (Huang et al., 2016). A meta-analysis conducted on 32 medical trials (with 1491 respondents) looking at the impacts of anthocyanin sources and anthocyanins on cardiometabolic health, on the other hand, reported that reduction in DBP and SBP does not achieve statistical importance (Yang et al., 2017).

In the same manner, meta-analysis of 19 studies indicated that anthocyanin supplementation had no meaningful effect on SBP or DBP (Daneshzad et al., 2019). Meta-analysis of six investigations with 472 participants indicated that anthocyanin supplementation had no meaningful effect on SBP or DBP. There was no significant impact of consuming blueberries to treat blood pressure in a meta-analysis of six trials considering 204 respondents (Zhu et al., 2017). All of the meta-analyses listed above included a mixed sample of women and men of various ages and from different locations, both with cardiovascular risk factors and healthy. Because of these inconsistent results, the theoretical data from clinical studies examining the impacts of anthocyanin-rich berries or anthocyanins on blood pressure and analysed the individual results to see whether any patterns could explain the wide range of blood pressure outcomes.

Multiple databases were searched for human chronic and acute intervention research using the keywords: "(blueberry* OR anthocyanin* OR raspberry* OR bilberry* OR lackberry* OR açai* OR blackcurrant* OR cherry* OR elderberry* OR Aronia* OR chokeberry*)" AND (diastolic OR systolic OR blood pressure). Sixty-six research studies were relevant, and their findings on blood pressure are shown in Tables 9.4 and 9.5.

TABLE 9.4
Summary of Single-Dose Interventions

Anthocyanin Source	Daily Anthocyanin Dosage	Effect on Blood Pressure	Reference
Blackcurrant extract drink	150, 300, or 600 mg	After 2 h =SBP, =DBP	Castro-Acosta et al. (2016)
Acai smoothie	493 mg	At 2 and 6 h =SBP, =DBP	Alquarashi et al. (2016)
Blueberry smoothie	348 mg	=SBP, =DBP	Del Bo' et al. (2014)
Blueberry drink plus smoking one cigarette	348 mg	= DBP, <SBP post-smoke	Del Bo' et al. (2014)
Blueberry drink with controlled sugar-matched drink plus smoking one cigarette	309 mg	It did not restore blood pressure	Del Bo' et al. (2014)
Blueberry drink with controlled sugar-matched drink	309 mg	=DBP, =SBP	Del Bo' et al. (2014)
Plum juce	369 mg	<SBP, <DBP, <MAP	Igwe et al. (2017)
Tart cherry juice	68 mg	=MAP, <SBP, =DBP	Keane et al. (2018)
Tart cherry juice	68 mg	<SBP	Keane et al. (2016)
Tart cherry juice	73.5 mg	=DBP, <MAP, <SBP	Keane et al. (2016)
Single dose of cherry juice or three dosages over 2 h	207 mg	<DBP, <SBP 2 h after consumption, when given in one dose	Kent et al. (2016)
Blackcurrant extract	17 mg/kg	=DBP, =SBP	Matsumoto et al. (2005)
Freeze-dried blueberry drink, blueberry baked product using baked control, or blueberry powder	196 mg in a baked product; 339 mg in glass	=DBP, =SBP	Rodriguez-Mateos et al. (2014)
Blueberry drink or nutrient-matched control	310 mg	=DBP, =SBP	Rodriguez-Mateos et al. (2013)

Source: Di Stefano et al. (2019).
Note: MAP, mean arterial pressure; DPB, diastolic blood pressure; SBP, systolic blood pressure; >, a statistically significant drop or increase; =, no statistically significant changes.

9.9.1.3   Single-Dose Interventions

Table 9.4 shows that six of 14 single-dose therapies were studied and identified substantially affected blood pressure. Every single-dose survey was conducted in healthy volunteers, except the Keane et al. research, which was carried out on 15 people with early hypertension. After 1, 2, 3, 4, 5, 6, 7, and 8 h, a single tart cherry juice serving was given containing 73.5 mg anthocyanin; MAP and SBP were considerably lower, however there was no change in DBP (Keane et al., 2016). Del Bo et al. studied the potential of serving blueberry juice with 300–350 mg anthocyanin in normalizing blood pressure in young, otherwise healthy smokers after smoking. In another experiment, blueberry juice was found to reduce the increase in SBP caused by smoke, but the result was not verified in the following analysis (Del Bo et al., 2017).

9.9.1.4   Long-Term Interventions

The findings of 52 long-term interventions are shown in Table 9.5. Only a few studies used fresh berries for evident logistical considerations and consistency. In contrast, the rest made use of processed and stable berries like concentrates, liquids, or freeze-dried powders. Just the phenolic

TABLE 9.5
Summary of Long-Term Interventions

ACN Source	Daily ACN Dosage	Effect on Blood Pressure	Reference
Sour cherry juice	720 mg	<DBP, <SBP	Ataie-Jafari et al. (2008)
Freeze-dried grape juice	35 mg	=DBP, <SBP	Barona et al. (2012)
Freeze-dried blueberry beverage plus control: water	742 mg	<DBP, <SBP	Basu et al. (2010)
Aronia extract	300 mg	<DBP, <SBP	Broncel et al. (2010)
Tart cherry juice plus control used: energy and sugar-matched drink	451 mg total phenolics	=DBP, <SBP	Chai et al. (2018)
Blackcurrant extract	105, 210, or 315 mg	=DBP, =SBP, <MAP with 210 and 315 =MAP=DBP=SBP at rest	Cook et al. (2017)
Blackcurrant extract	315 mg	<MAP, <DBP, <SBP, during isomeric condition	Cook et al. (2017)
ACN-rich elderberry extract	500 mg	=DBP, =SBP	Curtis et al. (2009)
"ACN-rich maqui berry extract	486 mg	=DBP, =SBP	Davinelli et al. (2015)
Tart cherry juice	540 mg	=DBP, =SBP post- or pre-exercise	Desai et al. (2018)
Grape and grape wine extracts	800 mg total phenolics	<DBP, <SBP with grape wine but not grape alone"	Draijer et al. (2015)
Berry mix	515 mg	=DBP, <SBP	Eruland et al. (2008)
Chokeberry and raspberry *Hibiscus sabdariffa* extract powder	19 mg	=DBP, =SBP in healthy Met's patients	Gurrola-Diaz et al. (2010)
Frozen blueberries (three times in 1 week)	456 mg	=DBP, =SBP	Habanova et al. (2016)
ACN capsule (extracted from blackcurrant and bilberry)	640 mg	Blood pressure during stress, 24-h ambulatory blood pressure, or =sitting supine	Hassellund et al. (2012)
Dried raspberry (black unripe) powder	2500 or 1500 mg powder	= DBP, <SBP with high dose	Jeong et al. (2016)
Dried raspberry (black unripe) powder	750 mg dry powder	=DBP, =SBP	Jeong et al. (2016)
Freeze-dried blueberry powder mixed with water	469 mg	>DBP, <SBP	Johnson et al. (2015)
Aronia juice enriched with glucomannan	25 mg	<DBP, <SBP	Kardum et al. (2014)
Aronia juice enriched with glucomannan	25 mg	=DBP, =SBP	Kardum et al. (2014)
Aronia juice	358 mg	<DBP, <SBP < average 24-h BP	Kardum et al. (2015)
ACN capsule extracted from blackcurrant and bilberry	300 mg	=DBP, =SBP	Karlsen et al. (2007)
Fresh sweet cherries	280 g	=DBP, =SBP after 1 month and at the end of the trial	Kelley et al. (2013)
Control used ACN plus cherry juice (free apple juice)	138 mg	=DBP, <SBP	Kent et al. (2017)
Blackcurrant juice, high- or low-dose	143 or 40 mg	=DBP, =SBP	Khan et al.2014
Bilberries (400 g fresh)	1381 mg	=DBP, =SBP	Kolehmainen et al. (2012)

(*continued*)

TABLE 9.5 (Continued)
Summary of Long-Term Interventions

ACN Source	Daily ACN Dosage	Effect on Blood Pressure	Reference
Mix of 18 berries (control: lifestyle intervention)	(Equal to 163 g fresh berries)	=DBP, =SBP	Lehtonen et al. (2011)
ACN-rich black soybean extract	31.45 mg	=DBP, =SBP	Lee et al. (2016)
ACN capsule (extracted from bilberry and blackcurrant)	320 mg	=DBP, <SBP	Li et al. (2015)
Oven-dried and cold-pressed *Aronia* powder	1024 mg	=SBP, < daytime DBP (recorded for 15 h)	Loo et al. (2016)
Tart cherry concentrate (control: energy-matched drink)	274.5 mg	=DBP, =SBP	Lynn et al. (2014)
Blackcurrant ACN extract	7.7 mg/kg	=DBP, =SBP after 30 min workload of typing	Matsumoto et al. (2005)
Blueberry 250 g		=DBP, =SBP, =ACE activity	McAnulty et al. (2005)
Freeze-dried blueberry powder, from 250 g berries		<Aortic systolic pressures, <DBP =DBP, <SBP, in subset of pre-hypertensive subjects (9 subjects)	McAnulty et al. (2014)
255 mg/day *Aronia* flavonoid extract	64 mg	<DBP, <SBP	Naruzzewicz et al. (2007)
Mixed berry drink (2/9 tomato, 1/9 blackcurrant, 1/3 blueberries, 1/9 strawberry, 1/9 lingonberry, 1/9 elderberry) or sugar control	248 mg	=DBP, =SBP	Nilsson et al.2017
ACN blackcurrant extract	50 mg	=DBP, =SBP, < intraocular pressure	Ohguro et al. (2013)
ACN blackcurrant extract	50 mg	=DBP, =SBP, < intraocular pressure,	Ohguro et al. (2012)
ACN capsule from blackcurrant and bilberry, 300 g frozen berries from cloudberries, strawberries, and raspberries	(Equivalent to 300 g fresh berries)	=DBP, =SBP	Puupponen-pimia et al. (2013)
Wild blueberry drink	400 mg	=DBP, =SBP	Riso et al. (2013)
Aronia juice 250 mL	90 mg	<SBP after 12 weeks, <DBP after 12 and 6 weeks	Skoczynka et al. (2007)
ACE inhibitors or *Aronia* extract supplements	60 mg	<DBP, <SBP, <ACE activity	Sikora et al. (2014)
Blueberry powder in smoothie and yoghurt	580 mg	=DBP, =SBP	Stull et al. (2010)
Blueberry powder in smoothie and yoghurt	580 mg	=DBP, =SBP	Stull et al. (2015)
ACN capsule	320 mg	=DBP, =SBP	Thompson et al. (2017)
Mixed juice of berry fruit (bilberry, cherry, *Aronia*, and red grape) or the same juice enriched with press residue of blackcurrant	210 or 43 mg	< SBP in both juices, more pronounced if high BP baseline value, =DBP	Tjelle et al. (2015)
Acai pulp (100g)	0.77 mg/mL	=DBP, =SBP	Udani et al. (2011)

TABLE 9.5 (Continued)
Summary of Long-Term Interventions

ACN Source	Daily ACN Dosage	Effect on Blood Pressure	Reference
Whole wild blueberry powder 1 or 2 g, or 200 mg extract	2.7, 5.4 or 14 mg	<SBP with extract, but not with powders, at 3 and 6 months	Whyte et al. (2018)
500 mg *Aronia* extract	45.1 mg	=DBP, =SBP	Xie et al. (2017)
ACN capsule (extracted from blackcurrant and bilberry)	320 mg	=DBP, =SBP	Zhang et al. (2015)
ACN capsules (extracted from blackcurrant and bilberry)	320 mg	=DBP, =SBP	Zhu et al. (2013)

Source: Di Stefano et al. (2019).
Note: ACN, anthocyanin; ACE, angiotensin-converting enzyme; MAP, mean arterial pressure; DPB, diastolic blood pressure; SBP, systolic blood pressure; >, a statistically significant drop or increase; =, no statistically significant changes.

extracts were employed in 11 investigations, while isolated anthocyanin-rich or anthocyanin extracts were used in 14 studies. Twenty-one of the 52 therapies studied had a meaningful impact on blood pressure.

9.10 CONCLUSION

Anthocyanins are pigments found abundantly in many fruits, vegetables, and cereals that occur naturally. Anthocyanins are colorful pigments found in plants with a broad range of health advantages. They are high-value colorants that can help prevent diseases like heart disease, cancer, diabetes, a variety of metabolic disorders, and microbial infection. These compounds are also neuroprotective and improve visual abilities. Tropical and subtropical fruits are high in anthocyanins and have health-promoting properties. Anthocyanidins and anthocyanins have been correlated to various mechanisms of action in disease prevention. Over the last decade, scientific understanding of anthocyanin metabolism, absorption, and bioavailability concerning CVD has dramatically expanded; however, much study remains to be done before definitive conclusions concerning the role of anthocyanins in CVD prevention can be drawn. There is an obvious need for more research in this area. Although laboratory research appears to show that anthocyanins can alter a variety of CVD biomarkers, epidemiological evidence is encouraging but lacking. Several studies have attributed anthocyanins and anthocyanin-rich berry consumption to a significant drop in blood pressure, demonstrating that an impact does occur.

REFERENCES

Agnihotri, N., Mishra, R., Goda, C., & Arora, M. (2012). Microencapsulation – A novel approach in drug delivery: A review. Indo Global Journal of Pharmaceutical Sciences, 2(1), 1–20.

Alqurashi, R.M., Galante, L. A., Rowland, I. R., Spencer, J. P., & Commane, D. M. (2016). Consumption of a flavonoid-rich açai meal is associated with acute improvements in vascular function and a reduction in total oxidative status in healthy overweight men. American Journal of Clinical Nutrition, 104, 1227–1235.

Ataie-Jafari, A., Hosseini, S., Karimi, F., & Pajouhi, M. (2008). Effects of sour cherry juice on blood glucose and some cardiovascular risk factors improvements in diabetic women: A pilot study. Nutrition Food Science, 38, 355–360.

Bansode, S. S., Banarjee, S. K., Gaikwad, D. D., Jadhav, S. L., & Thorat, R. M. (2010). Microencapsulation: A review. International Journal of Pharmaceutical Sciences Review and Research, 2(1), 38–43.

Barona, J., Aristizabal, J. C., Blesso, C. N., Volek, J. S., & Fernandez, M. L. (2012). Grape polyphenols reduce blood pressure and increase flow-mediated vasodilation in men with metabolic syndrome. Journal of Nutrition, 142, 1626–1632.

Basu, A., Rhone, M., & Lyons, T. J. (2010). Berries: Emerging impact on cardiovascular health. Nutrition Reviews, 68(3), 168–177.

Basu, A., Du, M., Leyva, M. J., Sanchez, K., Betts, N. M., Wu, M., Aston, C. E., & Lyons, T. J. (2010). Blueberries decrease cardiovascular risk factors in obese men and women with metabolic syndrome. Journal of Nutrition, 140, 1582–1587.

Bell, D. R., & Gochenaur, K. (2006). Direct vasoactive and vasoprotective properties of anthocyanin-rich extracts. Journal of Applied Physiology, 100, 1164–1170.

Berg, S., Bretz, M., Hubbermann, E. M., & Schwarz, K. (2012). The influence of different pectins on powder characteristics of microencapsulated anthocyanins and their impact on shellac coated granulate drug retention. Journal of Food Engineering, 108(1), 158–165.

Bridle, P., & Timberlake, C. F. (1997). Anthocyanins as natural food colours – selected aspects. Food Chemistry, 58(1), 103–109.

Broncel, M., Kozirog, M., Duchnowicz, P., Koter-Michalak, M., Sikora, J., & Chojnowska-Jezierska, J. (2010). *Aronia melanocarpa* extract reduces blood pressure, serum endothelin, lipid, and oxidative stress marker levels in patients with metabolic syndrome. Medical Science Monitoring, 16, CR28–CR34.

Bub, A., Watzl, B., Heeb, D., Rechkemmer, G., & Briviba, K. (2001). Malvidin-3-glucoside bioavailability in humans after ingestion of red wine, dealcoholized red wine and red grape juice. European Journal of Nutrition, 40(3), 113–120.

Bueno, J. M., Saez-Plaza, P., Ramos-Escudero, F., Jimenez, A. M., Fett, R., & Asuero, A. G. (2012). Analysis and antioxidant capacity of anthocyanin pigments. Part II: Chemical structure, color, and intake of anthocyanins. Critical Reviews in Analytical Chemistry, 42, 126–151.

Cassidy, A., O'Reilly, É. J., Kay, C., Sampson, L., Franz, M., Forman, J. P., Curhan, G., & Rimm, E. B. (2011). Habitual intake of flavonoid subclasses and incident hypertension in adults. American Journal of Clinical Nutrition, 93, 338–347.

Castaneda, O. A., Pacheco, H. M. L., Paez, H. M. E., Rodriguez, J. A., & Galan, C. A. (2009). Chemical studies of anthocyanins: A review. Food Chemistry, 113(4), 859–871.

Castro-Acosta, M. L., Smith, L., Miller, R. J., McCarthy, D. I., Farrimond, J. A., & Hall, W. L. (2016). Drinks containing anthocyanin-rich blackcurrant extract decrease postprandial blood glucose, insulin and incretin concentrations. Journal of Nutrition and Biochemistry., 38, 154–161.

Cavalcanti, R. N., Santos, D. T., & Meireles M. A. A. (2011). Non-thermal stabilization mechanisms of anthocyanins in model and food systems – An overview. Food Research International, 44(2), 499–509.

Chai, S. C., Davis, K., Wright, R. S., Kuczmarski, M. F., & Zhang, Z. (2018). Impact of tart cherry juice on systolic blood pressure and low-density lipoprotein cholesterol in older adults: A randomized controlled trial. Food Function, 9, 3185–3194.

Chen, F., Sun, Y., Zhao, G., Liao, X., Hu, X., Wu, J., & Wang, Z. (2007). Optimization of ultrasound-assisted extraction of anthocyanins in red raspberries and identification of anthocyanins in extract using high-performance liquid chromatography–mass spectrometry. Ultrasonics Sonochemistry, 14, 767–778.

Chen, X. Q., Nagao, N., Itani, T., & Irifune, K. (2012). Anti-oxidative analysis, and identification and quantification of anthocyanin pigments in different colored rice. Food Chemistry, 135, 2783–2788.

Chethana, S., Nayak, C. A., & Raghavarao, K. S. M. S. (2007). Aqueous two phase extraction for purification and concentration of betalains. Journal of Food Engineering, 81(4), 679–687.

Cicero, A. F. G., Fogacci, F., & Colletti, A. (2017). Food and plant bioactives for reducing cardiometabolic disease risk: An evidence based approach. Food Function, 8, 2076–2088.

Cicero, A. F. G., Grassi, D., Tocci, G., Galletti, F., Borghi, C., & Ferri, C. (2019). Nutrients and nutraceuticals for the management of high normal blood pressure: An evidence-based consensus document. High Blood Pressure Cardiovascular Prevention, 26, 9–25.

Cisse, M., Vaillant, F., Pallet, D., & Dornier, M. (2011). Selecting ultrafiltration and nanofiltration membranes to concentrate anthocyanins from roselle extract (*Hibiscus sabdariffa* L.). Food Research International, 44(9), 2607–2614.

Clark, J. L., Zahradka, P., & Taylor, C. G. (2015). Efficacy of flavonoids in the management of high blood pressure. Nutrition Review, 73, 799–822.

Clifford, M. N. (2000). Anthocyanins: Nature, occurrence and dietary burden. Journal of the Science of Food and Agriculture, 80(7), 1063–1072.

Cook, M. D., Myers, S. D., Gault, M. L., Edwards, V. C., & Willems, M. E. (2017). Cardiovascular function during supine rest in endurance-trained males with New Zealand blackcurrant: A dose–response study. European Journal of Applied Physiology, 117, 247–254.

Cook, M. D., Myers, S. D., Gault, M. L., & Willems, M. E. T. (2017). Blackcurrant alters physiological responses and femoral artery diameter during sustained isometric contraction. Nutrients, 9, 556.

Corona, G., Tang, F., Vauzour, D., Rodriguez-Mateos, A., & Spencer, J. (2011). Assessment of the anthocyanidin content of common fruits and development of a test diet rich in a range of anthocyanins. Journal of Berry Research, 1, 209–216.

Curtis, P. J., Kroon, P. A., Hollands, W. J., Walls, R., Jenkins, G., Kay, C. D., & Cassidy, A. (2009). Cardiovascular disease risk biomarkers and liver and kidney function are not altered in postmenopausal women after ingesting an elderberry extract rich in anthocyanins for 12 weeks. Journal of Nutrition, 139, 2266–2271.

Czank, C., Cassidy, A., Zhang, Q., et al. (2013). Human metabolism and elimination of the anthocyanin, cyanidin-3-glucoside: a 13C-tracer study. American Journal of Clinical Nutrition, 97(5), 995–1003.

Daneshzad, E., Shab-Bidar, S., Mohammadpour, Z., & Djafarian, K. (2019). Effect of anthocyanin supplementation on cardio-metabolic biomarkers: A systematic review and meta-analysis of randomized controlled trials. Clinical Nutrition, 38, 1153–1165.

Davinelli, S., Bertoglio, J. C., Zarrelli, A., Pina, R., & Scapagnini, G. (2015). A randomized clinical trial evaluating the efficacy of an anthocyanin-maqui berry extract (Delphinol) on oxidative stress biomarkers. Journal of the American College of Nutrition, 34, 28–33.

De Ferrars, R. M., Czank, C., Zhang, Q., Botting, N. P., Kroon, P. A., Cassidy, A., & Kay, C. D. (2014). The pharmacokinetics of anthocyanins and their metabolites in humans. British Journal of Pharmacology, 171, 3268–3282.

Del Bo', C., Porrini, M., Fracassetti, D., Campolo, J., Klimis-Zacas, D., & Riso, P. (2014). A single serving of blueberry (V. corymbosum) modulates peripheral arterial dysfunction induced by acute cigarette smoking in young volunteers: A randomized-controlled trial. Food Function, 5, 3107–3116.

Del Bo', C., Deon, V., Campolo, J., Lanti, C., Parolini, M., Porrini, M., Klimis-Zacas, D., & Riso, P. (2017). A serving of blueberry (V. corymbosum) acutely improves peripheral arterial dysfunction in young smokers and non-smokers: Two randomized, controlled, crossover pilot studies. Food Function, 2017, 8, 4108–4117.

Desai, T., Bottoms, L., & Roberts, M. (2018). The effects of Montmorency tart cherry juice supplementation and FATMAX exercise on fat oxidation rates and cardio-metabolic markers in healthy humans. European Journal of Applied Physiology, 118, 2523–2539.

Di Stefano, V., Pitonzo, R., Novara, M. E., Bongiorno, D., Indelicato, S., Gentile, C., ... & Melilli, M. G. (2019). Antioxidant activity and phenolic composition in pomegranate (Punica granatum L.) genotypes from south Italy by UHPLC–Orbitrap-MS approach. Journal of the Science of Food and Agriculture, 99(3), 1038–1045.

Diaz, F., Santos, E. M., Filardo, S., Villagómez, R., & Scheinvar, L. (2006). Colorant extraction from a red prickly pear (Opuntia lasiacantha) for food application. Journal of Environmental Agricture and Food Chemistry, 5, 1330–1337.

Donner, H., Gaob, Tb. L., & Mazzab, G. (1997). Separation and characterization of simple and malonylated anthocyanins in red onions, Allium cepa L. Food Research International, 30 (8), 63–74.

Draijer, R., de Graaf, Y., Slettenaar, M., de Groot, E., & Wright, C. I. (2015). Consumption of a polyphenol-rich grape-wine extract lowers ambulatory blood pressure in mildly hypertensive subjects. Nutrients, 7, 3138–3153.

Duangmal, K., Saicheua, B., & Sueeprasan, S. (2008). Color evaluation of freeze-dried roselle extract as a natural food colorant in a model system of a drink. Food Science and Technology, 41(8), 1437–1445.

Dubey, R., Shami, T. C., & Rao, K. U. B. (2009). Microencapsulation technology and applications. Defence Science Journal, 59(1), 82–95.

Durana, E. A., Giustib, M. M., Wrolstadc, R. E., & Gloria, M. B. A. (2001). Anthocyanins from Oxalis triangularis as potential food colorants. Food Chemistry, 75(2), 211–216.

Durge, A. V., Sarkar, S., & Singhal, R. S. (2013). Stability of anthocyanins as pre-extrusion coloring of rice extrudates. Food Research International, 50, 641–646.

Erlund, I., Koli, R., Alfthan, G., Marniemi, J., Puukka, P., Mustonen, P., Mattila, P., & Jula, A. (2008). Favorable effects of berry consumption on platelet function; blood pressure; and HDL cholesterol. American Journal of Clinical Nutrition, 87, 323–331.

Ersus, S., & Yurdagel, U. (2007). Microencapsulation of anthocyanin pigments of black carrot (*Daucuscarota* L.) by spray drier. Journal of Food Engineering, 80(3), 805–812.

Frank, T., Netzel, M., Strass, G., et al. (2003). Bioavailability of anthocyanidin-3-glucosides following consumption of red wine and red grape juice. Canadian Journal of Physiology and Pharmacology, 81(5), 423–435.

Galán-Saúco, V. (2002). Greenhouse cultivation of tropical fruits. Acta Horticulturae, 575, 727–735.

García-Conesa, M.T., Chambers, K., Combet, E., Pinto, P., Garcia-Aloy, M., Andrés Lacueva, C., de Pascual-Teresa, S., Mena, P., Konic Ristic, A., Hollands, W. J. et al. (2018). Meta-analysis of the effects of foods and derived products containing ellagitannins and anthocyanins on cardiometabolic biomarkers: Analysis of factors influencing variability of the individual responses. Int. J. Mol. Sci., 19, 694.

Garzona, G. A., & Wrolstad, R. E. (2001). The stability of pelargonidin-based anthocyanins at varying water activity. Food Chemistry, 75, 185–196.

Ghosh, D., & Konishi, T. (2007). Anthocyanins and anthocyanin-rich extracts: Role in diabetes and eye function. Asia Pacific Journal of Clinical Nutrition, 16(2), 200–208.

Giusti, M., & Wrolstad, R. E. (2003). Acylated anthocyanins from edible sources and their applications in food systems. Biochemical Engineering Journal, 14(3), 217–225.

Goldmann, I. L., Kader, A. A., & Heintz, C. (1999). Influence of production, handling, and storage on phytonutrient content of foods. Nutrition Review, 57, 46–52.

Gradinaru, G., Biliaderis, C. G., Kallithraka, S., Kefalas, P., & Garcia-viguera, C. (2003). Thermal stability of *Hibiscus sabdariffa* L anthocyanins in solution and in solid state: Effects of copigmentation and glass transition. Food Chemistry, 83, 423–436.

Gurrola-Díaz, C. M., García-López, P. M., Sánchez-Enríquez, S., Troyo-Sanromán, R., Andrade-González, I., & Gómez-Leyva, J. F. (2010). Effects of *Hibiscus sabdariffa* extract powder and preventive treatment (diet) on the lipid profiles of patients with metabolic syndrome (MeSy). Phytomedicine, 17, 500–505.

Habanova, M., Saraiva, J. A., Haban, M., Schwarzova, M., Chlebo, P., Predna, L., Gažo, J., & Wyka, J. (2016). Intake of bilberries (*Vaccinium myrtillus* L.) reduced risk factors for cardiovascular disease by inducing favorable changes in lipoprotein profiles. Nutrition Research, 36, 1415–1422.

Hassellund, S. S., Flaa, A., Sandvik, L., Kjeldsen, S. E., & Rostrup, M. (2012). Effects of anthocyanins on blood pressure and stress reactivity: A double-blind randomized placebo-controlled crossover study. Journal of Human Hypertension 26, 396–404.

He, K., Li, X., Chen, X., et al. (2011). Evaluation of antidiabetic potential of selected traditional Chinese medicines in STZ-induced diabetic mice. Journal of Ethnopharmacology, 137(3), 1135–1142.

Huang, H., Chen, G., Liao, D., Zhu, Y., & Xue, X. (2016). Effects of berries consumption on cardiovascular risk factors: A meta-analysis with trial sequential analysis of randomized controlled trials. Science Report, 6, 23625.

Idham, Z., Muhamad, I. I., & Sarmidi, M. R. (2012). Degradation kinetics and color stability of spray-dried encapsulated anthocyanins from *Hibiscus sabdariffa* L. Journal of Food Process Engineering, 35, 522–542.

Igwe, E. O., Charlton, K. E., Roodenrys, S., Kent, K., Fanning, K., & Netzel, M. E. (2017). Anthocyanin-rich plum juice reduces ambulatory blood pressure but not acute cognitive function in younger and older adults: A pilot crossover dose-timing study. Nutrition Research, 47, 28–43.

Jennings, A., Welch, A. A., Fairweather-Tait, S. J., Kay, C., Minihane, A. M., Chowienczyk, P., Jiang, B., Cecelja, M., Spector, T., Macgregor, A., et al. (2012). Higher anthocyanin intake is associated with lower arterial stiffness and central blood pressure in women. American Journal of Clinical Nutrition, 96, 781–788.

Jeong, H. S., Hong, S. J., Cho, J. Y., Lee, T. B., Kwon, J. W., Joo, H. J., Park, J. H., Yu, C. W., & Lim, D. S. (2016). Effects of *Rubus occidentalis* extract on blood pressure in patients with prehypertension: Randomized, double-blinded, placebo-controlled clinical trial. Nutrition, 32, 461–467.

Jeong, H. S., Kim, S., Hong, S. J., Choi, S. C., Choi, J. H., Kim, J. H., Park, C. Y., Cho, J. Y., Lee, T. B., Kwon, J. W., et al. (2016). Black raspberry extract increased circulating endothelial progenitor cells and improved arterial stiffness in patients with metabolic syndrome: A randomized controlled trial. Journal of Medicinal Food,, 19, 346–352.

Jing, P., Zhao, S.J., Ruan, S.Y., Xie, Z.H., Dong, Y., & Yu, L. (2012). Anthocyanin and glucosinolate occurrences in the roots of Chinese red radish (*Raphanus sativus* L.) and their stability to heat and pH. Food Chemistry, 133(4), 1569–1576.

Johnson, S. A., Figueroa, A., Navaei, N., Wong, A., Kalfon, R., Ormsbee, L. T., Feresin, R. G., Elam, M. L., Hooshmand, S., Payton, M. E., et al. (2015). Daily blueberry consumption improves blood pressure and arterial stiffness in postmenopausal women with pre- and stage 1-hypertension: A randomized, double-blind, placebo-controlled clinical trial. Journal of the Academy of Nutrition and Dietetics, 115, 369–377.

Ju, H., Chiang, F., & Wrolstad, R. E. (2005). Anthocyanin pigment composition of blackberries. Food Chemistry and Toxicology, 70, 198–202.

Jurikova, T., Mlcek, J., Skrovankova, S., Sumczynski, D., Sochor, J., Hlavacova, I., Snopek, L., & Orsavova, J. (2017). Fruits of black chokeberry *Aronia melanocarpa* in the prevention of chronic diseases. Molecules, 22, 944.

Kader, A. A. (2001). Importance of fruits, nuts, and vegetables in human nutrition and health. Perishables Handling Quarterly, 106, 4–6.

Kahkonen, M. P., Hopia, A. I., & Heinonen, M. (2001). Berry phenolics and their antioxidant activity. Journal of Agricultural and Food Chemistry, 49(8), 4076–4082.

Kardum, N., Koníc-Risti´c, A., Savikin, K., Spasíc, S., Stefanovíc, A., Ivaniševíc, J., & Miljkovíc, M. (2014). Effects of polyphenol-rich chokeberry juice on antioxidant/pro-oxidant status in healthy subjects. Journal of Medicinal Food, 17, 869–874.

Kardum, N., Petrovíc-Oggiano, G., Takic, M., Glibetíc, N., Zec, M., Debeljak-Martacic, J., & Koníc-Ristíc, A. (2014). Effects of glucomannan-enriched, aronia juice-based supplement on cellular antioxidant enzymes and membrane lipid status in subjects with abdominal obesity. Scientific World Journal,, 2014, 869250.

Kardum, N., Milovanovíc, B., Šavikin, K., Zduníc, G., Mutavdžin, S., Gligorijevíc, T., & Spasíc, S. (2015). Beneficial effects of polyphenol-rich chokeberry juice consumption on blood pressure level and lipid status in hypertensive subjects. Journal of Medicinal Food, 18, 1231–1238.

Karlsen, A., Retterstøl, L., Laake, P., Paur, I., Bøhn, S.K., Sandvik, L., & Blomhoff, R. (2007). Anthocyanins inhibit nuclear factor-kappaB activation in monocytes and reduce plasma concentrations of pro-inflammatory mediators in healthy adults. Journal of Nutrition, 137, 1951–1954.

Keane, K. M., George, T. W., Constantinou, C. L., Brown, M. A., Clifford, T.; & Howatson, G. (2016). Effects of Montmorency tart cherry (*Prunus cerasus* L.) consumption on vascular function in men with early hypertension. American Journal of Clinical Nutrition, 103, 1531–1539.

Keane, K. M., Bailey, S. J., Vanhatalo, A., Jones, A. M., & Howatson, G. (2018). Effects of Montmorency tart cherry (L. *Prunus cerasus*) consumption on nitric oxide biomarkers and exercise performance. Scandinavian Journal of Medicine,. Science and Sports, 28, 1746–1756.

Keane, K. M., Haskell-Ramsay, C. F., Veasey, R. C., & Howatson, G. (2016). Montmorency tart cherries (*Prunus cerasus* L.) modulate vascular function acutely, in the absence of improvement in cognitive performance. British Journal of Nutrition, 116, 1935–1944.

Kelley, D. S., Adkins, Y., Reddy, A., Woodhouse, L. R., Mackey, B. E., & Erickson, K. L. (2013). Sweet bing cherries lower circulating concentrations of markers for chronic inflammatory diseases in healthy humans. Journal of Nutrition, 143, 340–344.

Kelley, D. S., Adkins, Y., & Laugero, K. D. (2018). A review of the health benefits of cherries. Nutrients, 10, 368.

Kent, K., Charlton, K. E., Jenner, A., & Roodenrys, S. (2016). Acute reduction in blood pressure following consumption of anthocyanin-rich cherry juice may be dose-interval dependent: A pilot cross-over study. International Journal of Food Science and Nutrition, 67, 47–52.

Kent, K., Charlton, K., Roodenrys, S., Batterham, M., Potter, J., Traynor, V., Gilbert, H., Morgan, O., & Richards, R. (2017). Consumption of anthocyanin-rich cherry juice for 12 weeks improves memory and cognition in older adults with mild-to-moderate dementia. European Journal of Nutrition, 56, 333–341.

Khan, F ., Ray, S., Craigie, A. M., Kennedy, G., Hill, A., Barton, K. L., Broughton, J., & Belch, J. J. (2014). Lowering of oxidative stress improves endothelial function in healthy subjects with habitually low intake of fruit and vegetables: A randomized controlled trial of antioxidant- and polyphenol-rich blackcurrant juice. Free Radical Biology and Medicine, 72, 232–237.

Khoo, H. E., Azlan, A., Ismail, A., & Abas, F. (2012). Antioxidative properties of defatted dabai pulp and peel prepared by solid phase extraction. Molecules, 17(8), 9754–9773.

Kim, J. N., Han, S. N., Ha, T. J., & Kim, H. K. (2017). Black soybean anthocyanins attenuate inflammatory responses by suppressing reactive oxygen species production and mitogen activated protein kinases signaling in lipopolysaccharide-stimulated macrophages. Nutrition Reseach and Practice, 11, 357–364.

Kirca, A., Ozkan, M., & Cemeroglu, B. (2006). Stability of black carrot anthocyanins in various fruit juices and nectars. Food Chemistry, 97, 598–605.

Koide, T., Kamei, H., Hashimoto, Y., Kojima, T., & Hasegawa, M. (1996). Antitumor effect of hydrolyzed anthocyanin from grape rinds and red rice. Cancer Biotherapy and Radiopharmaceuticals, 11, 273–277.

Kolehmainen, M., Mykkänen, O., Kirjavainen, P. V., Leppänen, T., Moilanen, E., Adriaens, M., Laaksonen, D. E., Hallikainen, M., Puupponen-Pimiä, R., Pulkkinen, L., et al. (2012). Bilberries reduce low-grade inflammation in individuals with features of metabolic syndrome. Molecular Nutrition and Food Research, 56, 1501–1510.

Laleh, G. H., Frydoonfar, H., Heidary, R., et al. (2006). The effect of light, temperature, pH and species on stability of anthocyanin pigments in four *Berberis* species. Pakistan Journal of Nutrition, 5(1), 90–92.

Lee, M., Sorn, S. R., Park, Y., & Park, H. K. (2016). Anthocyanin rich-black soybean testa improved visceral fat and plasma lipid profiles in overweight/obese Korean adults: A randomized controlled trial. Journal of Medicinal Food, 19, 995–1003.

Lehtonen, H. M., Suomela, J. P., Tahvonen, R., Yang, B., Venojärvi, M., Viikari, J., & Kallio, H. (2011). Different berries and berry fractions have various but slightly positive effects on the associated variables of metabolic diseases on overweight and obese women. European Journal of Clinical Nutrition 65, 394–401.

Li, D., Zhang, Y., Liu, Y., Sun, R., & Xia, M. (2015). Purified anthocyanin supplementation reduces dyslipidemia; enhances antioxidant capacity; and prevents insulin resistance in diabetic patients. Journal of Nutrition, 145, 742–748.

Liazid, A., Guerrero, R. F., Cantos, E., Palma, M., & Barrosso, C. G. (2011). Microwave assisted extraction of anthocyanins from grape skin. Food Chemistry, 3, 1238–1243.

Loo, B. M., Erlund, I., Koli, R., Puukka, P., Hellström, J., Wähälä, K., Mattila, P., & Jula, A. (2016). Consumption of chokeberry (*Aronia mitschurinii*) products modestly lowered blood pressure and reduced low-grade inflammation in patients with mildly elevated blood pressure. Nutrition Research, 36, 1222–1230.

Lynn, A., Mathew, S., Moore, C. T., Russell, J., Robinson, E., Soumpasi, V., & Barker, M. E. (2014). Effect of a tart cherry juice supplement on arterial stiffness and inflammation in healthy adults: A randomised controlled trial. Plant Foods and Human Nutrition, 69, 122–127.

Ma, R., Xu, X., Zhao, C., Wang, Z., Chen, F., Hu, X., Li, Y., & Han, L. (2012). Effect of energy density and citric acid concentration on anthocyanins yield and solution temperature of grape peel in microwave-assisted extraction process. Journal of Food Engineering, 109, 274–280.

Manach, C., Gary, W., Morand, C., Scalbert, A., & Remesy, C. (2005). Bioavailability and bioefficacy of polyphenols in humans. I. Review of 97 bioavailability studies. American Journal of Clinical Nutrition, 81, 230–242.

Mantell, C., Rodriguez, M., & Martinez, E. (2003). A screening analysis of the high-pressure extraction of anthocyanins from red grape pomace with carbon dioxide and cosolvent. Engineering in Life Sciences, 3, 38–42.

Matera, R., Gabbanini, S., Nicola, G. R., Iori, R., Petrillo, G., & Valgimigli, L. (2012). Identification and analysis of isothiocyanates and new acylated anthocyanins in the juice of *Raphanus sativus* cv. Sango sprouts. Food Chemistry, 133(2), 563–572.

Matsumoto, H., Inaba, H., Kishi, M., et al. (2001). Orally administered delphinidin 3-rutinoside and cyanidin 3-rutinoside are directly absorbed in rats and humans and appear in the blood as the intact forms. Journal of Agriculture and Food Chemistry, 49(3), 1546–1551.

Matsumoto, H., Takenami, E., Iwasaki-Kurashige, K., Osada, T., Katsumura, T., & Hamaoka, T. (2005). Effects of blackcurrant anthocyanin intake on peripheral muscle circulation during typing work in humans. European Journal of Applied Physiology, 94, 36–45.

Mazza, G. (1995). Anthocyanins in grape and grape products. Critical Reviews in Food Science and Nutrition, 35(4), 341–371.

Mazza, G., & Francis, F. J. (1995). Anthocyanins in grapes and grape products. Critical Review in Food Sci Nutrition, 35(4), 341–371.

Mazza, G., & Miniati, E. (1993). Anthocyanins in Fruits, Vegetables and Grains. Boca Raton, FL: CRC Press; pp. 1–87.

McAnulty, S. R., McAnulty, L. S., Morrow, J. D., Khardouni, D., Shooter, L., Monk, J., Gross, S., & Brown, V. (2005). Effect of daily fruit ingestion on angiotensin converting enzyme activity, blood pressure, and oxidative stress in chronic smokers. Free Radical Research, 39, 1241–1248.

McAnulty, L. S., Collier, S. R., Landram, M. J., Whittaker, D. S., Isaacs, S. E., Klemka, J. M., Cheek, S. L., Arms, J. C., & McAnulty, S. R. (2014). Six weeks daily ingestion of whole blueberry powder increases natural killer cell counts and reduces arterial stiffness in sedentary males and females. Nutrition Research, 34, 577–584.

McGhie, T. K., & Walton, M. C. (2007). The bioavailability and absorption of anthocyanins: Towards a better understanding. Molecular Nutrition and Food Research, 51(6), 702–713.

Mink, P. J., Scrafford, C. G., Barraj, L. M., Harnack, L., Hong, C. P., Nettleton, J. A. & Jacobs, D. R., Jr. (2007). Flavonoid intake and cardiovascular disease mortality: A prospective study in postmenopausal women. American Journal of Clinical Nutrition, 85, 895–909.

Miyazawa, T., Nakagawa, K., Kudo, M., et al. (1999). Direct intestinal absorption of red fruit anthocyanins, cyanidin-3- glucoside and cyanidin-3, 5-diglucoside, into rats and humans. Journal of Agriculture and Food Chemistry, 47(3), 1083–1091.

Moskovitz, J., Yim, K. A., & Choke, P. B. (2002). Free radicals and disease. Archives of Biochemistry and Biophysics, 397, 354–359.

Mülleder, U., Murkovic, M., & Pfannhauser, W. (2002). Urinary excretion of cyanidin glycosides. Journal of Biochemical and Biophysical Methods, 53(1), 61–66.

Naruszewicz, M., Laniewska, I., Millo, B., & Dłuzniewski, M. (2007). Combination therapy of statin with flavonoid rich extract from chokeberry fruits enhanced reduction in cardiovascular risk markers in patients after myocardial infraction (MI). Atherosclerosis, 194, e179–e184.

Nayak, C. A., Srinivas, P., & Rastogi, N. K. (2010). Characterisation of anthocyanins from *Garcinia indica* Choisy. Food Chemistry, 118, 719–724.

Ngo, T., Wrolstad, R. E., & Zhao, Y. (2007). Color quality of Oregon strawberries impact of genotype, composition, and processing. Food Chemistry and Toxicology, 72, 25–32.

Nichenametla, S. N., Taruscio, T. G., Barney, D. L., & Exon, J. H. (2006). A review of the effects and mechanisms of polyphenolics in cancer. Critical Reviews in Food Science and Nutrition, 46(2), 161–183.

Nilsson, A., Salo, I., Plaza, M., & Björck, I. (2017). Effects of a mixed berry beverage on cognitive functions and cardiometabolic risk markers; A randomized cross-over study in healthy older adults. PLoS ONE, 12, e0188173.

Ohguro, H., Ohguro, I., Katai, M., & Tanaka, S. (2012). Two-year randomized, placebo-controlled study of black currant anthocyanins on visual field in glaucoma. Ophthalmologica, 228, 26–35.

Ohguro, H., Ohguro, I., & Yagi, S. (2013). Effects of black currant anthocyanins on intraocular pressure in healthy volunteers and patients with glaucoma. Journal of Ocular and Pharmacological Therapy, 29, 61–67.

Ordaz-Galindo, A., Wesche-Ebeling, P., Wrolstad, R. E., Rodriguez-Saona, L., & Argaiz Jamet, A. (1999). Purification and identification of Capulin (*Prunus serotina* Ehrh) anthocyanins. Food Chemistry, 65, 201–206.

Oren, S. M. (2009). Does anthocyanin degradation play a significant role in determining pigment concentration in plants? Plant Science, 177, 310–316.

Parichatikanond, W., Pinthong, D., & Mangmool, S. (2012). Blockade of the renin–angiotensin system with delphinidin, cyanin, and quercetin. Planta Medica, 78, 1626–1632.

Pazmino, D. A. E., Giusti, M. M., Wrolstad, R. E., & Gloria, M. B. A. (2001). Anthocyanins from *Oxalis triangularis* as potential food colorants. Food Chemistry, 75(2), 211–216.

Percival, S. S., Talcott, S. T., Chin, S. T., Mallak, A. C., Lound-Singleton, A., & Pettit-Moore, J. (2006). Neoplastic transformation of BALB/3T3 cells and cell cycle of HL-60 cells are inhibited by mango (*Mangifera indica* L.) juice and mango juice extract. Journal of Nutrition, 136, 1300–1304.

Puupponen-Pimiä, R., Seppänen-Laakso, T., Kankainen, M., Maukonen, J., Törrönen, R., Kolehmainen, M., Leppänen, T., Moilanen, E., Nohynek, L., Aura, A. M., et al. (2013). Effects of ellagitannin-rich berries on blood lipids; gut microbiota; and urolithin production in human subjects with symptoms of metabolic syndrome. Molecular Nutrition & Food Research, 57, 2258–2263.

Qin, Y., Xia, M., Ma, J., Hao, Y., Liu, J., Mou, H., Cao, L., & Ling, W. (2009). Anthocyanin supplementation improves serum LDL- and HDL-cholesterol concentrations associated with the inhibition of cholesteryl ester transfer protein in dyslipidemic subjects. American Journal of Clinical Nutrition, 90, 485–492.

Qin, Y., Zhai, Q., Li, Y., Cao, M., Xu, Y., Zhao, K., & Wang, T. (2018). Cyanidin-3-O-glucoside ameliorates diabetic nephropathy through regulation of glutathione pool. Biomedicine & Pharmacotherapy, 103, 1223–1230.

Rechkemmer, G. (2001). Funktionelle Lebensmittel-Zukunft de Ernahrung oder MarketingStrategie. Forschungereport Sonderheft, 1, 12–15 .

Rein, M. (2005). Copigmentation Reactions and Color Stability of Berry Anthocyanins. Helsinki: University of Helsinki; pp. 10–14.

Reyes, L. F., & Zevallos, L. C. (2007). Degradation kinetics and color of anthocyanins in aqueous extracts of purple and red-flesh potatoes (*Solanum tuberosum* L.). Food Chemistry, 100(3), 885–894.

Riso, P., Klimis-Zacas, D., Del Bo', C., Martini, D., Campolo, J., Vendrame, S., Møller, P., Loft, S., De Maria, R., & Porrini, M. (2013). Effect of a wild blueberry (*Vaccinium angustifolium*) drink intervention on markers of oxidative stress, inflammation and endothelial function in humans with cardiovascular risk factors. European Journal of Nutrition, 52, 949–961.

Rodriguez-Mateos, A., Rendeiro, C., Bergillos-Meca, T., Tabatabaee, S., George, T. W., Heiss, C., & Spencer, J. P. (2013). Intake and time dependence of blueberry flavonoid-induced improvements in vascular function: A randomized, controlled, double-blind, crossover intervention study with mechanistic insights into biological activity. American Journal of Clinical Nutrition, 98, 1179–1191.

Rodriguez-Mateos, A., Del Pino-García, R., George, T. W., Vidal-Diez, A., Heiss, C., & Spencer, J. P. (2014). Impact of processing on the bioavailability and vascular effects of blueberry (poly)phenols. Molecular Nutrition and Food Research, 58, 1952–1961.

Rogez, H., Pompeu, D. R., Akwie, S. N. T., & Larondelle, Y. (2011). Sigmoidal kinetics of anthocyanin accumulation during fruit ripening: A comparison between acai fruits (*Euterpe oleracea*) and other anthocyanin-rich fruits. Journal of Food Composition and Analysis, 24(6), 796–800.

Rosso, V. V. D., & Mercadante, A. Z. (2007). The high ascorbic acid content is the main cause of the low stability of anthocyanin extracts from acerola. Food Chemistry, 103, 935–943.

Selim, K. A., Khalil, K., Abdel-bary, M., & Abdel-azeim, N. (2008). Extraction, encapsulation and utilization of red pigments from roselle (*Hibiscus sabdariffa* L.) as natural food colorants. 5th Alexandria Conference of Food and Dairy Science and Technology. Alexandria, Egypt.

Shahidi, F., & Han, X. Q. (1993). Encapsulation of food ingredients. Critical Reviews in Food Science and Nutrition, 33(6), 501–547.

Sikora, J., Broncel, M., & Mikiciuk-Olasik, E. (2014). *Aronia melanocarpa* Elliot reduces the activity of angiotensin-converting enzyme – In vitro and ex vivo studies. Oxidative Medicine and Cellular Longevity, 2014, 739721.

Silva, F. M., Bailon, M. T. E., Alonso, J. J. P., Gonzalo, J. C. R., & Buelga, C. S. (2007). Anthocyanin pigment in strawberry. Lebensmittel Wissenschaft und Technologie, 40, 374–382.

Skoczynska, A., Jedrychowska, I., Por.eba, R., Affelska-Jercha, A., Turczyn, B., Wojakowska, A., Andrzejak, R., & Jedrychowska-Bianchi, I. (2007). Influence of chokeberry juice on arterial blood pressure and lipid parameters in men with mild hypercholesterolemia. Pharmacological Reports, 59, 177–182.

Stintzing, F. C., & Carle, R. (2004). Functional properties of anthocyanins and betalains in plants, food, and in human nutrition. Trends in Food Science and Technology, 15(1), 19–38.

Stoner, G. D., Sardo, C., Apseloff, G., Mullet, D., Wargo, W., Pound, V., Singh, A., Sanders, J., Aziz, R., Casto, B., et al. (2005). Pharmacokinetics of anthocyanins and ellagic acid in healthy volunteers fed freeze-dried black raspberries daily for 7 days. Journal of Clinical Pharmacology, 45, 1153–1164.

Strack, D., & Wray, V. (1989). Anthocyanins, In: Dey, P. M., & Harborne, J. B. (Ed.), Methods in Plant Biochemistry, Vol. 1, Plant Phenolics. London: Academic Press; pp. 325–356.

Stull, A. J., Cash, K. C., Johnson, W. D., Champagne, C. M., & Cefalu, W. T. (2010). Bioactives in blueberries improve insulin sensitivity in obese, insulin-resistant men and women. Journal of Nutrition, 140, 1764–1768.

Stull, A. J., Cash, K. C., Champagne, C. M., Gupta, A. K., Boston, R., Beyl, R. A., Johnson, W. D., & Cefalu, W. T. (2015). Blueberries improve endothelial function, but not blood pressure, in adults with metabolic syndrome: A randomized, double-blind, placebo-controlled clinical trial. Nutrients, 7, 4107–4123.

Tamura, H., & Yamagami, A. (1994). Antioxidative activity of monoacylated anthocyanins isolated from Muscat Bailey A grape. Journal of Agriculture and Food Chemistry, 42, 1612–1615.

Thompson, K., Hosking, H., Pederick, W., Singh, I., & Santhakumar, A. B. (2017). The effect of anthocyanin supplementation in modulating platelet function in sedentary population: A randomised, double-blind, placebo-controlled, cross-over trial. British Journal of Nutrition, 118, 368–374.

Tiwari, S., Goel, A., Jha, K. K., & Sharma, A. (2010). Microencapsulation techniques and its application: A review. The Pharma Research, 3, 112–116.

Tjelle, T. E., Holtung, L., Bøhn, S. K., Aaby, K., Thoresen, M., Wiik, S. Å., Paur, I., Karlsen, A. S., Retterstøl, K., Iversen, P. O., et al. (2015). Polyphenol-rich juices reduce blood pressure measures in a randomised controlled trial in high normal and hypertensive volunteers. British Journal of Nutrition, 114, 1054–1063.

Torronen, R., & Maatta, K. (2002). Bioactive substances and health benefits of strawberries. Acta Horticulturae, 567, 797–803.

Tsai, P. J., McIntosh, J., Pearce, P., Camden, B., & Jordan, B. R. (2002). Anthocyanin and antioxidant capacity in Roselle (*Hibiscus sabdariffa* L.) extract. Food Research International, 35, 351–356.

Tsuda, T., Shiga, K., Ohshima, K., Kawakishi, S., & Osawa, T. (1996). Inhibition of lipid peroxidation and the active oxygen radical scavenging effect of anthocyanin pigments isolated from *Phaseolus vulgaris* L. Biochemical Pharmacology, 52(7), 1033–1039.

Udani, J. K., Singh, B. B., Singh, V., & Barrett, M. L. (2011). Effects of Açai (*Euterpe oleracea* Mart.) berry preparation on metabolic parameters in a healthy overweight population: A pilot study. Nutrition Journal, 10, 45.

Vareed, S. K., Reddy, K., Schutzki, R. E., & Nair, M. G. (2006). Anthocyanins in *Cornus alternifolia, Cornus controversa, Cornus kousa* and *Cornus florida* fruits with health benefits. Life Sciences, 78, 777–784.

Veitch, N. C., & Grayer, R. J. (2008). Flavonoids and their glycosides, including anthocyanins. Natural Product Reports, 25, 555–611.

Vendrame, S., & Klimis-Zacas, D. (2015). Anti-inflammatory effect of anthocyanins via modulation of nuclear factor-κB and mitogen-activated protein kinase signaling cascades. Nutrition Review, 73, 348–358.

Vendrame, S. & Klimis-Zacas, D. (2019). Potential factors influencing the effects of anthocyanins on blood pressure regulation in humans: A review. Nutrients, 11(6), 1431.

Vendrame, S., Del Bo', C., Ciappellano, S., Riso, P., & Klimis-Zacas, D. (2016). Berry fruit consumption and metabolic syndrome. Antioxidants, 5, 34.

Wallace, T. C. (2011). Anthocyanins in cardiovascular disease. Advances in Nutrition, 2, 1–7.

Wani, A. A., Singh, P., Shah, M. A., Weisz, U. S., Gul, K., & Wani, I. A. (2012). Rice starch diversity: Effects on structural, morphological, thermal, and physiochemical properties – A review. Comprehensive Reviews in Food Science & Food Safety, 11, 417–436.

Whyte, A. R., Cheng, N., Fromentin, E., & Williams, C. M. (2018). A randomized, double-blinded, placebo-controlled study to compare the safety and efficacy of low dose enhanced wild blueberry powder and wild blueberry extract (ThinkBlue) in maintenance of episodic and working memory in older adults. Nutrients, 10, 660.

Wilska, J. J. (2007). Food colorants. In Sikorski, Z. E. (Ed.), Chemical and Functional Properties of Food Components. Boca Raton, FL: CRC Press; pp. 245–274.

Wu, X., Beecher, G. R., Holden, J. M., Haytowitz, D. B., Gebhardt, S. E., & Prior, R. L. (2006). Concentrations of anthocyanins in common foods in the United States and estimation of normal consumption. Journal of Agriculture and Food Chemistry, 54, 4069–4075.

Wu, X., Liang, L., Zou, Y., Zhao, T., Zhao, J., Li, F., & Yang, L. (2011). Aqueous two-phase extraction, identification and antioxidant activity of anthocyanins from mulberry (*Morus atropurpurea* Roxb). Food Chemistry, 129(2), 443–453.

Xie, L., Vance, T., Kim, B., Lee, S. G., Caceres, C., Wang, Y., Hubert, P. A., Lee, J. Y., Chun, O. K., & Bolling, B. W. (2017). Aronia berry polyphenol consumption reduces plasma total and low-density lipoprotein cholesterol in former smokers without lowering biomarkers of inflammation and oxidative stress: A randomized controlled trial. Nutrition Research, 37, 67–77.

Yahia, E.M., De Jesus Ornelas-Paz, J. & Gonzalez-Aguilar, G.A. (2011). Nutritional and health-promoting properties of tropical and subtropical fruits. In Elhadi, M. Yahia (Ed.), Postharvest Biology and Technology of Tropical and Subtropical Fruits. Sawston: Woodhead Publishing.

Yang, Z., & Zhai, W. (2010). Optimization of microwave-assisted extraction of anthocyanins from purple corn (*Zea mays* L.) cob and identification with HPLC–MS. Innovative Food Science and Emerging Technologies, 11, 470–476.

Yang, L., Ling, W., Du, Z., Chen, Y., Li, D., Deng, S., Liu, Z., & Yang, L. (2017). Effects of anthocyanins on cardiometabolic health: A systematic review and meta-analysis of randomized controlled trials. Advances in Nutrition, 8, 684–693.

Yi, L., Chen, C. Y., Jin, X., Mi, M.T., Yu, B., Chang, H., Ling, W.H. & Zhang, T. (2010). Structural requirements of anthocyanins in relation to inhibition of endothelial injury induced by oxidized low-density lipoprotein and correlation with radical scavenging activity. FEBS Letters, 584, 583–590.

Zhang, Z., Xuequn, P., Yang, C., Ji, Z., & Jiang, Y. (2004). Purification and structural analysis of anthocyanins from litchi pericarp. Food Chemistry, 84(4), 601–604.

Zhang, P. W., Chen, F. X., Li, D., Ling, W. H., & Guo, H. H. (2015). A CONSORT-compliant, randomized, double-blind, placebo-controlled pilot trial of purified anthocyanin in patients with nonalcoholic fatty liver disease. Medicine (Baltimore), 94, e758.

Zhu, Y., Ling, W., Guo, H., Song, F., Ye, Q., Zou, T., Li, D., Zhang, Y., Li, G., Xiao, Y., et al. (2013). Anti-inflammatory effect of purified dietary anthocyanin in adults with hypercholesterolemia: A randomized controlled trial. Nutrition, Metabolism and Cardiovascular Disease, 23, 843–849.

Zhu, Y., Bo, Y., Wang, X., Lu, W., Wang, X., Han, Z., & Qiu, C. (2016). The effect of anthocyanins on blood pressure: A PRISMA-compliant meta-analysis of randomized clinical trials. Medicine (Baltimore), 95, e3380.

Zhu, Y., Sun, J., Lu, W., Wang, X., Wang, X., Han, Z., & Qiu, C. (2017). Effects of blueberry supplementation on blood pressure: A systematic review and meta-analysis of randomized clinical trials. Journal of Human Hypertension, 31, 165–171.

Zou, T., Wang, D., Guo, H., Zhu, Y., Luo, X., Liu, F., & Ling, W. 2012. Optimization of microwave assisted extraction of anthocyanins from mulberry and identification of anthocyanins in extract using HPLC-ESI-MS. Journal of Food Science, 77(1), 46–50.

10 Anti-Inflammatory Properties of Natural Anthocyanins Obtained from Subtropical Fruits

M. Selvamuthukumaran

CONTENTS

10.1 INTRODUCTION

Infection of immune system may cause inflammation, which is a biological symptom that may occur in particular due to pathogen assault, when human tissues are injured (Mantovani et al., 2008; Brevetti et al., 2010; Hummasti and Hotamisligil, 2010; Kim et al., 2014). This inflammation can lead to diseases such as pulmonary disease, metabolic disease, and cancer (Christaki et al., 2010; Lambrecht and Hammad, 2010; Park et al., 2015). Fruits which are grown and harvested in a subtropical climate have potent anti-inflammatory properties (Table 10.1 and Figure 10.1) and these fruits are discussed below.

10.2 SUBTROPICAL FRUITS THAT POSSESS POTENT ANTI-INFLAMMATORY ACTIVITY

10.2.1 ORANGE

The blood red orange belonging to the sweet orange variety possesses ample health benefits. In particular, it contains high concentrations of anthocyanins, which are responsible for contributing the red color to the flesh. Anthocyanins present in this fruit exhibit significant anti-inflammatory activity, which can help to cure cancer, heart disease, and diabetes by scavenging the formation of free radicals and inflammation. Blood red oranges are part of the Mediterranean diet; their juice has been consumed from ancient times, when they were successfully used as a medicine because of their health-giving properties. The anti-inflammatory activity of blood red orange anthocyanins,

TABLE 10.1
Anti-Inflammatory Activities of Various Subtropical Fruits

Name of the Subtropical fruit	Active Constituent Responsible for Inducing Anti-Inflammatory Activities	References
Pineapple	Bromelain	Insuan et al. (2021)
Kiwi	Procyanidins	Eliseo et al. (2019).
Grape	Polyphenols	Estruch et al. (2004)
Pomegranate	Punicalagin, gallic acid, and ellagic acid	Bensaad et al. (2017)
Lychee	Flavonol	Yaminishi et al. (2014)
Passion fruit	Polyphenols	Park et al. (2018)
Red blood orange	Anthocyanins	Cardile et al. (2010)

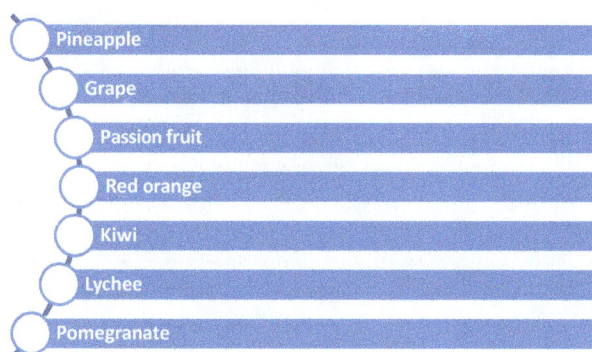

FIGURE 10.1 Subtropical fruits that possess potent anti-inflammatory properties.

i.e., *Citrus sinensis* of variety Sanguinello, Moro, and Tarocco was studied by Cardile et al. (2010) after they had been exposed to histamine and interferon-gamma on human keratinocyte line NCTC 2544. As a result of exposure, the authors observed immunomodulatory membrane molecule expression, i.e., intercellular adhesion molecule-1 (ICAM-1), and release of chemokines like interleukin-8 (IL-8) and monocyte chemoattractant protein-1 (MCP-1). The researchers also found that ICAM-1 modulated T-lymphocyte activation in the epidermis. They further concluded that the use of various concentrations of red blood orange anthocyanins along with histamine and interferon-gamma successfully induced the inhibition of MCP-1 and IL-8 release and expression of ICAM-1 in a dose-dependent manner, thereby significantly exhibiting potent anti-inflammatory activity.

Buscemi et al. (2012) observed the effect of intake of red orange juice on markers of inflammation in test subjects with enhanced cardiovascular risks. Their results show that the consumption of red orange juice for a period of 1 week successfully reduced inflammation to a great extent. This may be ascribed to diminishing concentrations of high-sensitivity C-reactive protein, which is an indicator of cardiovascular risk and inflammatory activation (Libby, 2006). It was further observed that red orange juice consumption also significantly reduced the inflammatory cytokine concentrations, i.e. tumor necrosis factor-α (TNF-α) and IL-6.

10.2.2 PASSION FRUIT

Passion fruit, scientifically known as *Passiflora foetida*, belongs to the family Passifloraceae, and is native to Mexico, Southwestern United States (Arizona, southern Texas) and the Caribbean islands.

From these countries the plant was introduced into Asia and Europe (Asadujjaman et al., 2014; Huang et al., 2015).

Passion fruit, which is being used to cure several diseases, was also recorded in ancient history. Park et al. (2018) studied the inflammatory action of methanolic extracts obtained from passion fruit. They studied the nuclear factor-κB (NF-κB) signaling and passion fruit methanolic extract involvement in inflammation regulation. They found that prostaglandin E_2 (PGE_2) production and inducible cyclooxygenase-2 (COX-2) in lipopolysaccharide (LPS)-induced macrophage cells expressions were successfully prevented. Further it was observed that release of pro-inflammatory cytokines was minimized. This extract suppressed the phosphorylation of MAPKs (ERK1/2, p38, and JNK) in LPS-induced RAW264.7 cells. The extract also prevented the activation of NF-κB induced by LPS. The authors' results project that methanolic extracts of passion fruit successfully prevented the LPS-induced inflammation, thereby exhibiting anti-inflammatory properties.

The progression of prostate cancer was correlated with inflammation markers. It was proposed that intake of polyphenols may delay prostate cancer, thereby improving the inflammatory response and antioxidant defenses. Baseggio et al. (2021) analyzed the effect of administration of yellow passion fruit (*Passiflora edulis* sp.) bagasse extract in transgenic mice (TRAMP) against inflammatory parameters which indicate the progression of cancer. They found that successful feeding of passion fruit bagasse extract may lead to a delay in the initial progression of cancer.

10.2.3 Lychee

Lychee fruit contains mainly anthocyanidins as flavonoids (Ververidis et al., 2007). Yaminishi et al. (2014) studied the effect of flavanol (flavan-3-ol)-rich lychee fruit extract on inflammation. Extracts obtained from lychee were the higher source of monomers, dimers, and trimers of flavonols. Therefore the food incorporated in such functional constituents derived from lychee fruit can suppress inflammation and prevent tissue damage. The effects of flavanol (flavan-3-ol)-rich lychee fruit extract were studied on inflammatory gene expression in IL-1β-treated rat hepatocytes. The authors found that lychee fruit extract successfully reduced the protein expression of the inducible nitric oxide synthase (iNOS) gene and mRNA, ultimately resulting in prohibition of production of nitric oxide (NO) induced by IL-1β. It also further reduced iNOS antisense transcript levels, which can stabilize the iNOS mRNA.

Flavanol (flavan-3-ol)-rich lychee fruit extract further reduced the activity of luciferase in the presence of IL-1β, which can suppress iNOS gene promoter activity at transcriptional level. The extract minimized the key regulator of nuclear transport and inhibited NF-κB phosphorylation. It also further minimized the mRNA levels of NF-κB target genes encoded with chemokines and cytokines. Inflammatory gene expression was suppressed and NF-κB activation was prohibited by this extract; this action which may be ascribed to flavonols which successfully exhibited significant anti-inflammatory activity.

10.2.4 Pomegranate

The pomegranate fruit has many therapeutic properties. In particular, it is highly successful at suppressing inflammation, exhibiting potent anti-inflammatory effects (Gonzalez-Trujano et al., 2015). This fruit can drastically reduce cancer (Lansky and Newman, 2007). Several research reports have shown that the ethanolic fractions of pomegranate fruit contains antioxidants like punicalagin A and B, gallic acid and ellagic acid, which have demonstrated significant anti-inflammatory effects in vitro (Fujihara et al., 2003; Kuroda and Yamashita, 2003; Laskin et al., 2011).

Bensaad et al. (2017) studied the anti-inflammatory activity of pomegranate fruit by extracting the fruit with ethyl acetate. The fractions obtained from pomegranate fruit were found to contain components like punicalagin, gallic acid, and ellagic acid and these compounds exhibited inhibitory activity on LPS, stimulated NO, PGE_2, IL-6, and COX-2, which are released from RAW264.7 cells. These compounds further inhibited the production of NO, PGE_2, and IL-6 induced by LPS, thereby exhibiting significant anti-inflammatory effects.

10.2.5 GRAPE

The grape contains several polyphenol complexes, especially flavonoids, which can suppress nuclear transcription factor B activation (Blanco-Colío et al., 2000) and minimize inflammatory biomarker concentration (Estruch et al., 2004). Castilla et al. (2006) studied the dietary supplementation of red grape juice concentrates against inflammatory biomarkers in human test subjects. Their results showed that after 3 weeks of feeding red grape juice concentrates, the MCP-1 concentration reduced to one-half of baseline value. The accumulation of inflammation plays an important role in the development of atherosclerosis (Ross, 1999; Libby, 2002). The elevated serum levels of MCP-1 in hemodialysis patients can lead to onset of atherosclerosis (Kusano et al., 2004). The researchers concluded that the dietary supplementation of concentrated red grape juice can successfully reduce serum concentrations of inflammatory biomarkers with additional effects like reducing cardiovascular risk and enhancing lipoprotein profile.

Greenspan et al. (2005) tested the anti-inflammatory activity of muscadine grape skins by extracting them with ethanol. The ethanol fractions obtained from this fruit skin were tested in two assays, i.e., superoxide release in phorbol myristate acetate-activated neutrophils and cytokine release (IL-6, IL-β, and TNF-α) by LPS-activated peripheral blood mononuclear cells. Their in vitro studies demonstrated that when the concentration of extract was increased, the release of cytokines and superoxide was inhibited. Superoxide release was inhibited to 60%, when extract dilution was used at a ratio of 1:100. IL-1β and TNF-α were reduced to 90 and 15% when the extract dilution was increased to a ratio of 1:200. The authors carried out an in vivo experiment with rats administered with diet containing 5% dried muscadine grape skins for 2 weeks followed by injection in the foot pad with carrageenan and found that after a 3-h period the rats administered with grape skin diet had less paw edema, i.e., less than 50%, which demonstrates that grape skin has potent anti-inflammatory properties both in vivo and in vitro.

10.2.6 KIWI

The kiwi fruit has considerable medicinal value. The fruit is native to China and its consumption can prevent several disorders, especially inflammatory diseases like cancer. Eliseo et al. (2019) obtained extracts from kiwi fruit peel, which is a rich source of polyphenols, and they explored anti-inflammatory activity by analyzing their capabilities to target the multiple pathways involved in monocyte-mediated inflammatory response. THP-1 monocytes stimulated by LPS were used as a model for human inflammation in vitro. The results showed that the extract obtained from kiwi fruit peel contained more procyanidins as polyphenols, representing around 92% w/w. The researchers found that this extract prohibited inflammatory molecule production, such as TNF-α pro-inflammatory cytokines, IL-1β, IL-6, granzyme B serine protease, and HMGB1 danger signal activated by monocytes. The inhibitory activity observed ranged from 60 to 81%. The extract also further prevented the activation of STAT3 and further promoted autophagy. The authors further concluded that kiwi fruit has a potent anti-inflammatory activity and can be successfully employed in nutraceutical and pharmaceutical formulations to counteract several inflammatory disorders.

10.2.7 PINEAPPLE

Pineapple fruit is widely consumed for its exotic flavor; in addition it has excellent medicinal properties. This fruit has bromelain as an active component, which has potent anti-inflammatory properties.

Insuan et al. (2021) conducted a study on the effect of bromelain obtained from crude as well as purified pineapple rhizome on LPS-induced inflammation in RAW 264.7 macrophage cells. They found that treating bromelain in a dose-dependent manner successfully reduced the pro-inflammatory cytokines and mediators induced by LPS and these are correlated with COX-2 expressions and of iNOS downregulations. Pure bromelain recorded the highest inflammation inhibitory activities compared to crude bromelain, thereby significantly suppressing MAPKs and NF-κB signaling pathways.

10.3 CONCLUSIONS

The consumption of subtropical fruits like kiwi, pineapple, red blood orange, lichee, and pomegranate can cure inflammation, thereby delaying or suppressing the progression of cancer. These fruits contain several flavonoids, especially anthocyanidin, a potent antioxidant, which can scavenge the formation of free radicals, prevent atherosclerosis, and reduce LDL cholesterol. Therefore, longevity can be enhanced by regular consumption of such fruit or including them as part of one's daily diet.

REFERENCES

Asadujjaman, M., Mishuk, A. U., Hossain, M. A., and Karmakar, U. K. (2014). Medicinal potential of *Passiflora foetida* L. plant extracts: Biological and pharmacological activities. Journal of Integrative Medicine 12: 121–126.

Baseggio, A. M., Kido, L. A., Viganó, J., Carneiro, M. J., Lamas, C. de A., Martínez, J., Sawaya, A. C. H. F., Cagnon, V. H. A., and Júnior, M. R. M. (2021). Systemic antioxidant and anti-inflammatory effects of yellow passion fruit bagasse extract during prostate cancer progression. Journal of Food Biochemistry, 46(3). https://doi.org/10.1111/jfbc.13885

BenSaad, L. A., Kim, K. H., Quah, C. C., Kim, W. R., and Shahimi, M. (2017). Anti-inflammatory potential of ellagic acid, gallic acid and punicalagin A&B isolated from *Punica granatum*. BMC Complementary and Alternative Medicine, 17: 47.

Blanco-Colío, L. M., Valderrama, M., Alvarez-Sala, L. A., et al. (2000). Red wine intake prevents nuclear factor-kappaB activation in peripheral blood mononuclear cells of healthy volunteers during postprandial lipemia. Circulation, 102: 1020–1026.

Brevetti, G., Giugliano, G., Brevetti, L., and Hiatt, W. R. (2010). Inflammation in peripheral artery disease. Circulation, 122: 1862–1875.

Buscemi, S., Rosafio, G., Arcoleo, G., Mattina, A., Canino, B., Montana, M., Verga, S., and Rini, G. (2012). Effects of red orange juice intake on endothelial function and inflammatory markers in adult subjects with increased cardiovascular risk. American Journal of Clinical Nutrition, 95: 1089–1095.

Cardile, V., Frasca, G., Rizza, L., Rapisarda, P., and Bonina F. (2010). Antiinflammatory effects of a red orange extract in human keratinocytes treated with interferon-gamma and histamine. Phytotherapy Research, 24(3): 414–418.

Castilla, P., Echarri, R., Dávalos, A., Cerrato, F., Ortega, H., Teruel, J. L., Lucas, M. F., Gómez-Coronado, D., Ortuño, J., and Lasunción, M. A. (2006). Concentrated red grape juice exerts antioxidant, hypolipidemic, and antiinflammatory effects in both hemodialysis patients and healthy subjects. American Journal of Clinical Nutrition, 84: 252–262.

Christaki, E., Opal, S. M., Keith, J. C. Jr, Kessinian, N., Palardy, J. E., Parejo, N. A., Lavallie, E., Racie, L., Mounts, W., Malamas, M. S., et al. (2010). Estrogen receptor beta agonism increases survival in experimentally induced sepsis and ameliorates the genomic sepsis signature: A pharmacogenomic study. Journal of Infectious Diseases 201: 1250–1257.

Eliseo, D. D., Pannucci, E., Bernini, R., Campo, M., Romani, A., Santi, L., and Velotti, F. (2019). *In vitro* studies on anti-inflammatory activities of kiwifruit peel extract in human THP-1 monocytes. Journal of Ethnopharmacology, 233: 41–46.

Estruch, R., Sacanella, E., Badía, E., et al. (2004). Different effects of red wine and gin consumption on inflammatory biomarkers of atherosclerosis: a prospective randomized crossover trial. Effects of wine on inflammatory markers. Atherosclerosis, 175: 117–1123.

Fujihara, M., Muroi, M., Tanamoto, K., Suzuki, T., Azuma, H., and Ikeda H. (2003). Molecular mechanisms of macrophage activation and deactivation by lipopolysaccharide:roles of the receptor complex. Pharmacology and Therapy, 100(2): 171–194.

Gonzalez-Trujano, M. E., Pellicer, F., Mena, P., Monerno, D. A., and Garcia-Viguera, C. (2015). Antinociceptive and anti-inflammatory activities of a pomegranate (*Punica granatum* L.) extract rich in ellagitannins. International Journal of Food Science and Nutrition, 66(4): 395–399.

Greenspan, P., Bauer, J. D., Pollock, S. H., Gangemi, J. D., Mayer, E. P., Ghaffar, A., Hargrove, J. L., and Hartle, D. K. (2005). Antiinflammatory properties of the muscadine grape (*Vitis rotundifolia*). Journal of Agriculture and Food Chemistry, 53(22), 8481–8484.

Huang, W. C., Wu, S. J., Tu, R. S., Lai, Y. R., and Liou, C. J. (2015). Phloretin inhibits interleukin-1β-induced COX-2 and ICAM-1 expression through inhibition of MAPK, Akt, and NF-κB signaling in human lung epithelial cells. Food Function, 6: 1960–1967.

Hummasti, S., and Hotamisligil, G. S. (2010). Endoplasmic reticulum stress and inflammation in obesity and diabetes. Circulation Research, 107: 579–591.

Insuan, O., Janchai, P., Thongchuai, B., Chaiwongsa, R., Saoin, S., Insuan, W., Pothacharoen, P., Apiwatanapiwat, W., Boondaeng, A., and Vaithanomsat, P. (2021). Anti-inflammatory effect of pineapple rhizome bromelain through downregulation of the NF-B- and MAPKs-signaling pathways in lipopolysaccharide (LPS)-stimulated RAW264.7 cells. Current Issues in Molecular Biology, 43(1), 93–106. https://doi.org/10.3390/cimb43010008

Kim, H. S., Park, J. W., Kwon, O. K., Kim, J. H., Oh, S. R., Lee, H. K., Bach, T. T., Quang, B. H., and Ahn, K. S. (2014). Anti-inflammatory activity of a methanol extract from *Ardisia tinctoria* on mouse macrophages and paw edema. Molecular Medicine Reports, 9, 1388–1394.

Kuroda, E., and Yamashita, U. (2003). Mechanisms of enchanced macrophage mediated prostaglandin E_2 production and its suppressive role in Th1 and Th2 dominant BALB/C mice. Journal of Immunology, 170, 757–764.

Kusano, K. F., Nakamura, K., Kusano, H., et al. (2004). Significance of the level of monocyte chemoattractant protein-1 in human atherosclerosis. Circulation Journal, 68, 671–676.

Lambrecht, B. N., and Hammad, H. (2010). The role of dendritic and epithelial cells as master regulators of allergic airway inflammation. Lancet, 376, 835–843.

Lansky, E. P., and Newman, R. A. (2007). *Punica granatum* (pomegranate) and its potential for prevention and treatment of inflammation and cancer. Journal of Ethnopharmacology, 109(2), 177–206.

Laskin, D. L., Sunil, V. R., Gardner, C. R., and Laskin, J. D. (2011). Macrophages and tissue injury: Agents of defence or destruction? Annual Review of Pharmacology and Toxicology, 51, 267–288.

Libby, P. (2002). Inflammation in atherosclerosis. Nature, 420, 868–874.

Libby, P. (2006). Inflammation and cardiovascular disease mechanisms. American Journal of Clinical Nutrution, 83, 456S–60S.

Mantovani, A., Allavena, P., Sica, A., and Balkwill, F. (2008). Cancer-related inflammation. Nature, 454, 436–444.

Park, J. W., Shin, I. S., Ha, U. H., Oh, S. R., Kim, J. H., and Ahn, K. S. (2015). Pathophysiological changes induced by *Pseudomonas aeruginosa* infection are involved in MMP-12 and MMP-13 upregulation in human carcinoma epithelial cells and a pneumonia mouse model. Infection and Immunity, 83, 4791–4799.

Park, J. W., Kwon, O. K., Ryu, H. W., Paik, J.-H., Paryanto, I., Yuniato, P., Choi, S. H., Oh, S. R., and Ah, K. S. (2018). Anti-inflammatory effects of *Passiflora foetida* L. in LPS-stimulated RAW264.7 macrophages. International Journal of Molecular Medicine, 41, 3709–3716.

Ross, R. (1999). Atherosclerosis – an inflammatory disease. New England Journal of Medicine, 340, 115–126.

Ververidis, F., Trantas, E., Douglas, C., Vollmer, G., Kretzschmar, G., et al. (2007). Biotechnology of flavonoids and other phenylpropanoid-derived natural products. Part I: Chemical diversity, impacts on plant biology and human health. Biotechnology Journal, 2, 1214–1234.

Yamanishi, R., Yoshigai, E., Okuyama, T., Mori, M., Murase, H., et al. (2014). The anti-inflammatory effects of flavanol-rich lychee fruit extract in rat hepatocytes. PLoS ONE, 9(4), e93818. doi:10.1371/journal.pone.0093818.

11 Antioxidative Effects of Subtropical Fruits Rich in Anthocyanins

Anil S. Nandane and Rahul C. Ranveer

CONTENTS

11.1 INTRODUCTION

Fruits are considered as an important source of food nutrients along with calories and therefore play a significant role in human nutrition. Essential minerals, vitamins, fatty acids, and dietary fiber are also available in sufficient amount and quality in fruits. The vitamins found in fruits include vitamin C, riboflavin, vitamin A, thiamine, niacin, pyridoxine, and fat-soluble vitamins. The minerals present in fruits are calcium, potassium, iron, and phosphorus. However, every fruit can be considered unique in its composition. Banana yields the highest calories per unit and takes the least time to digest. Cashew nut is an excellent source of protein (21.2%) whereas mango and papaya contain high amounts of vitamin A. Fruits are considered healthy foods due to low or no cholesterol, compared to several other food sources. The presence of a significant quantity of dietary fiber is another characteristic of several tropical and subtropical fruits. Dietary fiber helps prevent constipation, reduces cholesterol levels, improves bowel movement, and helps fight obesity, hypertension, and heart diseases. Fruits are a rich source of natural physiologically active components known as phytochemicals and antioxidants that have the ability to prevent oxidative chain reaction by removing free radicals and other mechanisms. The literature suggests that phytochemicals and nutrients found in fruits could slow the aging process, stimulate the immune system, and reduce the danger of many diseases.

Fruits and vegetables provide a variety of micronutrients, like minerals, fiber, and vitamins, and phytochemicals like polyphenols (Harborne and Williams, 2001; Grayer and Veitch, 2008;

DOI: 10.1201/9781003242598-11

Ziyatdinova and Budnikov, 2015). Phenolic compounds have attracted researchers for many years. This was initially because of their physiological importance to plants, principally about pigmentation and flavor and, recently, due to their radical scavenging capability, which helps provide antioxidative property and prevent cellular oxidation (Tirzitis and Bartosz, 2010; Martinis et al., 2016). Flavonoids represent the largest cluster of phenols and are thought to be responsible for the color and taste of many fruits and vegetables. Over 9000 flavonoid structures have been identified together with their formula, references, and biological information (Fraga, 2010; He and Giusti, 2010). These embody over 600 different anthocyanins that are spread among at least 27 families, 73 genera, and a number of species. It has been shown that, of the flavonoids studied, around 5000 have antioxidant activity (Martín Bueno et al., 2012a, b). Anthocyanins are typically identified as the largest and most vital group of water-soluble pigments in nature (Harborne, 1998). They provide shades of blue, purple, red, and orange colors in many fruits and vegetables. The important types and food sources of anthocyanin are shown in Figure 11.1.

The word anthocyanin is derived from two Greek words: *anthos,* meaning flowers, and *kianos,* meaning dark blue. Important sources of anthocyanins are blueberries, cherries, raspberries, strawberries, blackcurrants, purple grapes, and red wine (Mazza et al., 2002). They belong to the family of compounds referred to as flavonoids; however they are distinguished from other flavonoids because of their ability to create flavylium cations. They occur chiefly as glycosides of their corresponding aglycone anthocyanidin chromophores with the sugar molecule attached at the 3-position on the C-ring or the 5-position on the A-ring (Prior and Wu, 2006). There are nearly 17 anthocyanidins found in nature; however only six (cyanidin, delphinidin, petunidin, peonidin, pelargonidin, and malvidin, with cyanidin being the most common) are ubiquitously distributed and of importance in the human diet (Harborne, 1998; Jaganath and Crozier, 2010). Consistent with a review on the role of anthocyanins in the prevention of cancer, the daily intake of anthocyanins within the U.S. diet was found to be 180–215 mg, which is considerably higher than the intake of other dietary flavonoids like genistein, quercetin, and apigenin, that is only 20–25 mg/day (Wang and Stoner, 2008). The chemical structures of some of the commonly found anthocyanins are shown in Figure 11.2.

Some beverages, such as wine, tea, and coffee, have received greater attention thanks to their protective effects against oxidative damage associated with numerous chronic diseases, as well as cancer, reducing the chance of getting these diseases by 30–50% (Zhang and Tsao, 2016). The principal cause of death in the Western world is said to be chronic diseases like coronary cardiovascular disease or heart attack. Low blood levels of tocopherol and ascorbic acid have been shown to enhance the chance of angina pectoris among the population of Scotland (Wang et al., 1996; Cao et al., 1998). This is often largely attributed to the low consumption of foodstuffs containing vitamins and antioxidants.

Anthocyanins have antioxidative ability that is twice as great as other known antioxidants like catechin, vitamin E, butylated hydroxy anisole, and butylated hydroxy toluene, which may adversely affect human health (He and Giusti, 2010). The known ability of highly polarized anthocyanins to regenerate lipophilic antioxidants like tocopherol may well be a result of their properties similar to vitamin C, such as protecting cell membranes from peroxidation, by effectively trapping peroxyl free radicals.

11.2 BIOAVAILABILITY OF ANTHOCYANINS

Nowadays foods rich in anthocyanin are getting a lot of attention. Studies indicate that the consumption of anthocyanins lowers the chances of cardiovascular disease, diabetes, arthritis, and cancer, owing to their antioxidative and anti-inflammatory properties (Rechner and Kroner, 2005; Wang and Stoner, 2008). However, to attain medicinal benefits, these bioactive compounds should be bioavailable; that is, effectively absorbed from the intestine into the circulation and delivered to the suitable location to reach the target (McDougall et al., 2005). Oral administration of anthocyanin-rich

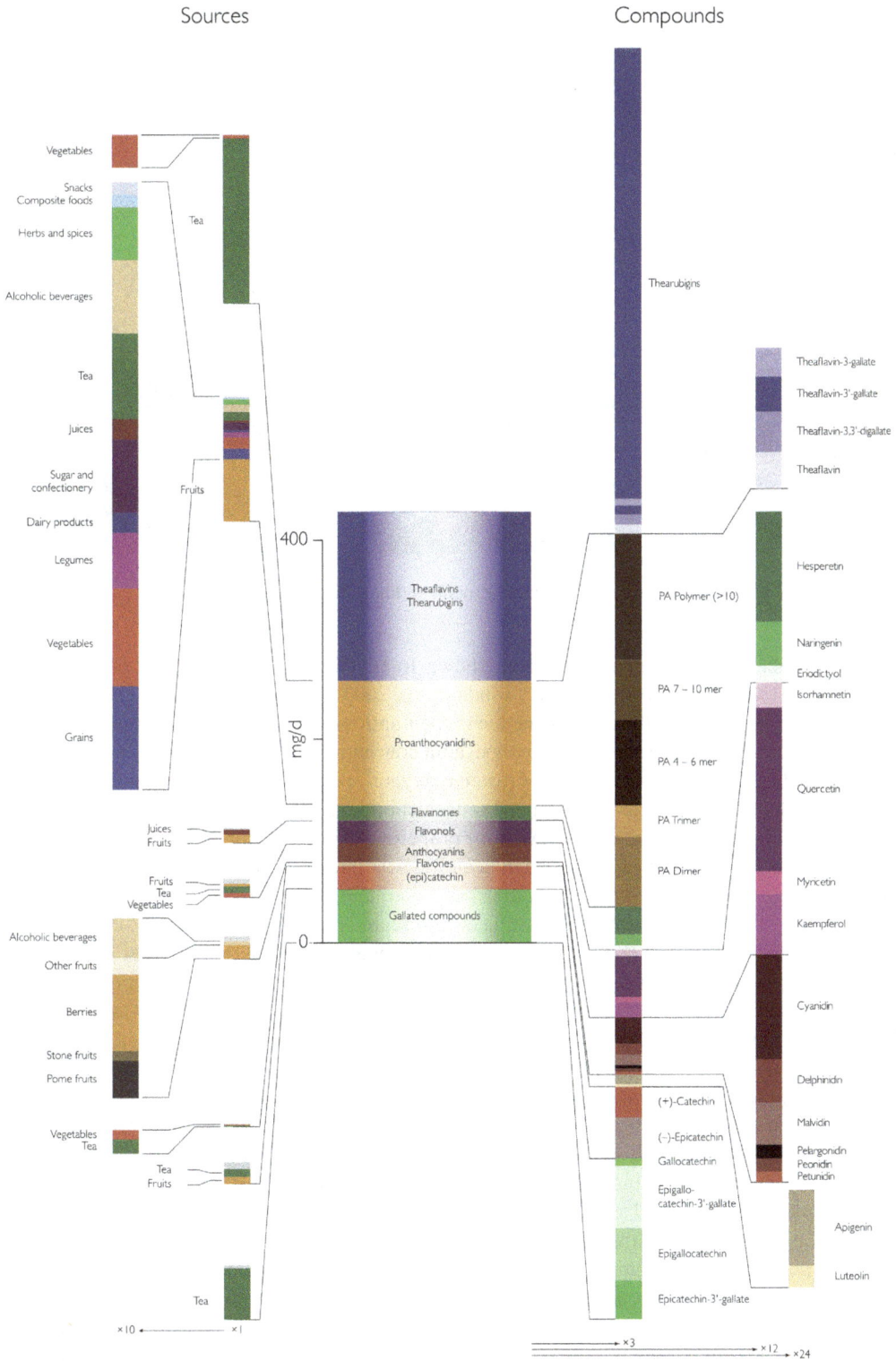

FIGURE 11.1 Important types and food sources of anthocyanin.

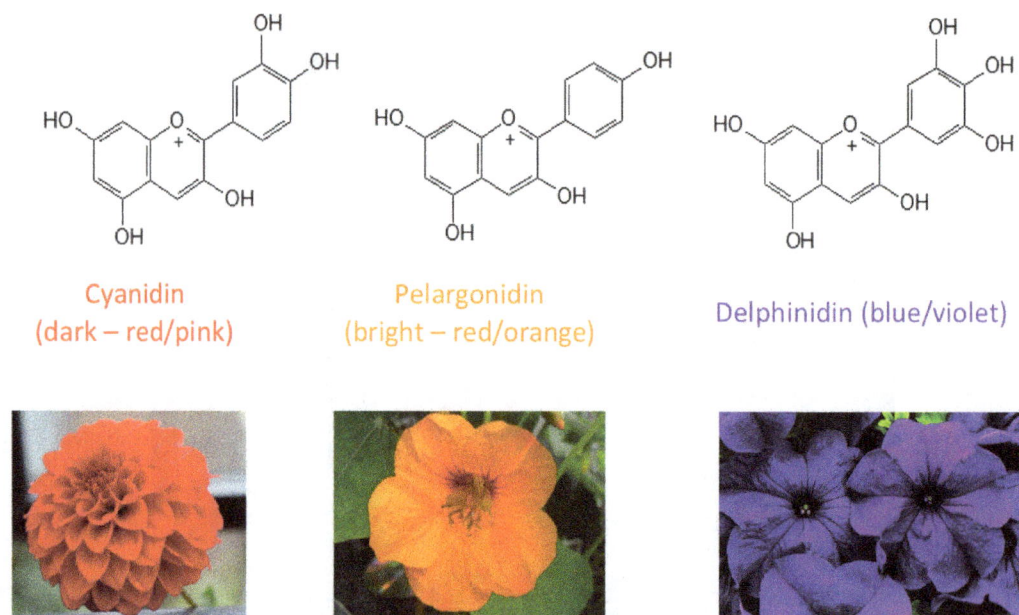

Cyanidin
(dark – red/pink)

Pelargonidin
(bright – red/orange)

Delphinidin (blue/violet)

FIGURE 11.2 Chemical structures of commonly found anthocyanins.

fruit extracts or pure compounds has been shown to be effective in preventing or suppressing various diseases (Ramirez-Tortosa, 2001; Tsuda et al., 2003; McDougall et al., 2005). Several of the activities reported for anthocyanins are due to their antioxidant activity and/or their metabolites. However, some studies have shown that anthocyanin concentrations were too low to make a significant contribution to in vivo scavenging of reactive oxygen species (ROS). However, this amount is sufficient to influence signal and gene signal transduction and gene expression pathways (Milbury et al., 2010). In vitro studies on the antioxidative properties of anthocyanins are numerous; however, after foods containing these pigments have been ingested they are likely to be metabolized, with possible change in their activity. Data on the absorption, metabolism, tissue and organ distribution and excretion of anthocyanins in human subjects are scarce because of complexity, cost, and time and generally have produced contradictory results (Kay et al., 2004; McDougall et al., 2005).

In vivo experiments with rats showed that malvidin-3-glucoside appeared in portal and systemic plasma after only 6 min and by that point steady state was reached. This finding helped the scientists conclude that this anthocyanin might permeate the gastric mucosa (Passamonti et al., 2003). Another study demonstrated that anthocyanin glycosides are quickly absorbed from the small intestine of rats followed by rapid metabolism and excretion into bile and urine as intact glycosides in addition to methylated forms and glucuronidated derivatives (Talavéra et al., 2004). Different studies have conjointly concluded that human subjects have the capability to metabolize cyanidin-3-glycosides into corrosponding glucuronide conjugates, as well as methylated and oxidized derivatives of cyanidin-3-galactoside and cyanidin glucuronide (Kay et al., 2004). Contrasting results were reported during rat experiments by other authors regarding raspberry anthocyanins which were poorly absorbed; significant amounts passed from small to large intestine (Borges et al., 2007). These authors conjointly stress the absence of information regarding either the metabolism or absorption of the pseudobases or the quinoidal base of anthocyanins within the digestive tract, mainly due to the absence of adequate analytical procedures. The same authors also reported that after anthocyanins leave the stomach, carbinol pseudobase, which is colorless, becomes the main form in the small intestine, resulting in limited absorption. As a result significant amounts pass into the large intestine where they are degraded by colonic bacteria. Microflora found in this part of the gastrointestinal tract may metabolize anthocyanins into phenolic compounds. The formation of protocatechuic acid (3,4-dihydroxybenzoic acid) takes place as a metabolite during bioconversion of cyanidin-3-rutinoside by human fecal microflora (Aura, 2005). Another group Vitaglione et al. (2007) also reported this metabolite in humans after ingestion of cyanidin-3-glucoside, accounting for about 40% of the ingested cyanidin-3-glucoside 6 h post-consumption in the blood stream. It was found in feculent samples. This phenolic compound is also responsible for many biological properties of anthocyanins, such as anti-obesity, cardiovascular-protective, and anti-inflammatory activities.

After cleavage of the 3-glycosidic moiety, free cyanidin is extremely unstable under physiological conditions and may be metabolized by the microorganism or degraded by chemical reaction (Galvano et al., 2007). Some other phenolics can be obtained as a result of biotransformation of anthocyanin by gut microflora. Examples are 3-O-methylgallic acid, syringic acid, and 2,4,6-trihydroxy-benzaldehyde when incubated in the large intestine of freshly slaughtered pigs of an extract from red-wine grape anthocyanins, containing delphinidin-3-glucoside, petunidin-3-glucoside, peonidin-3-glucoside, and malvidin-3-glucoside (Forester and Waterhouse, 2008). Some authors propose that the many health advantages attributed to dietary anthocyanin consumption are due to the presence of the phenolic metabolites (Vitaglione et al., 2007; Forester and Waterhouse, 2008).

Espín et al. (2007) mentioned that anthocyanins, their aglycones, and both methylated and glucuronide derivatives of anthocyanins may be detected in tissues like abdomen, bowel, liver, bile, kidney, respiratory organ, and eye. McGhie and Wlaton (2007) have compiled the results obtained by numerous authors on the processes of absorption and metabolism of anthocyanins. On the basis of this they reported that anthocyanin glycosides will be speedily absorbed from the stomach after ingestion and that they enter the circulation by passing through the liver. Methylation and glucuronidation reactions can take place in these organs and a few of the metabolites are transported to the intestine as bile. Unabsorbed anthocyanin glycosides from the stomach pass to the small intestine and due to the pH are converted in the form of chalcone and quinonoidal. Absorbed anthocyanins then enter the circulation system by passing though liver and can be metabolized. Anthocyanins that reach the colon are exposed to gut microflora which convert them to sugar and phenolic compounds, which later are degraded by disruption of the C-ring to yield phenolic acids and aldehydes.

11.3 PROCESS OF OXIDATION

The process of oxidation has been studied for several years, because of its importance for organisms as well as foodstuffs. In living organisms, aerobic metabolism is essential for survival of the cells. Oxidation is the process of energy generation utilizing carbohydrates, lipids, and proteins and is also related to the detoxification of many xenobiotics and to the immune response through free radicals. Oxygen is vital for life and is the driving force for the overall metabolism of living organisms. At the same time, it is involved in the formation of partially reduced mediators with high reactivity that are known as ROS. Most are in the form of free radicals containing a lone pair of electrons, and therefore have paramagnetic properties and high reactivity (Ali et al., 2016) The systems of antioxidant protection need to act on the substrates vulnerable to oxidation in a controlled way to maintain the physiological equilibrium of the organism. The protective action of some enzymes, like the enzyme superoxide dismutase, and glutathione oxidase, can begin once excess free radical is created. If this excess cannot be neutralized it will result in oxidation of the lipidic membrane, low-density lipoprotein (LDL), the protein cellular components, deoxyribonucleic acid, and enzymes. Arteriosclerosis, which is considered to be chronic inflammation of the vascular system, is triggered by an inflammatory agent like oxidized LDL. LDLs are tiny particles composed of lipids, cholesterol, and proteins, whose main role is to transport cholesterol and lipids from the blood to the adipose and muscular tissues as well as to all cells of body. However, LDLs will be oxidized by the free radicals, affecting the molecules of cholesterol and fatty acids of LDL. Oxidized LDLs are associated with pathogenesis of coronary heart disease. The oxidative stress may be the result of an imbalance in favor of oxidation caused by environmental, dietary, or physiological factors. However it is debatable whether oxidation and in particular oxidative stress is either a primary cause or a side effect of many chronic diseases and the process of aging. Therefore, several efforts and resources are dedicated to the search for the role of oxidants in reducing the extent of oxidation, leading to either the interference or retardation of oxidative stress.

An excessive production of ROS, notably hydroxyl radicals, will simply initiate oxidation of the LDLs. In turn, they contribute to a greater or lesser degree to the onset of coronary heart disease, arthritis, inflammatory disease, cancer, renal disease, pancreatitis, Parkinson's disease, cataracts, diabetes, and respiratory organ disorders. Intake of dietary antioxidants, that is exogenous antioxidants, is essential, and a few compounds of this family, like vitamin E, β-carotene, and phenolic compounds, are exclusively synthesized by plants (Zhang and Tsao, 2016). Therefore, it is necessary to create an optimum balance between oxidants and antioxidants. Keep in mind that with age the balance tilts towards oxidants.

In foods, oxidation will be one amongst many causes of alterations resulting in rancidity, deterioration, and loss of nutritional, marketing, and organoleptic quality. For this reason, the food-processing industry is making efforts to prevent and slow down food deterioration with process modification and use of antioxidants. However, in keeping with studies administered in vivo over the last 20 years, free radicals and ROS are not only considered to be harmful factors but also messengers responsible for intracellular signaling. So, there has been a considerable shift in in the conception of these processes in normal as well as pathological conditions. The concept regarding the role of free radicals within the functioning of cells and organisms has been renewed with evolution of a new concept of redox equilibrium. Oxidative stress is then considered to be the effect of thiol redox reactions chiefly involved in signaling pathways. On these lines it can be reported that non-radical oxidants play a basic role within the oxidation of thiols for the sake of signaling, and the formation of free-radical intermediates may not be necessary.

11.4   ANTIOXIDATIVE PROPERTY OF ANTHOCYANINS

All living organisms rely on the reduction oxidation system to maintain normal healthy life. Free radicals are considered essential for maintaining the living state of the cells and organisms. Some free radicals like nitric oxide, superoxide ion, and related ROS and/or reactive nitrogen species (RNS) contribute to the signaling processes of cells (Droge, 2002; Jing, 2006). But there is the possibility of imbalance with respect to redox homeostasis. Oxidative stress is the situation representing an off-balance state where an excessive amount of ROS/RNS dominates endogenous antioxidant capability, resulting in oxidation of enzymes, proteins, DNA, and lipids. Oxidative stress is mainly responsible for the development of chronic degenerative diseases like coronary heart disease, cancer, and aging (Dai and Mumper, 2010).

Antioxidants are compounds that may delay, inhibit, or prevent the oxidation of oxidizable materials by quenching free radicals and reducing oxidative stress. The mode of action of antioxidants can be by: (1) preventing chain initiation by scavenging ROS/RNS; (2) reducing localized oxygen concentrations; (3) binding metal ions so that they do not generate reactive species; (4) chain breaking, i.e., scavenging intermediate radicals, like peroxyl and alkoxyl radicals, to stop continued hydrogen abstraction; and (5) degrading peroxides by changing them to non-radical components, like alcohol (Dai and Mumper, 2010; Miguel, 2010). The antioxidative properties of various plant extracts, food, and beverages have been assessed with the help of in vitro assays.

11.5 ANTIOXIDANT BIOASSAYS FOR ANTHOCYANINS

Antioxidant assays in foods and biological systems can be categorized into two groups: those based on the measurement of lipid peroxidation, and those based on measurement of free radical scavenging ability (Miguel, 2010). While assessing lipid peroxidation, many lipid substrates may be used and antioxidant activity in these systems may be detected by quantification of substrate and oxidant consumption or by measurements of the intermediates or final products formed. On the other hand methods based on the measurement of free radical scavenging ability can be further grouped into two categories depending on the type of chemical reaction involved: hydrogen or electron transfer. Additionally, analytical tests based on assessment of effectiveness against ROS/ RNS may be needed.

On the basis of the source and nature of anthocyanins, extraction becomes a prerequisite before bioassay. The process used for extraction may be an important step during bioassay of antioxidants. The major challenges faced during extraction can be low stability of anthocyanins and their tendency to stay bound to the matrix of the sample. From the literature it was found that the important factors affecting extraction efficiency are temperature, pH, solvent system used, solvent-to-solid ratio, and number of extractions. After extraction of anthocyanin in liquid form its antioxidant ability can be assessed using any suitable bioassay method. The antioxidant activity of anthocyanin can be explained by two mechanisms: hydrogen atom donor (HAT) and single-electron transfer (SET). In the HAT mechanism, the free radical R• removes a hydrogen atom from the antioxidant (AH^+), changing the free radical to a more stable form. In the case of the SET mechanism, the antioxidant (AH+) donates an electron to the free radical, reducing the oxidized intermediates into a stable form (Ali et al., 2016).

However, it is very difficult to distinguish between the mechanisms of HAT and SET. Most of the time the two mechanisms operate simultaneously and are dependent on factors such as structure of antioxidant, partition coefficient, solvent polarity, and solubility (Guo et al., 2016). The antioxidant ability of anthocyanins can be determined using different bioassays; a comparison of commonly used bioassays is given in Table 11.1.

Each of these methods gives information concerning the antioxidant activity of the anthocyanins under a specified set of conditions. In order to provide additional information concerning the reaction mechanism, an additional comprehensive description of the procedure followed in each of those bioassays is given below.

11.5.1 DIPHENYL-1-PICRYLHYDRAZYL (DPPH) ASSAY

DPPH assay is a spectrophotometric technique that may be applied to solid as well as liquid samples and it is not specific for any antioxidant. Thus, it may be applied to evaluate the total antioxidant capability of the sample. The basic principle of this assay is the ability of the DPPH* free radical to react with the hydrogen donor (AH^+). DPPH* is supposed to have maximum absorption within the UV-vis spectral region at 515 nm; this absorbance is decreased when the free radical is reduced. The antioxidant activity of anthocyanins for DPPH has been assessed by several authors (Jacob et al., 2012; Aguilera et al., 2016; Joseph et al., 2016; Masisi et al., 2016; Shahidi and Alasalvar, 2016). Reliable results may be obtained with this technique in a very quick and easy manner. Currently, the option of colorimeter is being explored where a spectrophotometer is not available so that application of this assay can be widened. What is more, because the free radical is stable, its production is not necessary each time that analysis is done. One of the major disadvantages of this assay is the complexity of analysis, which becomes further complicated when other compounds present with sample also have absorption at the specified wavelength.

TABLE 11.1

Comparison of Bioassays Used to Determine the Antioxidant Ability of Anthocyanins

Sr. No.	Bioassay Name	Method	Reaction	Detection Principle
1	Diphenyl-1-picrylhydrazyl (DPPH)	Spectrophotometric or colorimetric	DPPH* free radical reacts with hydrogen donor (AH⁺)	Decrease in absorbance at 515 nm
2	Oxygen radical absorbance capacity (ORAC)	Fluorescence spectroscopy	Sample (AH⁺) reacts with fluorescent compound, phycoerthrin (β-PE) with a generation of free radicals, 2,2′-azobis(2-amidino-propane) dihydrochloride (AAPH)	Decrease in fluorescence
3	Total peroxyl radical trapping antioxidant parameter (TRAP)	Chemiluminescence	Azo-radical initiator (e.g., 2AAPH, which produces radicals (R$^{\bullet}$) that react rapidly with O_2 to give peroxyl radicals (ROO$^{\bullet}$)	Decrease in luminescence
4	Ferric thiocyanate (FTC)	Spectrophotometric	Oxidation of ferrous chloride to ferric ion by reaction with peroxide, which then combines with ammonia thiocyanate, forming red-colored ferric thiocyanate	Increase in absorbance at 500 nm
5	Ferric reducing antioxidant power (FRAP)	Colorimetric	Reduction of complexes of 2,4,6-tripyridyl-s-triazine (TPTZ) with ferric chloride hexahydrate ($FeCl_3.6H_2O$) in acidic conditions	Increase in absorbance at 593 nm
6	Cupric ion reducing antioxidant capacity (CUPRAC)	Spectrophotometric	Reaction with the CUPRAC reagent (cupric neocuproine) leading to formation of a chromophore Cu(I)-neocuproine	Increase in absorbance at 450 nm
7	2,2′-Azinobis (3-ehtylbenzothiazoline-6-sulfonic acid) (ABTS) diamonium salt	Colorimetric	Production of a stable radical, blue/green ABTS chromophore, by the reaction of ABTS with potassium persulfate	Decrease in absorbance at 415 nm

11.5.2 OXYGEN RADICAL ABSORBANCE CAPACITY (ORAC) ASSAY

ORAC assay is a type of fluorescence technique based on the principle of combination of the sample (AH⁺) with fluorescent compound, protein phycoerthrin (β-PE) and with a generation of free radicals, 2,2′-azobis(2-amidino-propane) dihydrochloride (AAPH). The mechanism of this assay involves loss of the fluorescent compound once it has been oxidized by the generated free radicals. When antioxidants (AH⁺) are involved in the reaction, the fluorescent compound is protected against oxidative degradation. The fluorescence signal is monitored for 1 h at λ em = 565 nm and λ ex = 540 nm respectively. The antioxidant activity of anthocyanins has been assessed by ORAC assay by several authors (Wang et al. 1996; Cao et al. 1998; Aguilera et al. 2016). This assay is considered to be suitable for determining the antioxidative capability of hydrophilic and hydrophobic samples, because it is adaptable to different samples changing the generator of free radicals. One drawback of this assay is the non-specificity of fluorescent compounds reacting with the sample, losing fluorescence when the free radical generator is not added.

11.5.3 TOTAL PEROXYL RADICAL TRAPPING ANTIOXIDANT PARAMETER (TRAP) ASSAY

TRAP assay is a chemiluminescence technique that involves the following components: (1) thermo-labile azo-radical initiator (e.g., AAPH, which produces radicals (R$^{\bullet}$) that react rapidly with O_2

to give peroxyl radicals (ROO$^{\bullet}$); (2) oxidizable compounds with chemiluminescence properties to monitor the reaction progress (e.g., luminol); and (3) the sample with the antioxidant properties (AH^{+}).

This assay can be used for determination of almost all known chain-breaking antioxidants. However, a major disadvantage is that it is extremely difficult to compare results from different laboratories. This issue is mainly attributed to a range of end points considered by different laboratories. The method is comparatively more complex, long, and needs a high degree of expertise.

11.5.4   FERRIC THIOCYANATE (FTC) ASSAY

The FTC assay is a spectrophotometric methodology that involves oxidation of ferrous chloride to ferric ion by reaction with peroxide. The peroxides are generated when the sample is mixed with ethanol, water, phosphate solution, and linoleic acid. Then, acidic solution of ferrous chloride in hydrochloric acid is added to the reaction, yielding ferric ions. The ferric ions then combine with ammonia thiocyanate to form red-coloured ferric thiocyanate. Absorbance of the sample is measured at 500 nm until the maximum value is reached. This assay is suitable for measuring the quantity of peroxide produced during the initial stage of oxidation. The methodology can be employed to determine the antioxidative ability of anthocyanin without peroxidation of polyunsaturated fatty acid. It is easy and reproducible. However results are unreliable when compounds in the sample absorb around 500 nm.

11.5.5   FERRIC REDUCING ANTIOXIDANT POWER (FRAP) ASSAY

FRAP assay is a colorimetric method that can be used for the determination of the total antioxidant activity of anthocyanins. The principle involved in this method is reduction of complexes of 2,4,6-tripyridyl-s-triazine (TPTZ) with ferric chloride hexahydrate (FeCl$_3$.6H$_2$O) in acidic conditions. After reduction the solution turns slightly brownish, forming ferrous complexes. Absorbance is measured at 593 nm against the blank. This method has been widely used for determining the antioxidant activity of anthocyanins in commodities like elderberry *Hibiscus sabdariffa* extract; raspberries, blackberries, redcurrants, gooseberries, and carnelian cherries; or in carrots, cabbage, cauliflower, potatoes, onions, asparagus, and eggplant. Fast and reproducible results are obtained with this method; however, it does have limitations. Firstly the sample should be aqueous and secondly, some compounds without antioxidative properties can reduce Fe^{3+} to Fe^{2+}, resulting in over-estimation of the antioxidant activity of the sample.

11.5.6   CUPRIC ION-REDUCING ANTIOXIDANT CAPACITY (CUPRAC) ASSAY

CUPRAC assay is a spectrophotometric method similar to FRAP. The method relies on the reaction of anthocyanins with the CUPRAC reagent (cupric neocuproine) leading to formation of a chromophore Cu(I)-neocuproine that has maximum absorbtion at 450 nm. The main advantages of this method are related to the desirable characteristics of the CUPRAC reagent, such as its availability and easy accessibility, rapidity, stability, lower cost, sensitivity towards thiol-type antioxidants, unlike FRAP, and responsiveness to both hydrophilic and lipophilic antioxidants.

11.5.7   2,2'-AZINOBIS (3-ETHYLBENZOTHIAZOLINE-6-SULFONIC ACID) (ABTS) DIAMMONIUM SALT ASSAY

The most recent ABTS assay method is based on decolorization techniques. It consists of the production of a stable radical, blue/green ABTS chromophore, by the reaction of ABTS with potassium persulfate. This stable radical has a maximum absorption at 415 nm. The reaction between radical and antioxidant results in a decrease in absorbance of this compound. Antioxidant activity is calculated relative to the reactivity of Trolox standard under similar conditions. This assay is normally used in combination with DPPH assay for the determination of the antioxidant activity of anthocyanins. The method is simple and require small sample volume. However, a standard solution is required to obtain accurate results.

REFERENCES

Aguilera, Y., Martin-Cabrejas, M. A., & de Mejia, E. G. Phenolic compounds in fruits and beverages consumed as part of the Mediterranean diet: Their role in prevention of chronic diseases. Phytochem. Rev. 2016;15(3):405–423.

Ali, H. M., Almagribi, W., & Al-Rashidi, M. N. Antiradical and reductant activities of anthocyanidins and anthocyanins, structure–activity relationship and synthesis. Food Chem. 2016;194:1275–1282.

Aura, A.-M. In vitro Digestion Models for Dietary Phenolic Compounds. PhD thesis. 2005. Espoo, Finland. pp. 63–67.

Borges, G., Roowi, S., Rouanet, J.-M., Duthie, G. G., Lean, M. E. J., & Crozier A. The bioavailability of raspberry anthocyanins and ellagitannins in rats. Mol. Nutr. Food Res. 2007;51:714–725.

Cao, G., Booth, S. L., Sadowski, J. A., & Prior, R. L. Increases in human plasma antioxidant capacity following consumption of controlled diets high in fruits and vegetables. Am. J. Clin. Nutr. 1998;68(5):1081–1087.

Dai, J., & Mumper, R. J. Plant phenolics: extraction, analysis and their antioxidant and anticancer properties. Molecules 2010;15:73137352.

Dröge, W. Free radicals in the physiological control of cell function. Physiol. Rev. 2002;82:47–95.

Espín, J. C., García-Conesa, M. T., & Tomás-Barberán, F. A. Nutraceuticals: Facts and fiction. Phytochemistry 2007;68:2986–3008.

Forester, S. C., & Waterhouse, A. L. Identification of Cabernet Sauvignon anthocyanin gut microflora metabolites. J. Agric. Food Chem. 2008;56:9299–9304.

Fraga, C. G. (ed.) Plant Phenolics and Human Health: Biochemistry, Nutrition and Pharmacology. New York: Wiley, 2010.

Galvano F., Fauci L., Vitaglione P., Fogliano V., Vanella L., & Felgines C. Bioavailability, antioxidant and biological properties of the natural free-radical scavengers cyanidin and related glycosides. Ann. Ist. Super. Sanità. 2007; 43: 382–393.

Guo, X., Yang, B., Tan, J., Jiang, J., & Li, D. Association of dietary intakes of anthocyanins and berry fruits with risk of type 2 diabetes mellitus: A systematic review and meta-analysis of prospective cohort studies. Eur. J. Clin. Nutr. 2016;70(12):1360–1367.

Harborne, J. B. Phenolic Compounds in Phytochemical Methods – A Guide to Modern Techniques of Plant Analysis. Third edition. New York: Chapman & Hall, 1998; pp. 66–74.

Harborne, J. B, & Williams, C. A. Anthocyanins and other flavonoids. Nat. Prod. Rep. 2001;18(3):310–333.

He, J., & Giusti, M. M. Anthocyanins: Natural colorants with health-promoting properties. Annu. Rev. Food Sci. Technol. 2010;1(1):163–187.

Jacob, J. K., Tiwari, K., Correa-Betanzo, J., Misran, A., Chandrasekaran, R., & Paliyath, G. Biochemical basis for functional ingredient design from fruits. Annu. Rev. Food Sci.Technol. 2012;3:79–104.

Jaganath, I. B., & Crozier, A. Dietary flavonoids and phenolic compounds. In Fraga, C. G. (ed.), Plant Phenolics and Human Health: Biochemistry, Nutrition, and Pharmacology. Hoboken, NJ: John Wiley ; 2010.

Jing, P. Purple Corn Anthocyanins: Chemical Structure, Chemopreventive Activity and Structure / Function Relationships. PhD thesis, 2006. The Ohio State University, USA; pp. 5–90.

Joseph, S. V., Edirisinghe, I., & Burton-Freeman, B. M. Fruit polyphenols: A review of antiinflammatory effects in humans. Crit. Rev. Food Sci. Nutr. 2016;56(3):419–444.

Kay, C. C., Mazza, G., Holub, B. J., & Wang, J. Anthocyanin metabolites in human urine and serum. Br. J. Nutr. 2004;91:933–942.

Martín Bueno, J., Ramos-Escudero, F., Sáez-Plaza, P., Muñoz, A. M., Navas, M. J., & Asuero, A. G. Analysis and antioxidant capacity of anthocyanin pigments. Part I: General considerations concerning polyphenols and flavonoids. Crit. Rev. Anal. Chem. 2012a;42(2):102–125.

Martín Bueno, J., Sáez-Plaza, P., Ramos-Escudero, F., Jímenez, A. M., Fett, R., & Asuero, A. G. Analysis and antioxidant capacity of anthocyanin pigments. Part II: Chemical structure, color, and intake of anthocyanins. Crit. Rev. Anal. Chem. 2012b;42(2):126–151.

Martins, N., Barros, L., & Ferreira, I. C. F. R. In vivo antioxidant activity of phenolic compounds: Facts and gaps. Trends Food Sci. Technol. 2016;28:1–12.

Masisi, K., Beta, T., & Moghadasian, M. H. Antioxidant properties of diverse grain cereals: A review on in vitro and in vivo studies. Food Chem. 2016;196:90–97.

Mazza, G., Kay, C. D., Cottrell, T., et al. Absorption of anthocyanins from blueberries and serum antioxidant status in human subjects. J. Agric. Food Chem. 2002;50:7731–7737.

McDougall, G. J., Dobson, P., Smith, P., Blake, A., & Stewart, D. Assessing potential bioavailability of rasp-berry anthocyanins using an in vitro digestion system. J. Agric. Food Chem. 2005;53:5896–5904.

McGhie, T. K., & Walton, M. C. The bioavailability and absorption of anthocyanins: Towards a better understanding. Mol. Nutr. Food Res. 2007;51 :702–713.

Miguel, M. G. Antioxidant activity of medicinal and aromatic plants. A review. Flavour Fragr. J. 2010;25:291–312.

Milbury, P. E., Vita, J. A., & Blumbery, J. B. Anthocyanins are bioavailable in humans following an acute dose of cranberry juice. J. Nutr. 2010;140:1099–1104.

Passamonti, S., Vrhovsek, U., Vanzo, A., & Mattivi F. The stomach as a site for anthocyanins absorption from food. FEBS Lett. 2003;544:210–213.

Prior, R. L., & Wu, X. Anthocyanins: structural characteristics that result in unique metabolic patterns and bio-logical activities. Free Rad. Res. 2006;40(10):1014–1028.

Ramirez-Tortosa, C., Andersen, O. M., Gardner, P. T., Morrice, P. C., Wood, S. G., Duthie, S. J., Collins, A. R., & Duthie, G. G. Anthocyanin-rich extract decreases indices of lipid peroxidation and DNA damage in vitamin E-depleted rats. Free Radical Biol. Med. 2001;31:1033–1037.

Rechner, A. R., & Kroner, C. Anthocyanins and colonic metabolites of dietary polyphenols inhibit platelet function. Thromb. Res. 2005;116:327–334.

Shahidi, F., & Alasalvar, C. Handbook of Functional Beverages and Human Health. Boca Raton, FL: CRC Press, 2016.

Talavéra, S., Felgines, C., Texier, O., Besson, C., Manach, C., Lamaison, J.-L., & Rémésy C. Anthocyanins are efficiently absorbed from the small intestine in rats. J. Nutr. 2004;134:2275–2279.

Tirzitis, G., & Bartosz, G. Determination of antiradical and antioxidant activity: Basic principles and new insights. Acta Biochim. Pol. 2010;57(1):139–142.

Tsuda, T., Horio, F., Uchida, K., Aoki, H., & Osawa, T. Dietary cyanidin 3-O-beta-D-glucoside-rich purple corn color prevents obesity and ameliorates hyperglycemia in mice. J. Nutr. 2003;133:2125–2130.

Veitch, N. C., & Grayer, R. J. Flavonoids and their glycosides including anthocyanins. Nat. Prod. Rep. 2008;25(3):555–611.

Vitaglione, P., Donnarumma, G., Napolitano, A., Galvano, F., Gallo, A., Scalfi, L., & Fogliano, V. Protocatechuic acid is the major human metabolite of cyanidin-glucosides. J. Nutr. 2007;137:2043–2048.

Vogiatzoglou, A., Mulligan, A. A., Lentjes, M. A. H., Luben, R. N., Spencer, J. P. E., Schroeter, H., et al. Flavonoid intake in European adults (18 to 64 years). PLoS ONE 2015;10(5):e0128132. doi:10.1371/journal.pone.0128132

Wang, L.-S., & Stoner, G. D. Anthocyanins and their role in cancer prevention. Cancer Lett. 2008;269:281–290.

Wang, H., Cao, G., & Prior, R. L. Total antioxidant capacity of fruits. J. Agric. Food Chem. 1996;44(3):701–705.

Zhang, H., & Tsao, R. Dietary polyphenols, oxidative stress and antioxidant and anti-inflammatory effects. Curr. Opin. Food Sci. 2016;8:33–42.

Ziyatdinova, G. K., & Budnikov, H. C. Natural phenolic antioxidants in bioanalytical chemistry: State of the art and prospects of development. Russ. Chem. Rev. 2015;84(2):194–224.

12 Bioavailability Efficacy of Anthocyanins from Various Subtropical Fruits

Rohini Dhenge, Asmita Acharya, Shraddha Bhat, and Ajay Chinchkar

CONTENTS

12.1 INTRODUCTION

A mature fruit is a "natural feast" that is full of nutritional food and natural health-protecting formula. Various ancient documents mention the existence of this natural meal for mankind's nutritional security, as well as the vague narrative of human civilization's origin. Fruits have always been a source of fascination for royal families, but scientific research on them began only in the nineteenth century. In recent years, Bose (2001) has documented the nutritional composition and usage of all known edible fruits. In some situations, the therapeutic value and health well-fare of certain fruits have been demonstrated (Block et al., 1992; Heber & Bowerman, 2001; Liu, 2003), while in others, it has been postulated or is under study. Tropical, subtropical, and temperate fruits are the three main classifications of fruits based on their climatic requirements for production. Since the mid-1990s, subtropical fruits have grown in importance in worldwide trade.

Especially, subtropical fruit is the most important food in human nutrition since it is an ample natural supply of key nutritional components and calories. It provides appropriate quantities and quality of essential vitamins, fatty acids, minerals, and dietary fiber. When compared to sources of food or other types, fruit is special in a healthy diet because it is cholesterol-free. Another important factor is the amount of high-quality fiber in fruit. It helps battle obesity by preventing constipation, lowering cholesterol levels, improving bowel movement, and rejuvenating the digestive system.

Fruits are also said to provide mental health benefits. Fruit decreases sadness and anxiety, boosts focus, and minimizes violent behavior when consumed on a regular basis.

Subtropical fruits are high in natural bioactive molecules known as phytochemicals and antioxidants, which can halt oxidative chain reactions by eliminating free radical intermediates and even being oxidized. According to a study, phytochemicals, when paired with nutrients found in fruits, can help delay the aging process, prevent cell damage, increase immunity, and protect against a variety of chronic diseases, including cardiovascular disease and cancer. However, the makeup of each fruit varies (Mitra et al., 2011).

Over the years, there has been a steady shift in interest from cultivating subtropical fruits primarily for nostalgic reasons or as horticultural oddities to investigating their potential as commercially viable crops which are now a familiar part of all over the world. It is the most recent of many subtropical fruits that have been exploited to acquire a high level of economic relevance, such as citrus fruits, banana, papaya, litchi, tree tomato, guava, plum, strawberry, grapes, pear, avocado, dates, figs, kiwi, olive, pomegranate. A couple of these fruits are produced in tropical climates as well. In order to have any chance of developing into a commercial crop, fruit must meet certain requirements:

1. It must be well adapted to the area in which one proposes to grow it.
2. It must have ready consumer acceptance.
3. It must lend itself to marketing as a high-quality product, either fresh or in processed form.
4. It must be producible with reasonable profit to the grower in competition with other domestic and foreign products.

The fact that a subtropical fruit plant species is currently consigned to ornamental status or small-scale cultivation does not rule it out as a possible crop plant. India, the Philippines, China, Indonesia, Bangladesh, Thailand, Brazil, Pakistan, Colombia, and Mexico are the top ten producers of subtropical fruits (FAO-FORCAST, 2020–2025). Table 12.1 gives details of the common and scientific names of subtropical fruit.

Phytochemicals and/or bioactive compounds can take many different forms, and some subtropical fruits have higher concentrations of one bioactive ingredient than others. Phenolic compounds, carotenoids, phytosterols, dietary fiber (soluble and insoluble), and fructo-oligosaccharide are only a few of the thousands of bioactive components present in subtropical fruits. Typically, these compound categories do not appear in a single fruit piecemeal, but rather in a mixture. Bioactive chemicals found in fruits provide health benefits in addition to basic nourishment (Kaur and Kapoor, 2001). Fruit consumption is connected to a lower risk of oxidative stress-related illnesses (Haminiuk et al., 2012).

The main category is phenolic compounds which occur naturally in fruits, but this category comprises numerous other subcompounds such as polyphenols. Polyphenols are chemical compounds found in plants that have more than one benzene ring, a variable number of hydroxyls, carbonyls, and carboxylic acid groups, and one or more sugar moieties. Polyphenol concentration varies greatly across and among fruits of the same species in subtropical fruits, and it is well recognized that they are good sources of polyphenols. Flavonoids are the most frequent type of polyphenol. Flavonoids are classified as catechins, anthocyanins, chalcones, flavonols, flavonones, flavanols, isoflavones, and flavanes, based on structural characteristics. Anthocyanins are a type of flavonoid commonly found in tropical fruits.

12.2 ANTHOCYANIN

Bioactive substances having antioxidant capabilities such as anthocyanins, are glycosylated anthocyanidins, called flavonoids. They are water-soluble and are responsible for the orange, red,

TABLE 12.1
The Scientific and Common Names of Subtropical Fruits

Common Name	Scientific Name	Regions Where the Commodities Are Produced
Avocado	*Persea americana*	Central America, Caribbean
Banana	*Musa paradisiaca* var. *sapientum*	Eastern Africa, Middle Africa, Central and South America, Caribbean, south and south-eastern Asia, Melanesia, Micronesia, Polynesia
Citrus fruits		
Lemon	*Citrus limon*	Central America
Lime	*Citrus aurantifolia*	Central America
Orange	*Citrus sinensis*	Northern and southern Africa, North, central, and south America, Caribbean, Southern Europe
Date	*Phoenix dactylifera*	Northern Africa, Western Asia
Fig	*Ficus carica*	Turkey, Egypt, Algeria, Iran, and Morocco
Guava	*Psidium guajava*	Central America, Caribbean, south and south-eastern Asia, Polynesia
Kiwi	*Actinidia deliciosa*	Australia and New Zealand
Lychee (litchi)	*Litchi chinensis* or *Nephelium litchi*	China, India, Southern Asia, Madagascar, South Africa, Taiwan, Thailand, Vietnam
Olive	*Olea europea*	Northern Africa, western Asia, Southern Europe
Papaya	*Carica papaya*	Asia and South America, including Brazil and Mexico, India, Indonesia, Africa
Pomegranate	*Punica granatum*	India, Afghanistan, China, Greece, Iran, Turkey, Spain, Tunisia, Morocco, Japan, France, Armenia, Cyprus, Egypt, Italy
Tamarillo (tree tomato)	*Cyphomandra betacea*	Peru, Colombia, Australia, and the United States, New Zealand, Ecuador, Rwanda

Source: FAO (2020); ISHS (2020).

and blue colors found in many fruits and vegetables. Cyanidin, pelargonidin, delphinidin, peonidin, petunidin, and malvidin are the only six anthocyanidins known to occur commonly in plants (Wu and Prior, 2005a, b).

Anthocyanins are water-soluble colored pigment compounds that give fruits, vegetables, and ornamental crops their red, blue, and purple colors. Plants have around 200 distinct anthocyanins. Around 70 of these have been found in fruits. Fruits, such as pink and green guavas, can be distinguished by their composition and number of anthocyanins (Mazza and Miniati, 2018). Bright red or purple pigments are seen in many subtropical fruits, indicating the presence of anthocyanins, while reports of precise anthocyanin concentration in tropical fruits are scarce (De Brito et al., 2007). The color of anthocyanin-containing media depends on the structure and concentration of the pigment, pH, temperature, light, co-pigments, enzymes, oxygen, metallic ions, sulfur dioxide, and sugar. Color changes in the fruit are caused by the breakdown of chlorophyll and the creation of anthocyanins and carotenoids, albeit few of these pigments may be synthesized in the green tissues and they are "hidden" by the chlorophyll until its degradation during ripening. Peroxidase has a variety of functions in the ripening and post-ripening processes, according to Miesle et al. (1991), including alterations in cell wall flexibility and anthocyanin degradation.

Fruits contain anthocyanins in large quantities. The anthocyanin profile has been shown to be influenced by environmental factors like light and temperature (Giusti et al., 1998). Pomegranate juice contains between 161.9 and 387.4 mg/L of total anthocyanins, making it a high-anthocyanin fruit. Gil et al. (2000) found that cyanidin-3-glucoside was the most common anthocyanin in pomegranate juice. In frozen litchi fruit, the anthocyanin concentration was found to be 5.21×10^2 unit/

g in control samples and 7.8×10^2 unit/g in samples treated with 1% hydrochloric acid by Jiang et al. (2004). Strawberry cultivars from various subtropical regions exhibited lower acidity than the acidity values described in the literature, according to Curi et al. (2016). After freezing two raspberry cultivars for 1 year, De Ancos et al. (2000) found a 5–17% increase in total anthocyanins in two cultivars.

12.3   EFFECT OF PROCESSING

During processing and subsequent storage, the stability of anthocyanins is affected by the food's content, as well as temperature, light, and organic compounds like ascorbic acid, and product contamination with metal ions (Markakis, 1982; Macheix et al., 2018). In the presence of oxygen, anthocyanins break down into brown-colored compounds. The temperature and duration of heat treatment affect anthocyanin degradation. The preservation of anthocyanins is improved by using high-temperature, short-time procedures. Thermal treatments that use high temperatures for a long time, such as sterilization and concentration, on the other hand, promote anthocyanin destruction (Mazza and Miniati, 2018). High-pressure processing has no effect on pigments that contribute to fruit color (chlorophyll, carotenoids, anthocyanins, and so on) (Ahmed and Ramaswamy, 2006). Due to the existence of bioactive substances with antioxidant qualities, fruit consumption has been linked to a lower incidence of chronic degenerative disorders (van't Veer et al., 2000). As a result of the interest in these compounds, as well as the seasonal nature of fruit, study on their stability during processing and storage has sprung up (Garcia-Perez et al., 2010).

12.4   SUBTROPICAL FRUITS

12.4.1   Guava

Guava is said to have originated in a region stretching from southern Mexico through Central America, though this is debatable. It is now grown in most tropical and subtropical locations of the world, as its perfect temperature and humidity for fruiting are 25–30°C and 50–80% humidity, respectively (GonzagaNeto and Soares, 1995).

Dalla Nora et al. (2014) analyzed that the anthocyanin malvidin-3-glucoside accounted for the majority of the total anthocyanin content in fresh guabiju samples, samples frozen for 30 days, and samples dried by hot air, accounting for around 60%, 52%, and 57% of total in the fruit, respectively. As documented by Liaqat et al. (2011) in raspberries, Basiouny (1995) in blackberries, and Lo Piero et al. (2005) in red oranges, the elevation in guabiju anthocyanins during freezing could be ascribed to continuous production in the fruit during storage or increased availability of free anthocyanins. In contrast Dalla Nora et al. (2014) observed that red guava contains three anthocyanins: cyanidin chloride, malvidin-3-glycoside, and cyanidin-3-glycoside, with a total of 685.8 g/g fruit (dry-weight basis). Only the red-skin cultivar extract was found to contain anthocyanins (Biegelmeyer et al., 2011). With respect to anthocyanins, the anthocyanin content steadily reduced during the storage time among the blending of jamun and guava juice (Sridhar et al., 2017).

12.4.2   Lychee

Lychee (*Litchi chinensis* Sonn.), generally known as litchi, is a tropical to subtropical fruit native to southern China and northern Vietnam. It is a member of the Sapindaceae family (Huang et al., 2005). Temperature and humidity have a substantial impact on lychee flower bud differentiation, flowering, fruit set, fruit quality, and flavor development (Dabral and Misra, 2007); however, the plant has evolved well in the subtropics, where summers are hot and humid and winters are dry and cool.

The existence of anthocyanins encircling the succulent edible aril and a single seed in the middle causes the fruit to have a tough, indehiscent crimson pericarp (skin) (Fuchs et al., 1993; Huang, 1995). The presence of anthocyanins is the reason why lychee skin exhibits a bright red color. With decreased chlorophyll concentration and rising anthocyanin production, the pericarp color of lychee fruit interchanges from green to reddish pink as it matures (Huang, 1995; Wang et al., 2005). The presence of anthocyanins in pericarp tissues, primarily cyanindin-3-glucoside, pelargonidin-3-glucoside, cyanindin-3-galactoside and pelargonidin-3,5-diglucozide, gives lychee fruit its bright red color (Zhang et al., 2000, 2004). Polyphenol oxidase oxidizes polyphenols and degrades anthocyanins, causing skin browning (Jiang, 2000; Zhang et al., 2001; Liu et al., 2010).The primary anthocyanin discovered was cyanidin-3-rutinoside. Lee and Wicker (1991) also discovered cyanidin-3-glucoside and malvidin-3-acetylglucoside.

Anthocyanins and phenols in lychee fruit skin are thought to degrade with time, and browning is caused by the development of polymeric brown pigments (Jiang et al., 2006b). The amounts of key anthocyanin chemicals in lychee fruit pericarp tissues decreased during storage, according to Zhang et al. (2000, 2001). Underhill and Critchley. (1994) reported that anthocyanin pigments decolorized over time in ambient storage, presumably due to pericarp pH alterations. Other authors reported that the action of polyphenol oxidase has been implicated in the degradation of anthocyanin pigments in lychee via condensation with quinones generated from endogenous phenolics (Wesche-Ebeling and Montgomery, 1990). Polyphenol oxidase and peroxidase were found to be involved in anthocyanin degradation and pericarp browning of lychee fruit by anthocyanase (Zhang et al., 2003).

Anthocyanin concentration, polyphenol oxidase action, pH value, and membrane permeability were studied in relation to post-harvest browning of litchi fruit affected by water loss. With pericarp desiccation, total anthocyanin concentrations dropped. Fruit held at 90% relative humidity (RH) lost the least total anthocyanins over time, whereas total anthocyanin content of fruit stored at 60 and 70% RH significantly reduced (Jiang and Fu, 1999). Anthocyanin was destroyed by some food additives, such as metallic Cu^{2+} and vitamin C, but not by Mg^{2+}, Zn^{2+}, Fe^{2+}, benzoic acid, hexadienic acid, and β-cyclodextrin.

12.4.3 PLUM

Plum (*Prunus domestica*), an attractively colored fruit, is consumed both fresh and processed. Some of the most frequent processed goods include plum purée, paste, sauce, juice concentrate, and prunes. In developed countries, 50% of the produce is used for processing, whereas in developing countries, commercial use of plums is minimal (Chang et al., 1994).

Anthocyanidins, a class of glycosidic pigments, are responsible for plum's red color (Shrikhande and Francis, 1976). The main source of cyanidin and peonidin derivatives is plum. Cyanidin-3-O-rutinoside and cyanidin-3-O-glucoside are the two main anthocyanins found in plums, although the corresponding peonidin derivatives are only found in trace amounts. Additionally reported components include cyanidin galactoside, xyloside, malonyl-glycoside, and (6"-acetoyl) glycoside. The majority of cultivars had the same anthocyanins, but there was significant variance in their abundance. Quantitative changes have been tracked. Interesting anthocyanin profiles were found in red and black plums (Wu & Prior, 2005). The predominant polyphenols found in plums are anthocyanins such as cyanidin-3-glucoside and cyanidin-3-rutinoside (primarily rutinoside derivatives) (Kim et al., 2003).

12.4.4 PASSION FRUIT

Passiflora is the largest genus in the Passifloraceae family, with approximately 400 species found in tropical and subtropical America, Asia, and Africa. About 50–60 of these species produce edible fruits, but only a few are commercially important, with many only available in local markets in South and Central America and the Caribbean. Passion fruit is commercially produced from the purple species *Passiflora edulis* Sims and the yellow variation *Passiflora edulis* f. *flavicarpa* Degener. The purple hue of *P. edulis* rind was discovered to be due to cyanidin-3-glucoside, which made up about 97% of the total anthocyanin content (Kidoy et al., 1997). There were also trace levels of cyanidin-3-(6-malonylglucoside) (2%) and pelargonidin-3-glucoside (1%).

12.4.5 DATE

Date palms (*Phoenix dactylifera* L.) grow in hot, arid climates around the world, and their fruit is sold as a high-value sweet fruit crop all over the world. Dates are an important subsistence crop in most desert places across the world (Al-Shahib and Marshall, 2003). Carotenoids, anthocyanins, flavones, flavonoles, lycopene, flavoxanthin, and lutein are among the pigments found in fresh Egyptian dates. (Barreveld, 1993,).

12.4.6 FIG

The fig tree (*Ficus carica* L.) is thought to have originated in western Asia and has been spread throughout the Mediterranean by humans, with traces discovered in excavations dating back to at least 5000 BC (Morton, 1987). In comparison to fig types with lighter skin, dark-skinned fig cultivars have higher quantities of polyphenols, anthocyanins, and flavonoids, as well as stronger antioxidant activity (Solomon et al., 2006). Rutin and cyanidin-3-rutinoside have been identified as the predominant flavonol and anthocyanin in both types of commercial figs (Sarilop and Bursa siyahi), which come in various colors (yellow and purple), respectively (Kamiloglu & Capanoglu, 2015). In the anthocyanin family, cyanidin-3-glucoside and cyanidin-3-rutinoside were discovered. When compared to the pulp, the skin of purple figs was the tissue that contributed the most anthocyanins, with levels of cyanidin-3-glucoside and cyanidin-3-rutinoside that were eight and ten times greater, respectively. However, despite the skin of the yellow fig being light in color, cyanidin-3-glucoside was not found there. According to Dueñas et al. (2008), figs also contain the anthocyanins cyanidin-3,5-diglucoside, pelargonidin-3-glucoside, and peonidin-3-glucoside.

12.4.7 POMEGRANATE

The pomegranate (*Punica granatum* L.) is an ancient subtropical fruit that has become increasingly popular in recent years. In contrast to prior records claiming that the pomegranate was native to Iran and/or northern India (Morton, 1987), the pomegranate most likely originated in northern Turkey, as evidenced by the presence of pomegranates near the late-14th-century BCE Uluburun shipwreck near Kas, Turkey.

Delphinidin-3-glucosides, delphinidin-3–5-glucosides, cyanidin-3-glucosides, cyanidin-3–5-glucosides, pelargonidin-3-glucosides, pelargonidin-3–5-glucosides, and cyanidin 3-arabinose are anthocyanins found in pomegranate juice, and their concentrations vary. Cold storage has been shown to increase anthocyanin concentrations (Gil et al., 1995), but not controlled atmospheric storage (Holcroft et al., 1998).

12.4.8 DRAGON FRUIT

In tropical and subtropical regions, dragon fruit and durian are grown in a variety of countries. The pitaya, or dragon fruit, is a tropical/subtropical fruit endemic to Mexico and Central/South America (*Hylocereusundatus* [Haw]) (Haber, 1983; Wichienchot et al., 2010). Three anthocyanins present in the skin of the dragon fruit were partially identified in the extracts by high-performance liquid chromatography; they were: cyanidin 3-O-glucoside, cyanidin 3,5-O-glucoside, and pelargonidin 3,5-O-glucoside (Vargas et al., 2013). According to E. A Saati (2009), dragon fruit rind that could be extracted with water as the solvent contains 1.1 mg/100 mL anthocyanin.

12.5 BIOACCESSIBILITY AND BIOAVAILABILITY OF ANTHOCYANINS FROM FRUITS

Despite their positive features, anthocyanins' success in preventing or treating a variety of diseases depends on their bioavailability (Faria et al., 2009). The consumption of anthocyanin-rich foods is related to reduced risk of developing cardiovascular disease and cancer (Braga et al., 2018). As a result, there are certain spaces in the literature addressing anthocyanins' mode of action, which has limited their use. Only a small percentage of these pigments are absorbed by humans, according to many studies.

Bioaccessibility and bioavailability research could provide answers to this dilemma, but the analytical methodologies remain a major challenge. Temperature, light, solvents, chemical structure, and pH changes are all known to degrade anthocyanin pigments (Welch et al., 2008). The human digestion operation is made up of multiple phases, each of which causes a significant pH change. Several writers have overlooked some of these details, confounding our knowledge of the overall anthocyanin pathway. The study of anthocyanin bioavailability has been made easier because of new digestion models produced by science and technology. Anthocyanins exist in the form of glycosides in foods. These chemicals go through the gastrointestinal tract after being consumed. The anthocyanin fraction not absorbed by the stomach is absorbed by the small intestine. The carbinol pseudobase is anticipated to predominate once anthocyanins reach more basic conditions in the small intestine. Unlike flavonoids, anthocyanin glycosides are absorbed quickly and effectively in the small intestine (Talavera et al., 2004). Only a small percentage of these pigments are absorbed by humans, according to many studies. Are anthocyanins, then, to blame for these positive outcomes?

12.6 CONCLUSION

The purpose of this chapter was to provide extensive information on anthocyanin bioaccessibility and bioavailability in order to comprehend them better. As a result, this chapter also covers the most widely used methodologies and demonstrates the critical function of gut microbiota in anthocyanin metabolism, which should not be overlooked when designing a standardized approach to assess anthocyanin bioavailability. The literature study was evaluated using two main methods. To begin, the following keywords were used to search the ISI Web of Knowledge and Medline databases: anthocyanins, anthocyanidins, bioavailability, bioaccessibility, saliva, absorption, stability, metabolism, gut, faeces, micro flora, plasma, blood, urine, gastrointestinal, digestion, "tissue uptake," "intestinal contents," Caco-2, cell, glycosylation, and acylation. A directed manual search was performed in a second technique by inspecting the references.

Anthocyanin bioavailability can be investigated using in vivo or ex vivo techniques, as well as in vitro studies that imitate human behavior. According to a review of the literature, studies on the bioaccessibility and bioavailability of anthocyanins were originally carried out using in vivo and ex vivo methodologies, assessing blood and urine anthocyanin concentrations after consuming a high-anthocyanin meal (Bub et al., 2001; Mullen et al., 2006). Table 12.2 summarizes the bioaccessibility and bioavailability of anthocyanins from fruits. It is only feasible to assume anthocyanin components from fruits and vegetables. Although anthocyanin concentration varies widely among fruits, it is often significantly higher in fruits than in vegetables. Figure 12.1 represents the lowest and highest anthocyanin content per 100 g of fresh weight.

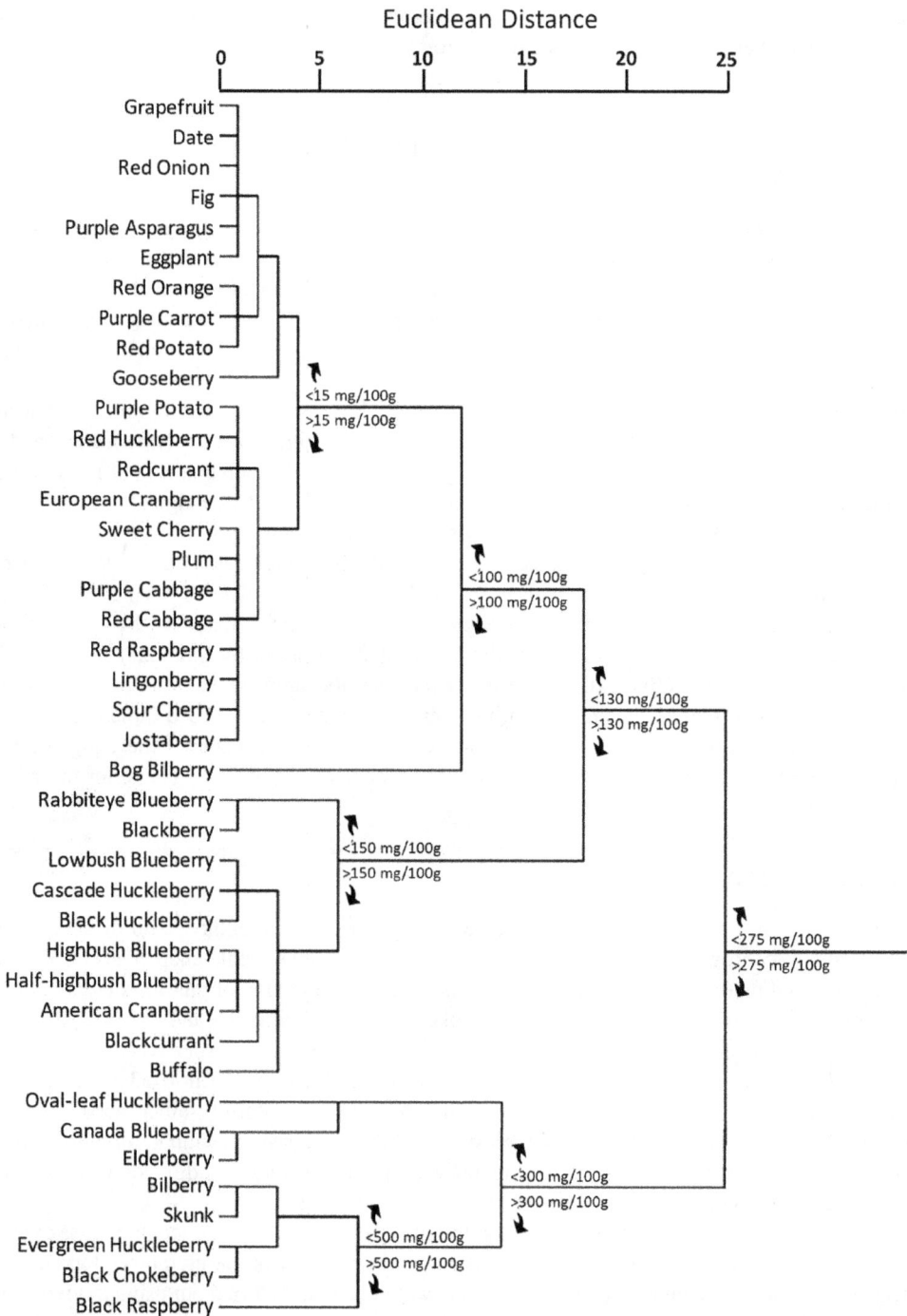

FIGURE 12.1 According to the Phenol-Explorer database, anthocyanin cluster distribution in plants depends on total anthocyanin concentration (Neveu et al., 2010; Rothwell et al., 2012, 2013).

TABLE 12.2
The Retention and Effect on Bioaccessibility of Anthocyanins from Fruits in Different Models

Technology and Processing Conditions/Methods/Model	Food	Retention	Effect on Bioaccessibility	References
No technology – is applied on human	Bilberries	8-h passage through the upper intestine	Low bioavailability of anthocyanins in plasma and urine	Mueller et al. (2017)
	Strawberries		Pelargonidin-3-glucoside absorbed by the intestine	Kosińska-Cagnazzo et al. (2015)
Rat stomach and intestine using in situ models	Red-orange		A high proportion (about 20%) of red-orange anthocyanins was absorbed from the stomach and again no anthocyanin metabolite was observed in the stomach after 30 min of incubation	Felgines et al. (2006)
Pig (diet supplemented with fruit)	Blueberries		Detected in the plasma or urine of the fasted animals; intact anthocyanins were detected in the liver, eye, cortex, and cerebellum	Kalt et al. (2008)
Anesthetized rats	*Vitis vinifera* grapes	Stomach for 10 min	Detected in the plasma (176.4 ± 50.5 ng/mL, mean $\pm$ SEM) but also in the brain (192.2 ± 57.5 ng/g)	Passamonti et al. (2005)
Male Wistar rats	Blackberry (14 and 750 µmol/L) and bilberry (88 µmol/L)	The stomach of rats for 30 min (absorption)	A high proportion ($\sim 25\%$) of anthocyanin monoglycosides (glucoside or galactoside) was absorbed from the stomach, whereas absorption of cyanidin-3-rutinoside was lower. Bilberry anthocyanins were also efficiently absorbed, but absorption varied greatly (19–37%)	Talavera et al. (2003)
Rat in situ models	Freeze-dried strawberry (feeding 8 days)		Low amounts of total anthocyanins were recovered in 24-h urine. Glycosides (about 53%) and glucuronidated metabolites (about 47%) were found. Pelargonidin-3-glucoside was rapidly absorbed from both stomach and small intestine	Felgines et al. (2007)
Male Fischer 344 rats	Black raspberry		Total anthocyanins in the gastric lumen and tissue steadily decreased during the 180 min whereas anthocyanin contents in the small intestinal lumen and tissue were highest at 120 min before decreasing	He et al. (2009)

(continued)

TABLE 12.2 (Continued)
The Retention and Effect on Bioaccessibility of Anthocyanins from Fruits in Different Models

Technology and Processing Conditions/Methods/Model	Food	Retention	Effect on Bioaccessibility	References
Female volunteers	Orange juice (600 mL/day for 21 days)		Significant increase in plasma vitamin C, cyanidin-3-glucoside, β-cryptoxanthin, and zeaxanthin (moderately improving the antioxidant defense system)	Riso et al. (2005)
Omnivorous subjects (5 males and 1 female)	Meal (purée + cereals) consisting of 50 g of oat cereal and 50 g of berry purée containing bilberries (*Vaccinium myrtillus* L.) and lingonberries (*Vaccinium vitis-idaea* L.) in a ratio of 3:1 (w/w) Purée + cereals contained 1435 μmol of anthocyanins and 339 μmol of phenolic acids		The urinary excretion of measured 18 phenolic acids increased 241 μmol during the 48-h follow-up after the purée + cereal supplementation. The excretion peak of dietary phenolic acids was observed at 4–6 h after the purée + cereal supplementation and 2 h earlier after the supplementation of the purée alone. Homovanillic and vanillic acids were the most abundant metabolites, and they were partly produced from anthocyanins	Nurmi et al. (2009)

REFERENCES

Ahmed, J., & Ramaswamy, H. S. (2006). Changes in colour during high pressure processing of fruits and vegetables. *Stewart Postharvest Review*, *2*(5), 1–8.

Al-Shahib, W., & Marshall, R. J. (2003). The fruit of the date palm: Its possible use as the best food for the future? *International Journal of Food Sciences and Nutrition*, *54*(4), 247–259.

Barreveld, W. H. (1993). Date Palm Products. www. fao.org/docrep/t0681E/t0681e12.htm

Basiouny, F. M. (1995). Ethylene evolution and quality of blackberry fruit as influenced by harvest time and storage intervals. *Postharvest Physiology of Fruits 398*, 195–204.

Biegelmeyer, R., Andrade, J. M. M., Aboy, A. L., Apel, M. A., Dresch, R. R., Marin, R., ... & Henriques, A. T. (2011). Comparative analysis of the chemical composition and antioxidant activity of red (*Psidium cattleianum*) and yellow (*Psidium cattleianum* var. *lucidum*) strawberry guava fruit. *Journal of Food Science*, *76*(7), C991–C996.

Block, G., Patterson, B., & Subar, A. (1992). Fruit, vegetables, and cancer prevention: A review of the epidemiological evidence. *Nutrition and Cancer*, *18*(1), 1–29.

Bose, T. K. (2001). Fruit: tropical and subtropical. *Naya Yug*, 1, 721.

Braga, A. R. C., Murador, D. C., de Souza Mesquita, L. M., & de Rosso, V. V. (2018). Bioavailability of anthocyanins: Gaps in knowledge, challenges and future research. *Journal of Food Composition and Analysis*, *68*, 31–40.

Bub, A., Watzl, B., Heeb, D., Rechkemmer, G., & Briviba, K. (2001). Malvidin-3-glucoside bioavailability in humans after ingestion of red wine, dealcoholized red wine and red grape juice. *European Journal of Nutrition*, *40*(3), 113–120.

Chang, T. S., Siddiq, M., Sinha, N. K., & Cash, J. N. (1994). Plum juice quality affected by enzyme treatment and fining. *Journal of Food Science*, *59*(5), 1065–1069.

Curi, P. N., de Sousa Tavares, B., de Almeida, A. B., Pio, R., Peche, P. M., & de Souza, V. R. (2016). Influence of subtropical region strawberry cultivars on jelly characteristics. *Journal of Food Science*, *81*(6), S1515–S1520.

Dabral, M., & Misra, K. K. (2007). Studies on flowering and fruiting in some litchi cultivars. *Indian Journal of Horticulture*, *64*(2), 207–210.

Dalla Nora, C., Jablonski, A., Rios, A. D. O., Hertz, P. F., de Jong, E. V., & Flôres, S. H. (2014). The characterisation and profile of the bioactive compounds in red guava (*Psidium cattleyanum* Sabine) and guabiju (*Myrcianthes pungens* (O. Berg) (D. Legrand). *International Journal of Food Science & Technology*, *49*(8), 1842–1849.

Dalla Nora, C., Müller, C. D. R., de Bona, G. S., de Oliveira Rios, A., Hertz, P. F., Jablonski, A., ... & Flôres, S. H. (2014). Effect of processing on the stability of bioactive compounds from red guava (*Psidium cattleyanum* Sabine) and guabiju (*Myrcianthes pungens*). *Journal of Food Composition and Analysis*, *34*(1), 18–25.

de Ancos, B., Ibanez, E., Reglero, G., & Cano, M. P. (2000). Frozen storage effects on anthocyanins and volatile compounds of raspberry fruit. *Journal of Agricultural and Food Chemistry*, *48*(3), 873–879.

De Brito, E. S., De Araújo, M. C. P., Alves, R. E., Carkeet, C., Clevidence, B. A., & Novotny, J. A. (2007). Anthocyanins present in selected tropical fruits: Acerola, jambolão, jussara, and guajiru. *Journal of Agricultural and Food Chemistry*, *55*(23), 9389–9394.

Dueñas, M., Pérez-Alonso, J. J., Santos-Buelga, C., & Escribano-Bailón, T. (2008). Anthocyanin composition in fig (*Ficus carica* L.). *Journal of Food Composition and Analysis*, *21*(2), 107–115.

FAO-FORCAST (2020–2025). International Society for Horticultural Science.

Faria, A., Pestana, D., Azevedo, J., Martel, F., de Freitas, V., Azevedo, I., ... & Calhau, C. (2009). Absorption of anthocyanins through intestinal epithelial cells: Putative involvement of GLUT2. *Molecular Nutrition & Food Research*, *53*(11), 1430–1437.

Felgines, C., Texier, O., Besson, C., Lyan, B., Lamaison, J. L., & Scalbert, A. (2007). Strawberry pelargonidin glycosides are excreted in urine as intact glycosides and glucuronidated pelargonidin derivatives in rats. *British Journal of Nutrition*, *98*(6), 1126–1131.

Fuchs, Y., Zauberman, G., Ronen, R., Rot, I., Weksler, A., & Akerman, M. (1993). The physiological basis of litch fruit pericarp color retention. *Physiological Basis of Postharvest Technologies, 343*, 29–33.

Gil, M. I., García-Viguera, C., Artés, F., & Tomás-Barberán, F. A. (1995). Changes in pomegranate juice pigmentation during ripening. *Journal of the Science of Food and Agriculture*, *68*(1), 77–81.

Gil, M. I., Tomás-Barberán, F. A., Hess-Pierce, B., Holcroft, D. M., & Kader, A. A. (2000). Antioxidant activity of pomegranate juice and its relationship with phenolic composition and processing. *Journal of Agricultural and Food Chemistry*, *48*(10), 4581–4589.

Giusti, M. M., Rodríguez-Saona, L. E., Baggett, J. R., Reed, G. L., Durst, R. W., & Wrolstad, R. E. (1998). Anthocyanin pigment composition of red radish cultivars as potential food colorants. *Journal of Food Science*, *63*(2), 219–224.

Gonzaga Neto, L., & Soares, J. M. (1995). A cultura da goiaba. *Área de Informação da Sede-Col Criar Plantar ABC 500P/500R Saber (INFOTECA-E)*.

Haber, W. A. (1983). Hylocereuscostaricensis (*Pitahaya silvestre*), wild pitahaya. *Costa Rican Natural History*, 252–253.

Haminiuk, C. W., Maciel, G. M., Plata-Oviedo, M. S., & Peralta, R. M. (2012). Phenolic compounds in fruits – An overview. *International Journal of Food Science & Technology*, *47*(10), 2023–2044.

He, J., Wallace, T. C., Keatley, K. E., Failla, M. L., & Giusti, M. M. (2009). Stability of black raspberry anthocyanins in the digestive tract lumen and transport efficiency into gastric and small intestinal tissues in the rat. *Journal of Agricultural and Food Chemistry*, *57*(8), 3141–3148.

Heber, D., & Bowerman, S. (2001). Applying science to changing dietary patterns. *The Journal of Nutrition*, *131*(11), 3078S–3081S.

Holcroft, D. M., Gil, M. I., & Kader, A. A. (1998). Effect of carbon dioxide on anthocyanins, phenylalanine ammonia lyase and glucosyltransferase in the arils of stored pomegranates. *Journal of the American Society for Horticultural Science*, *123*(1), 136–140.

Huang, H. B. (1995). Advances in fruit physiology of the arillate fruits of litchi and longan. *Annual Review of Horticultural Science*, *1*, 107–120.

Huang, X., Subhadrabandhu, S., Mitra, S. K., Ben-Arie, R., & Stern, R. A. (2005). Origin, history, production and processing. *Litchi and Longan*, 25–34.

Jiang, Y. (2000). Role of anthocyanins, polyphenol oxidase and phenols in lychee pericarp browning. *Journal of the Science of Food and Agriculture*, *80*(3), 305–310.

Jiang, Y. M., & Fu, J. R. (1999). Postharvest browning of litchi fruit by water loss and its prevention by controlled atmosphere storage at high relative humidity. *LWT-Food Science and Technology*, *32*(5), 278–283.

Jiang, Y., Li, Y., & Li, J. (2004). Browning control, shelf life extension and quality maintenance of frozen litchi fruit by hydrochloric acid. *Journal of Food Engineering*, *63*(2), 147–151.

Jiang, Y. M., Wang, Y., Song, L., Liu, H., Lichter, A., Kerdchoechuen, O., ... & Shi, J. (2006). Postharvest characteristics and handling of litchi fruit – An overview. *Australian Journal of Experimental Agriculture*, *46*(12), 1541–1556.

Kalt, W., Blumberg, J. B., McDonald, J. E., Vinqvist-Tymchuk, M. R., Fillmore, S. A., Graf, B. A., ... & Milbury, P. E. (2008). Identification of anthocyanins in the liver, eye, and brain of blueberry-fed pigs. *Journal of Agricultural and Food Chemistry*, *56*(3), 705–712.

Kamiloglu, S., & Capanoglu, E. (2015). Polyphenol content in figs (*Ficus carica* L.): Effect of sun-drying. *International Journal of Food Properties*, *18*(3), 521–535.

Kaur, C., & Kapoor, H. C. (2001). Antioxidants in fruits and vegetables – The millennium's health. *International Journal of Food Science & Technology*, *36*(7), 703–725.

Kidøy, L., Nygård, A. M., Andersen, Ø. M., Pedersen, A. T., Aksnes, D. W., & Kiremire, B. T. (1997). Anthocyanins in fruits of *Passiflora edulis* and *P. suberosa*. *Journal of Food Composition and Analysis*, *10*(1), 49–54.

Kim, D. O., Jeong, S. W., & Lee, C. Y. (2003). Antioxidant capacity of phenolic phytochemicals from various cultivars of plums. *Food Chemistry*, *81*(3), 321–326.

Kosińska-Cagnazzo, A., Diering, S., Prim, D., & Andlauer, W. (2015). Identification of bioaccessible and uptaken phenolic compounds from strawberry fruits in in vitro digestion/Caco-2 absorption model. *Food Chemistry*, 170, 288–294. http://dx.doi.org/ 10.1016/j.foodchem.2014.08.070.

Lee, H. S., & Wicker, L. (1991). Anthocyanin pigments in the skin of lychee fruit. *Journal of Food Science*, *56*(2), 466–468.

Liaqat, A., Svensson, B., Alsanius, B. W., & Olsson, M. E. (2011). Late season harvest and storage of *Rubus* berries – Major antioxidant and sugar levels. *Scientia Horticulturae*, *129*(3), 376–381.

Liu, R. H. (2003). Health benefits of fruit and vegetables are from additive and synergistic combinations of phytochemicals. *The American Journal of Clinical Nutrition*, *78*(3), 517S–520S.

Liu, L., Cao, S. Q., Xu, Y. J., Zhang, M. W., Xiao, G. S., Deng, Q. C., & Xie, B. J. (2010). Oxidation of (–)-epicatechin is a precursor of litchi pericarp enzymatic browning. *Food Chemistry,* 118, 508–511.

Lo Piero, A. R., Puglisi, I., Rapisarda, P., & Petrone, G. (2005). Anthocyanins accumulation and related gene expression in red orange fruit induced by low temperature storage. *Journal of Agricultural and Food Chemistry,* 53(23), 9083–9088.

Macheix, J. J., Fleuriet, A., & Billot, J. (2018). *Fruit Phenolics.* CRC Press.

Markakis, P. (1982). Stability of anthocyanins in foods. *Anthocyanins as Food Colors,* 163, 180.

Mazza, G., & Miniati, E. (2018). *Anthocyanins in Fruits, Vegetables, and Grains.* CRC Press.

Miesle, T. J., Proctor, A., & Lagrimini, L. M. (1991). Peroxidase activity, isoenzymes, and tissue localization in developing highbush blueberry fruit. *Journal of the American Society for Horticultural Science,* 116(5), 827–830.

Mitra, S. K., Devi, H. L., & Debnath, S. (2011). Tropical and subtropical fruits and human health. In *International Symposium on Tropical and Subtropical Fruits 1024* (pp. 39–47).

Morton, J. F. (1987). Fig. In: *Fruits of Warm Climates.* Winterville, NC: Creative Resource Systems. www. hort. purdue.edu/newcrop/morton/fig.html.

Morton, J. F. (1987). *Fruits of Warm Climates.* JF Morton.

Mueller, D., Jung, K., Winter, M., Rogoll, D., Melcher, R., & Richling, E. (2017). Human intervention study to investigate the intestinal accessibility and bioavailability of anthocyanins from bilberries. *Food Chemistry,* 231, 275–286.

Mullen, W., Edwards, C. A., & Crozier, A. (2006). Absorption, excretion and metabolite profiling of methyl-, glucuronyl-, glucosyl-and sulpho-conjugates of quercetin in human plasma and urine after ingestion of onions. *British Journal of Nutrition,* 96(1), 107–116.

Neveu, V., Perez-Jiménez, J., Vos, F., Crespy, V., du Chaffaut, L., Mennen, L., Knox, C., Eisner, R., Cruz, J., Wishart, D., et al. Phenol-Explorer: An online comprehensive database on polyphenol contents in foods. *Database* 2010, *2010*, bap024.

Nurmi, T., Mursu, J., Heinonen, M., Nurmi, A., Hiltunen, R., & Voutilainen, S. (2009). Metabolism of berry anthocyanins to phenolic acids in humans. *Journal of Agricultural and Food Chemistry,* 57(6), 2274–2281.

Passamonti, S., Vrhovsek, U., Vanzo, A., & Mattivi, F. (2005). Fast access of some grape pigments to the brain. *Journal of Agricultural and Food Chemistry,* 53(18), 7029–7034.

Riso, P., Visioli, F., Gardana, C., Grande, S., Brusamolino, A., Galvano, F., ... & Porrini, M. (2005). Effects of blood orange juice intake on antioxidant bioavailability and on different markers related to oxidative stress. *Journal of Agricultural and Food Chemistry,* 53(4), 941–947.

Rothwell, J. A., Perez-Jimenez, J., Neveu, V., Medina-Remón, A., M'hiri, N., García-Lobato, P., Manach, C., Knox, C., Eisner, R., Wishart, D. S., et al. (2012). Phenol-Explorer 3.0: A major update of the Phenol-Explorer database to incorporate data on the effects of food processing on polyphenol content. *Database,* 2013, *2013*, bat070.

Rothwell, J. A., Urpi-Sarda, M., Boto-Ordoñez, M., Knox, C., Llorach, R., Eisner, R., Cruz, J., Neveu, V., Wishart, D., Manach, C., et al. (2013). Phenol-Explorer 2.0: A major update of the Phenol-Explorer database integrating data on polyphenol metabolism and pharmacokinetics in humans and experimental animals. *Database,* 2012, *2012*, bas031.

Saati, E. A. (2009). Identification and quality test of red dragon fruit skin pigment (*Hylocereus costaricensis*) at multiple age save with different type of solvent. Directorate of Research and Community Service, UMM. Poor.

Shrikhande, A. J., & Francis, F. J. (1976). Anthocyanins in foods. *Critical Reviews in Food Science & Nutrition,* 7(3), 193–218.

Solomon, A., Golubowicz, S., Yablowicz, Z., Grossman, S., Bergman, M., Gottlieb, H. E., ... & Flaishman, M. A. (2006). Antioxidant activities and anthocyanin content of fresh fruits of common fig (*Ficus carica* L.). *Journal of Agricultural and Food Chemistry,* 54(20), 7717–7723.

Sridhar, D., Prashanth, P., Kumar, M. R., & Jyothi, G. (2017). Studies on the effect of blending of jamun juice and guava juice on sensory quality and storage. *International Journal of Pure and Applied Biosciences,* 5(4), 1089–1096.

Talavera, S., Felgines, C., Texier, O., Besson, C., Lamaison, J. L., & Rémésy, C. (2003). Anthocyanins are efficiently absorbed from the stomach in anesthetized rats. *The Journal of Nutrition,* 133(12), 4178–4182.

Talavera, S., Felgines, C., Texier, O., Besson, C., Manach, C., Lamaison, J. L., & Rémésy, C. (2004). Anthocyanins are efficiently absorbed from the small intestine in rats. *The Journal of Nutrition, 134*(9), 2275–2279.

Underhill, S., & Critchley, C. (1994). Anthocyanin decolorisation and its role in lychee pericarp browning. *Australian Journal of Experimental Agriculture, 34*(1), 115–122.

van't Veer, P., Jansen, M. C., Klerk, M., & Kok, F. J. (2000). Fruits and vegetables in the prevention of cancer and cardiovascular disease. *Public Health Nutrition, 3*(1), 103–107.

Vargas, M. D. L. V., Cortez, J. A. T., Duch, E. S., Lizama, A. P., & Méndez, C. H. H. (2013). Extraction and stability of anthocyanins present in the skin of the dragon fruit (*Hylocereus undatus*). *Food and Nutrition Sciences, 4*(12), 1221.

Wang, H. C., Huang, X. M., Hu, G. B., Yang, Z. Y., & Huang, H. B. (2005). A comparative study of chlorophyll loss and its related mechanism during fruit maturation in the pericarp of fast-and slow-degreening litchi pericarp. *Scientia Horticulturae, 106*(2), 247–257.

Welch, C. R., Wu, Q., & Simon, J. E. (2008). Recent advances in anthocyanin analysis and characterization. *Current Analytical Chemistry, 4*(2), 75–101.

Wesche-Ebeling, P. E. D. R. O., & Montgomery, M. W. (1990). Strawberry polyphenoloxidase: Its role in anthocyanin degradation. *Journal of Food Science, 55*(3), 731–734.

Wichienchot, S., Jatupornpipat, M., & Rastall, R. A. (2010). Oligosaccharides of pitaya (dragon fruit) flesh and their prebiotic properties. *Food Chemistry, 120*(3), 850–857.

Wu, X., & Prior, R. L. (2005a). Systematic identification and characterization of anthocyanins by HPLC-ESI-MS/MS in common foods in the United States: Fruits and berries. *Journal of Agricultural and Food Chemistry, 53*(7), 2589–2599.

Wu, X., & Prior, R. L. (2005b). Identification and characterization of anthocyanins by high-performance liquid chromatography–electrospray ionization–tandem mass spectrometry in common foods in the United States: Vegetables, nuts, and grains. *Journal of Agricultural and Food Chemistry, 53*(8), 3101–3113.

Zhang, D., Quantick, P. C., & Grigor, J. M. (2000). Changes in phenolic compounds in litchi (*Litchi chinensis* Sonn.) fruit during postharvest storage. *Postharvest Biology and Technology, 19*(2), 165–172.

Zhang, Z., Pang, X., Ji, Z., & Jiang, Y. (2001). Role of anthocyanin degradation in litchi pericarp browning. *Food Chemistry, 75*(2), 217–221.

Zhang, Z. Q., Pang, X. Q., Duan, X. W., & Ji, Z. L. (2003). The anthocyanin degradation and anthocyanase activity during the pericarp browning of lychee fruit. *Scientia Agricultura Sinica, 36*(8), 945–949.

Zhang, Z., Xuequn, P., Yang, C., Ji, Z., & Jiang, Y. (2004). Purification and structural analysis of anthocyanins from litchi pericarp. *Food Chemistry, 84*(4), 601–604.

Index

Note: Page numbers in *italics* refer to figures; page numbers in **bold** refer to tables.

For Product Safety Concerns and Information please contact our EU
representative GPSR@taylorandfrancis.com
Taylor & Francis Verlag GmbH, Kaufingerstraße 24, 80331 München, Germany

www.ingramcontent.com/pod-product-compliance
Lightning Source LLC
Chambersburg PA
CBHW082034230326
41598CB00081B/6345